THE WOODY PLANTS OF OHIO

This work is a project of the Ohio Flora Committee of the Ohio Academy of Science, sponsored by the Academy and the National Science Foundation.

THE WOODY PLANTS OF OHIO

TREES, SHRUBS, AND WOODY CLIMBERS
NATIVE, NATURALIZED, AND ESCAPED

By E. Lucy Braun

A contribution toward
THE VASCULAR FLORA OF OHIO

A Sandstone Book

OHIO STATE UNIVERSITY PRESS
COLUMBUS

PREFACE

Soon after the publication of the eighth edition of *Gray's Manual of Botany*, by M. L. Fernald, a group of Ohio botanists, who were taking part in a field trip, conceived the idea of preparing an Ohio Flora, which would bring the nomenclature up-to-date and show, by means of maps, the distribution of Ohio's vascular plants. Later, Edward S. Thomas, then president of the Ohio Academy of Science, appointed a committee to consider the advisability of preparing such a flora, and to determine its scope. The report of this committee, which was presented to and approved by the council of the Ohio Academy of Science, resulted in the establishment of the Ohio Flora Committee.

During the formulation of plans for the project and early work on it, it became evident that the preparation of the flora as a whole would be a long-range undertaking. *The Woody Plants of Ohio* is a contribution toward the major project *The Vascular Flora of Ohio*.

This book has two principal objectives: (1) to give information as to what species occur in Ohio and to show by means of maps the distribution of these species; and (2) to give to amateurs, students, and fieldworkers in the natural sciences a ready means of identifying woody plants at any season of the year. In addition, data on variation within the species are included, particularly when of geographic significance or when not in accord with the usual "manual" concepts. It has seemed desirable, also, to include some general information concerning economic plants, relationships, and broad geographic distribution of genera and families.

Technical descriptions of species have been reduced to a minimum—such characters are included in the keys. Recognition characters which may serve in the field and characters which distinguish similar species have been emphasized. Some such characters, useful in Ohio, may not suffice elsewhere.

This book should prove useful, not only to residents of Ohio, but also to those of neighboring states. Ohio's flora contains many species at the margins of their ranges— northeastern species occur locally in northern Ohio, and southern species in a few of the southern counties. The inclusion of all species, whether rare and local, or widespread, will obviate the usual difficulties encountered in the use of books where only the commoner things are selected.

Flowers (which are present only for a short time) have usually been omitted from the illustrations, unless, as in the case of papaw, neither leaf characters nor winter-twig characters can be used for determination at flowering season. Many who know our shrubs and trees when in bloom will wish to recognize them at other seasons and will find that they can use leaves and fruits (mature or immature) for identification.

Illustrations of winter-twigs have been made from fresh specimens collected in late February and March. Fresh material has been used whenever possible for the drawings of fruiting specimens. Herbarium specimens and specimens collected especially for illustrations have been used for many of the leaf drawings.

I wish to express my gratitude to the National Science Foundation for a three-year grant which made it possible to include the many illustrations in this book, and enabled me to travel about collecting fresh specimens to be used in the preparation of the illustrations and to visit herbaria for study; and to the Ohio Academy of Science for its support of the initial phases of the Flora work. My thanks are also extended to the members of the Ohio Flora Committee for their suggestions and comments, and for their reading of all or parts of the manuscript; to Clara Weishaupt, curator of the herbarium of Ohio State University, for her co-operation during my work in the herbarium and for the loan of hundreds of specimens; to those of other institutions who loaned specimens; to the many individuals who contributed one or a few fresh specimens for use in preparing the illustrations— especially to J. Arthur Herrick who furnished many specimens from northeastern Ohio; to those whose collecting, since the publication, in 1953, of the preliminary list of woody plants, has made the distribution maps more complete—especially to Ervin M. Herrick who has added many hundreds of county records during his travels throughout Ohio; and, finally, to my sister, Annette F. Braun, for her aid in the final revision of the manuscript, checking the distribution maps, and arranging the plates.

All illustrations were made by Mrs. Elizabeth Dalvé, assisted by Mrs. Elizabeth King. Mrs. Dalvé's interest in the work and her careful observation of details is attested by the drawings themselves.

To Ernest J. Palmer, acknowledgement is made of his contribution of the text of the genus *Crataegus*, and his identification of all Ohio specimens; to the late Carleton R. Ball for his identification of all specimens of *Salix* on which distribution maps are based; and to Wayne E. Manning for his identification of many specimens of *Carya*.

E. LUCY BRAUN

Cincinnati, Ohio
June, 1959

TABLE OF CONTENTS

INTRODUCTION

This work includes all woody plants of Ohio—trees, shrubs, and woody climbers—that are indigenous (native), naturalized, or escaped.

Woody plants are those whose above-ground stems persist over winter and are woody or somewhat woody. There is no sharp line of demarcation between woody and non-woody plants, and some which are included here should, perhaps, be omitted. However, the inclusion or omission of borderline plants has usually been decided by other features. Certain members of the Ericaceae, as wintergreen, are almost herbaceous; their inclusion is desirable, however, for by so doing, all members of the family which occur in Ohio are in this text. In the genus *Cornus*, *C. canadensis* is included to make the treatment of *Cornus* complete for Ohio. Some plants which in warm latitudes are slightly woody near the base are omitted. *Clematis virginiana*, for example, is not included (except in the key to vines), for this would require that the family Ranunculaceae be included; usually, *Xanthorhiza* (not a member of the Ohio flora) is considered the only woody member of the family in the "manual range"; it has escaped near a Rhododendron planting in Hocking County.

Native or indigenous plants are those which occur (or did occur within the period represented by Ohio herbaria) naturally in Ohio. They grow where they do because of natural forces. Some species, whose occurrence in Ohio is known from herbarium specimens, are now extinct or may soon become extinct. This is particularly true of some rare northern species found in bogs or along shores; their habitats have been largely destroyed by man's activities. Others have long since disappeared from some localities (and counties) but still persist in a few spots. It has not been feasible to determine which species still occur in the counties recorded and which are now extinct. Our maps show the known distribution, by counties, of all native members of the woody flora of Ohio, whether extinct or still growing in the counties indicated. They do not include (at least not intentionally) the extension of range which has resulted from planting. Native species are distinguished in the text by boldface type. It is possible that a few species thus indicated are not truly native to Ohio, and that a few native species have been omitted. With few exceptions, all native species are illustrated; both summer and winter characters of most of them are shown.

Naturalized species have entered our flora from other parts of the United States or from abroad either by intentional or by accidental introduction; they have become established and they reproduce freely, so that in some instances they appear to belong. Some native species have greatly extended their ranges in historic time, so that in parts of their present range they may actually be introduced and naturalized. This is true of black locust and trumpet-creeper; information on their original ranges is insufficient to make possible the mapping of only native occurrences.

Escaped plants are also introduced; they occur sporadically and may become established (naturalized) or may disappear. A few species "spreading from cultivation" are included because specimens are sent to herbaria more or less frequently. Most of such species are so obviously garden plants that they have no place in a treatment of Ohio's flora. A few (for example *Akebia*) have been included because of the difficulty in identifying them in

1

the usual manuals. Others, and a number of exotic species, are mentioned in the description of family or genus, so as to give a clue to the identity of specimens found in parks or arboreta.

Naturalized and escaped species are not mapped; their names appear in the text in small capitals.

RECORDS

For a taxonomic work, herbarium specimens are the only acceptable records of the occurrence of species. Within Ohio, there are a large number of herbaria, some large, some small. Information concerning what is (or is said to be) in various herbaria has been provided in the form of "herbarium slips;" our slip is a form 3 by 5 inches, which, when filled out, gives the name of the herbarium, the name of the plant (as it appears on the herbarium sheet), the county and the exact locality in which it was found, the date of collection, and (often) data on habitat. These herbarium slips were used to compile a tentative list of Ohio's woody plants and to ascertain where specimens for study were located. They have given much information about frequency of occurrence and variation in habitats. They have not been used in making the distribution maps, however. No species has been shown to occur in a county unless a specimen of that species from the county has been studied and its identification verified. In most cases, the sheet and corresponding slip are stamped

RECORDED FOR OHIO FLORA

E. L. B.

If identification has been made by a specialist in the genus, this fact has been noted and the herbarium sheet has not been stamped. By no means have all of the many thousands of herbarium specimens in Ohio been examined. The file of herbarium slips, now housed with the herbarium at Ohio State University, constitutes a permanent catalogue; it consists of approximately twenty-five thousand entries. It will furnish monographers with information concerning the number of specimens of a given genus that are available and where they are located. It has been an invaluable aid in the present work.

Because of changing concepts of species, recent studies in natural hybridization and introgression, and the great variation displayed by many species, sight records and published records without specimens cannot be used. Some of our commonest "species" can no longer be thought of as simple entities (see *Fagus grandifolia*, *Acer saccharum*, and *A. nigrum*). Specimens and field studies are necessary for modern work.

A few species that have been reported as occurring in Ohio are included in the text although efforts to locate the specimens on which such reports are based have been unsuccessful.

LIST OF HERBARIA

Herbarium slips received from the following herbaria are filed in the catalogue located in the herbarium of Ohio State University:

Antioch College, Yellow Springs, Ohio
Athens High School, Athens, Ohio
Baldwin-Wallace College, Berea, Ohio
Bowling Green State University, Bowling Green, Ohio

Butler University, Indianapolis, Indiana
Carnegie Museum, Pittsburgh, Pennsylvania
Defiance College, Defiance, Ohio
Denison University, Granville, Ohio
Kent State University, Kent, Ohio
Marietta College, Marietta, Ohio
Muskingum College, New Concord, Ohio
Oberlin College, Oberlin, Ohio
Ohio Agricultural Experiment Station, Wooster, Ohio
Ohio State University, Columbus, Ohio
Ohio University, Athens, Ohio
Ohio Wesleyan University, Delaware, Ohio
Franz Theodore Stone Laboratory, Put-in-Bay, Ohio
University of Cincinnati, Cincinnati, Ohio

E. Lucy Braun, Cincinnati, Ohio
Arthur S. Brooks, Van Wert, Ohio
F. W. Buchanan, Amsterdam, Ohio
Charles R. Goslin, Lancaster, Ohio
J. Guccion, Cleveland, Ohio
R. C. McCafferty, Wadsworth, Ohio
Almon N. Rood, Phalanx Station, Ohio

PREVIOUS FLORAS AND LISTS

Many papers and books have been written about the vascular flora and plant communities of Ohio; the descriptions of the vegetation often contain lists of species growing in the area studied.

State-wide floras or lists include the entire vascular flora, or are confined to the spring flora, the woody plants, the trees, the shrubs, or the pteridophytes. These range in date from 1860 to 1932, when the last list was published.

The six catalogs of the Ohio flora are:

1860. J. S. NEWBERRY. Catalogue of the flowering plants and ferns of Ohio. 14th Ann. Rept. Ohio State Board Agr. 1859:235–273.

1878. H. C. BEARDSLEE. Catalogue of the plants of Ohio, including flowering plants, ferns, mosses and liverworts. 32nd Ann. Rept. Ohio State Board Agr. 1877:335–363.

1893. W. A. KELLERMAN and W. C. WERNER. Catalogue of Ohio plants. Rept. Geol. Surv. Ohio 7:56–406.

1899. W. A. KELLERMAN. The fourth state catalogue of Ohio. Bull. Ohio State Univ. Ser. 4, 10:1–65.

1914. J. H. SCHAFFNER. Catalog of Ohio vascular plants. Ohio Biol. Surv. Bull. 2.

1932. J. H. SCHAFFNER. Revised catalog of Ohio vascular plants. Ohio Biol. Surv. Bull. 25.

In addition to the above lists, Schaffner's *Field Manual of the Flora of Ohio and Adjacent Territory* (R. G. Adams & Co., Columbus), 1928, is state-wide in scope.

In 1882, an annotated list of "Woody plants of Ohio," by Warder, James, and James, appeared (*36th Ann. Rept. Ohio State Board Agr.* 1881:73–112). In 1905, Schaffner published a "Key to the genera of Ohio woody plants in the winter condition" and a "Key to the genera of Ohio woody plants based on leaf and twig characters" (*Ohio Nat.* 5:277–286, 364–373).

Publications on trees alone are numerous, among them a number of keys and lists and at least two small books or pamphlets devoted to the more common Ohio trees, including some planted species. One list of shrubs has been published. The first publication on Ohio trees was by John Hussey in 1873, where, in "Forest distribution" he gives a "list of trees found growing indigenously in Ohio" (*27th Ann. Rept. Ohio State Board Agr.* 1872: 23–40). Other publications on Ohio trees and shrubs are:

1886. ADOLPH LEUE. The forestal relation of Ohio. This "describes forests and gives lists of principal trees for each county of the State." 1st Ann. Rept. Ohio State Forestry Bureau 1885:15–64.

1887. ADOLPH LEUE. Catalogue of forest trees growing in Ohio. 2nd Ann. Rept. Ohio State Forestry Bureau 1886:16–60.

1895. W. A. KELLERMAN. Ohio forest trees containing brief descriptions of all native species and a key for their identification based on the leaves and fruit. 16 pp. The author. Columbus.

1906. J. H. SCHAFFNER. Check list of Ohio trees. Ohio Nat. **6**:457–461.

1907. J. H. SCHAFFNER. Check list of Ohio Shrubs. Ohio Nat. **8**:205–209.

1909. J. H. SCHAFFNER. Trees of Ohio and surrounding territory. Proc. Ohio Acad. Sci. Spec. Paper 15.

1914, '22, '26. J. H. SCHAFFNER. Field Manual of Trees. (Covers the "manual range" and thus includes Ohio.) Adams & Co., Columbus.

1927. J. S. ILLICK. Common Trees of Ohio. Amer. Tree Assn., Washington, D. C.

——. F. W. DEAN and L. C. CHADWICK. Ohio Trees. Ohio State Univ. Press, Columbus.

A great many county or local lists have added to the knowledge of Ohio's flora. These range in date from 1834, when John L. Riddell's "Catalogue of the plants, growing spontaneously in Franklin County, central Ohio; excluding grasses, mosses, lichens, fungi, etc." (*West Med. Gaz.* **2**:116–120, 154–159) appeared, and 1849, when T. G. Lea's *Catalogue of plants, native and naturalized, collected in the vicinity of Cincinnati, Ohio, during the years 1834–1844* (Philadelphia) was published, to 1955, when the most recent county list, "The vascular flora of Clinton County, Ohio," by E. E. Terrell, appeared (*Ohio Jour. Sci.* **55**: 215–240). Lea's herbarium of 1,056 sheets representing 714 species formed the basis for his list; it is deposited in the Academy of Natural Sciences of Philadelphia. A later list (Braun, 1934, *Amer. Midl. Nat.* **15**:1–75) covered a comparable area and made use of Lea's habitat notes to reconstruct something of a picture of what the flora had been before urban growth completely destroyed Lea's favorite collecting sites.

Titles and places of publication of the many lists, additions to lists, notes, and so on, can be found in Miller's "Bibliography of Ohio botany" (*Ohio Biol. Surv. Bull.* **27**, 1932).

NOMENCLATURE

Unfortunately, scientific names sometimes change. Many names used in this book differ from those used by Schaffner in his catalogue of Ohio plants (1932). At that time, two codes of nomenclature were in use, the American Code and the International Code. Schaffner followed the first of these. Now, the International Code is accepted by all botanists. Even so, there are instances in which a species appears under different names, even in recent books. This may be due to differences in interpretation of a particular species; or it may be a result of very recent research which has discovered incorrect usage, or which necessitates subdivision of a former species. Whatever the cause, it always results in some confusion. The most difficult instances are those in which a specific name has been transferred from one plant to another, and perhaps back to the first again. Rock chestnut oak is an example: it was known as *Quercus prinus*; then study of the type suggested to certain students that the name should be applied to the swamp chestnut oak, which then became *Q. prinus* instead of *Q. michauxii*. Rock chestnut oak was then called *Q. montana*, a very fitting name as far as distribution is concerned, and one which is used in the *Yearbook of Agriculture: Trees* (Little, 1949). In the same manner, the name of red oak has shifted back and forth—from *rubra* to *borealis* and back to *rubra* (in some manuals, though not in Gleason, 1952).

With few exceptions, the nomenclature in this book accords with that in *Gray's Manual of Botany*, eighth edition (Fernald, 1950). Synonyms are given when names used here differ from those used by Schaffner (1932), by Fernald (1950), or by Gleason (1952), and in a few other cases.

A rule adopted by a recent International Botanical Congress (1950) states that a specific name applies to the species as a whole, whether it includes named varieties or is indivisible. Even though the typical variety of a species does not occur in our area, we are still correct in referring to the plants by the specific name. Thus we may refer to pussy willow as *Salix discolor*, disregarding the fact that within the species there are two varieties in our area, the typical which is var. *discolor* (repeating specific name) and another, var. *latifolia*. And, although the typical variety of *Chimaphila umbellata* does not occur in Ohio, we may correctly refer to plants by the specific name; but to be more exact we should add var. *cisatlantica*. Forms (forma, formae) are lower in rank than varieties; species may include formae, but if first divisible into varieties, the formae are then divisions of a variety. Variations may be pronounced and still not be designated by names if they intergrade freely or intermingle.

SCIENTIFIC AND COMMON NAMES

The scientific name of a plant—genus and species and, sometimes, variety and/or form—is universally (geographically speaking) used and applies to but one taxon. Common names are often provincial, and the same name may apply to more than one species; thus wahoo is *Euonymus atropurpureus* (or *E. americanus*) or *Ulmus alata*! White oak is *Quercus alba*; but red oak may be *Q. borealis* (*Q. rubra*) or *Q. falcata* (*Q. digitata*).

Generic names are usually derived from the Greek and Latinized, though sometimes they are taken from the Latin; specific names are Latin or Latinized. Generic and specific names agree in gender. Apparent exceptions to this are seen among the names of trees, but since most Latin names of trees are feminine, the generic name with a masculine ending, such as *Quercus*, may be followed by a specific name with a feminine ending, such as *alba*. An anglicized pronunciation of the Latin names is generally used.

Generic names should always be capitalized. Most specific names should not be capitalized; those derived from proper nouns—the names of persons and genera—may be capitalized or not as desired: *Ulmus Thomasi* or *Ulmus thomasi*. Specific names of geographic origin are never capitalized—*canadensis, virginiana*, and so on.

The spelling of scientific names may be puzzling in a few instances. The specific names of *Prunus pensylvanica* and *Acer pensylvanicum* are correctly spelled with one *n*; at the time these names were given, this was the customary spelling of the state name. The original spelling must be maintained.

Abbreviations following a scientific name refer to the authors of the names. Thus *Quercus alba* L. means that Linnaeus named this species. Sometimes two authors are cited, one name in parentheses, as *Rhododendron nudiflorum* (L.) Torr.; in this case, Linnaeus first used this specific name in conjunction with the generic name *Azalea—A. nudiflora* L.— and Torrey transferred it to *Rhododendron*.

Common names sometimes appear to be translations of the scientific ones; but unless they are in common use, such manufactured names may seem unsatisfactory. Sometimes they are descriptive; sometimes they are misleading. The common name "ash," for example, refers to all species of *Fraxinus*. Wafer-ash is not an ash, but a *Ptelea*, hence the common name is hyphenated. For the same reason, we write blue-beech (for *Carpinus*) and witch-hazel (for *Hamamelis*), using the hyphen to show that the former is not a *Fagus* (beech), and the latter not a *Corylus* (hazel). In some instances where euphony permits, the two words are run together, as in pineapple and tuliptree. Such usage is a recommendation of the U.S. Forest Service, and is followed in Forest Service publications (see Little, 1953, p. 15). The use of a hyphen between two parts of a common name when both are nouns, as in post oak, is not recommended. The hyphen is used only when the common name of a genus is combined to form a name for a plant that is not a member of that genus.

PURPOSE AND USE OF KEYS

A key is a sort of outline of characters so arranged that its use simplifies identification of plants. Because woody plants may be observed in the field in any season, keys are provided for both summer and winter use.

No key to families is included in this book. The classification of families depends largely on floral characters, characters which are often difficult to see and which can be found for only a limited time. Although genera within the family may be best distinguished by flower (and fruit) characters, it is usually possible to recognize genera by other characters. The keys provided here are built largely on vegetative characters; they do not associate related genera, but are entirely artificial.

For effective use of keys, particularly of winter keys, a good hand lens is essential; one with 10× magnification is most satisfactory. Since the metric system is used for all measurements, a ruler graduated to millimeters will be needed.

Keys are not always arranged in the same manner; some are indented, some not. The former are usually easier to follow and have the advantage of indicating obvious grouping of genera or species with similar characters. Some indented keys depend entirely on differences in the amounts that succeeding subdivisions are indented; others number or letter at least the principal divisions. Some use single letters, as *a* and *a* for the two main divisions; others designate these as *a* and *aa*. The latter system is followed in this book.

The Synopsis of Keys (p. 33), which is itself a brief key, will enable the reader to select the proper key for identification of the specimen at hand. Within each key, a dichotomous arrangement, an either-or sort of plan, is used (except that three choices are offered occasionally, especially in some ultimate divisions). Your first choice will always be between *a* and *aa*; for example, if it is summer and your plant is a vine, you will have selected Key II (p. 35). The first point to be decided is whether the leaves are opposite (*a*) or alternate (*aa*). If the latter, you then choose the "fork of the road" *aa*, and henceforth must stay on this route, choosing successive forks in turn: Are the leaves simple (*b*) or compound (*bb*)? If simple, you must again make a choice, this time between *c* and *cc*. Since a considerable number of characters are given for *c*, read all of the points, for you may have a small piece of *Smilax* that has no tendrils, and yet all the other characters are present in your specimen. Turn to *Smilax* in the text and check your identification. Or your specimen may not fit the several points in *c*, but only one of them, the cordate base of leaves; in that case, look further in the key to *cc*, which has some but not the whole combination of characters given for *c*. Then turning to the next letter, decide how your plant climbs—if by tendrils, you will choose *dd*, and then *ee*; but you will still have a final choice to make. Check your determination by turning to the text and figures. If your plant is a grape (*Vitis*), there you will find a key to species (p. 263) arranged in the same way as the key you have been using.

If when using a key, you are uncertain at any place along the route concerning which choice to make (perhaps because your specimen is not complete enough), remember which choice you make—which turn in the route you follow, so to speak—so that if your first choice turns out to be a blind lead, you can go back to that point and try again.

A considerable number of characters are given in many of the divisions of the keys—

7

more, perhaps, than your specimen illustrates, and more, perhaps, than seem needed since one of the characters may appear to be sufficient. Another specimen collected a few months later, however, may show other characters or lack the one that the earlier specimen had. As Fernald has said, a "one-character key . . . often fails to turn in the lock."

In some cases, you may choose to identify your specimen by turning the pages and looking at illustrations. Having done so, you may want to pick out the characters which distinguish your plant, some of which may not be present in your specimen. In this case, the text may not suffice, for few of the characters used in the keys are repeated in the text. It is possible, however, to read a key backward, so to speak. Let us return to the vine you were identifying a while back. By looking at illustrations, you have decided it is a grape (*Vitis*). By looking in the "Index to Genera in Keys" (p. 55), you will find that *Vitis* is on p. 36 of the keys. Let us assemble its characters by reading the divisions of the key you would have to select to get to *Vitis*, namely, the first choice under *ee*, *ee* under *dd*, *dd* under *cc*, *cc* under *b*, and *b* under *aa*. Reading these pieces of the key, you have these characters for *Vitis*: leaves alternate, simple; stems climbing by tendrils arising opposite the leaves; leaves palmately veined; bark shredding (a character you will see in the field but which your specimen will not show), no lenticels; leaves coarsely toothed and often lobed, cordate. These characters can be re-assembled to group those pertaining to leaves and to stems. No other genus in our Ohio flora has this combination of characters.

Upon your first attempt to use the keys, it will be evident that you must become familiar with the terms used. These are all defined in the Glossary; many of them, grouped as to organ, are more fully explained or illustrated in the section "Descriptive Terms," immediately following.

DESCRIPTIVE TERMS

The Glossary (p. 337) defines all the more or less technical (descriptive) terms used in the keys or text. It is, of necessity, arranged alphabetically, hence terms relating to a single structure are separated. Because definitions often fail to convey adequately a picture of concrete features—such as leaf-shape, leaf-margin, fruit-type, and so on—groups of terms used for such features are explained here. Some terms—such as thorn, fruit, and berry—have technical meanings which do not coincide exactly with popular usage.

The dictionary defines a tree as "a woody perennial plant having a single main axis or stem (trunk), commonly exceeding 10 feet in height," and a shrub as "a low, usually several-stemmed, woody plant." We have found it impracticable to separate trees from shrubs. Although most trees have only one trunk, this is not a constant character. Black willow (*Salix nigra*), gray birch (*Betula populifolia*), umbrella magnolia (*Magnolia tripetala*), and many others commonly have several trunks, yet we always call them trees. Height is also an unsatisfactory means of distinction, for it is affected by age and environment. Low, much branched shrubs will never be mistaken for trees, but intermediate forms might be called either tree or shrub.

Whether the plant is a tree or a shrub, the trunk or trunks branch. The ultimate divisions are known as **branchlets** or **twigs**. In this book, branchlet is used to designate stems with leaves; a branchlet is a leafy branchlet. Twig is used for the winter condition, the winter-twig.

The places where leaves are attached (and where leaf-scars are found on winter-twigs) are **nodes**; the part of the stem between two nodes is an **internode**. Leaves (and leaf-scars) may be one at a node, in which case leaves are said to be **alternate**; two at a node, on opposite sides of the branchlet, or **opposite**; or three or more at a node, or **whorled**.

Alternate leaves are arranged spirally and may be two-ranked, three-ranked, five-ranked, or even more complexly spiral. If they are two-ranked, the first leaf is on one side of the stem, the next is on the opposite side, and the third is directly above the first; if three-ranked, the fourth leaf is directly above the first; and if five-ranked, the sixth leaf is above the first, but the spiral goes around twice. When viewed from the end of the branchlet, the leaves appear to be arranged in longitudinal rows along the stem—if two-ranked, there are two rows; if three-ranked, three rows, and so on. This spiral arrangement is easily demonstrated with a piece of string: fasten it to the stem at a low node, then pass each successively higher leaf (or leaf-scar); in the two- and three-ranked arrangements, the string will encircle the stem only once before it comes back to a position directly above the first leaf; in the five-ranked arrangement, it will go around twice before a leaf is encountered which is in line vertically with the first.

At the base of the current year's growth of a branch (usually distinguished by greener color) are narrow, crowded, ringlike markings, the scars of scales of the terminal bud. Since these scars usually persist for several years, the annual increase in the length of the branch and the age of the branch can be ascertained.

Examination of a winter-twig will reveal a number of structures and types of markings (Fig. 1). Buds are present, sometimes all of one kind, sometimes of more than one kind. Buds may be alternate, opposite, or whorled, just as are leaves. The buds at stem-nodes

9

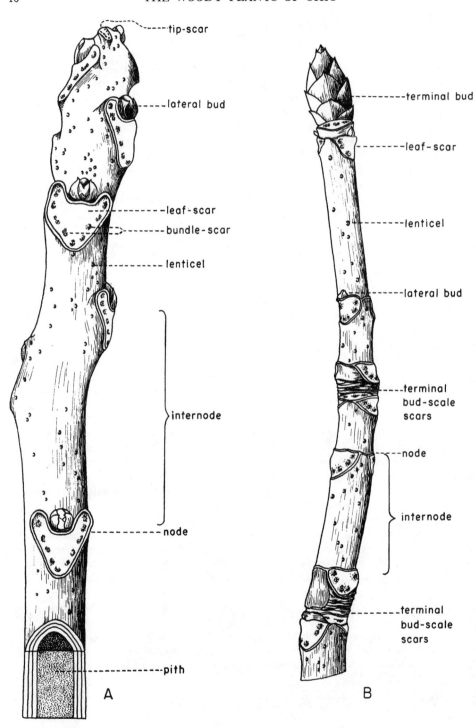

A

B

are **lateral buds.** In a large proportion of species, a **terminal bud** is located at the tip of the twig; a true terminal bud is lacking in many species, and in its place is a **tip-scar,** which is sometimes difficult to see. In such plants, the uppermost lateral bud functions as a terminal bud, its axis continuing stem elongation.

Lateral buds are usually axillary in position, placed immediately above a leaf-scar. **Accessory buds** may be present: they are **collateral** if side by side (see *Lindera*, p. 148), **superposed** if one above another (see *Asimina*, p. 148, and *Gymnocladus*, p. 224). Buds may be **leaf-buds** (also called vegetative buds), **flower-buds**, or **mixed buds**. Flower-buds differ in shape from leaf-buds and are often larger; they are seen only on plants whose flowers are born on old wood. The flowers of plants with mixed buds are produced on leafy shoots of the current season's growth.

Buds may be **naked** (see *Hamamelis*, p. 157, and *Juglans*, p. 92) or covered with **bud-scales.** Bud-scales are arranged in opposite pairs in opposite-leaved species, or spirally in alternate-leaved species. The number of exposed bud-scales (visible bud-scales) varies in different genera—only one in *Salix*, p. 80, many in *Quercus*, p. 122. In a few genera, what appear to be bud-scales are actually stipules, which will enlarge and show their true character after growth is initiated (*Liriodendron*, p. 146; *Magnolia*, p. 142).

Leaf-scars are the marks left after leaf-fall. Their shape corresponds exactly to the shape of the base of the petiole. In most species a leaf-scar is just below a lateral bud, or it partly or completely surrounds it (see *Platanus*, p. 160). Within each leaf-scar are small scars which often look like mere dots. Each dot is the scar left by the severance of a vascular bundle passing from the stem into the leaf. The number and arrangement of these **bundle-scars** are useful characters in identifying trees and shrubs in winter.

More or less distinct, long or short linear scars project laterally from the upper angles of a leaf-scar in some species; these are **stipule-scars** (see *Fagus*, p. 114; *Magnolia*, p. 142).

Lenticels are small, circular or elongated structures which dot the otherwise even surface of the twig. In most species they become inconspicuous after a few years, but in others they elongate laterally, giving a characteristic appearance to the bark, as in cherry and birch.

Branchlets or twigs, and branches as well, sometime bear pungent processes; that is, they are "thorny." But thorn cannot correctly be used for all such processes, for some are prickles. The fundamental difference is the presence or absence of vascular tissue. A **thorn** is a small, sharp-pointed, usually leafless stem (*Crataegus*, p. 175) or a greatly modified leaf (*Berberis*) or stipule (*Robinia*, p. 228), all of which contain vascular tissue. It has been suggested that **spine** should be used for the modified leaf structures and that thorn be reserved for modified stems. However, the terms **spine-tipped** or **spinescent** branchlet (as of *Pyrus coronaria*, p. 163) are firmly established and cannot readily be replaced. Such branchlets frequently bear leaves. A **prickle** is a sharp outgrowth of epidermis or cortex (as in *Aralia*, p. 276, *Rubus*, p. 204, *Rosa*, p. 212, and *Smilax*, p. 70). Rose "thorns" are really prickles.

If a branchlet or twig is cut lengthwise, a series of unlike tissues are disclosed. In the center, is a soft region of **pith.** The pith may be **continuous** (longitudinally as in Fig. 1, A), **chambered** (with cavities separated by thin plates, as in *Juglans*, p. 92), or **interrupted** at the nodes by woody diaphragms (as in *Vitis*, p. 264). Surrounding the

---

EXPLANATION OF FIGURE 1

FIG. 1.—Winter-twigs of Ailanthus, *A*, with alternate leaf-scars, and buckeye, *B*, with opposite leaf-scars.

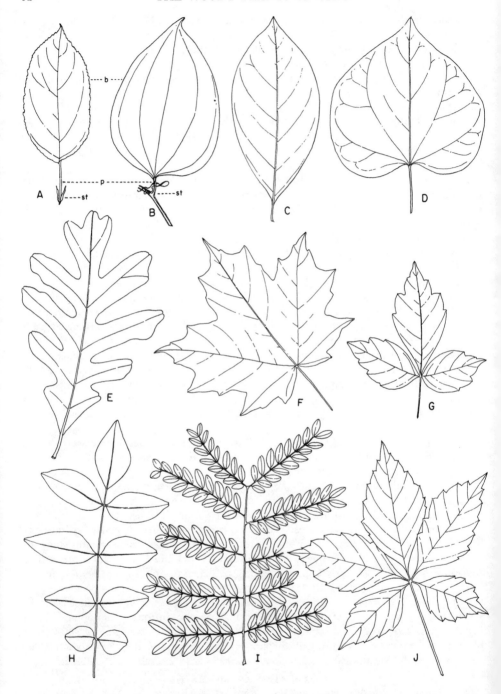

pith is a hard cylinder, the **wood-cylinder**; and surrounding it is a very thin cylinder, the **cambium**, which is only one or a few cells thick. A cylinder of **bark** surrounds these. The bark is made up of a number of tissues: adjacent to the cambium is a cylinder of phloem; then (if the twig is young) there are cylinders of pericycle, cortical chlorenchyma, cork, and epidermis. The **vascular cylinder** is the wood-cylinder and phloem-cylinder together. Within a few years the outer layers of bark slough off, until finally the bark on a tree-trunk consists only of phloem and cork in layers that develop annually. The thickness of the bark varies greatly in different species. If the bark is torn from a tree, an essential part of the conducting (vascular) system is destroyed. In young stems **vascular bundles** branch off from the vascular cylinder and enter leaves where they are known as **veins**.

A "complete leaf" consists of blade, petiole, and stipules (Fig. 2, A, B). The **petiole**, which is largely of vascular tissue, may be long or short; or it may be absent, in which case the leaf is **sessile**. In some leaves there is, at the base of the petiole, a pair of appendages, the **stipules**, which may be persistent (as in *Rosa*, p. 212) or early deciduous (as in many willows). Stipules may be large or small, or variously modified (the tendrils of *Smilax*, p. 70, the stipular spines of *Robinia*, p. 228). Within the blade and arising as branches of the vascular bundle or bundles of the petiole are the **veins**. The type of branching and the course of the principal veins of a leaf are useful characters in identification. With few exceptions, leaves of our woody plants are either **pinnately** or **palmately** veined (Fig. 2, C, D). **Dichotomous** or **forked venation** is illustrated by the Asiatic *Ginkgo* (and by maidenhair fern); **parallel venation** is characteristic of most monocots (*Yucca* and *Arundinaria*, p. 68, among our woody plants).

Leaves are either simple or compound. A **simple leaf** consists of a single blade which is uncut or variously parted but in which all parts are continuous (Fig. 2, C-F). In a **compound leaf**, the blade is divided into distinct parts, the **leaflets**, which are separately attached. Compound leaves may be **pinnately compound** or **pinnate** (Fig. 2, H), in which case the leaflets are attached to a central axis or **rachis**; or they may be **palmately compound** or **digitate** (Fig. 2, J), with the leaflets all attached to the end of the petiole. Leaves with only three leaflets are **ternate** or **3-foliate** (Fig. 2, G). Pinnate leaves may be again divided, **bipinnate** (*Gymnocladus*, p. 224, and *Gleditsia*, Fig. 2, I). Pinnate leaves are **odd-pinnate** if a terminal leaflet is present, as in *Ailanthus* (p. 232) or the sumacs (p. 234); **abruptly pinnate** if there is no single terminal leaflet, as in honey-locust (Fig. 2, I, and p. 226).

Leaf-shape and the shape of the apex and the base of a leaf are sometimes specific characters. The leaf-shapes most commonly mentioned are illustrated diagrammatically (Fig. 3). Intermediate forms are designated by compounding the terms for standard outlines; that is, ovate-lanceolate is broader than typical lanceolate, and so on. The forms of apexes and bases of leaves are also illustrated diagrammatically (Fig. 3). **Leaf-margins** may be entire, toothed, or lobed. The terms applied to types of leaf-margins are best defined by illustrations (Fig. 3). Again, intermediate forms may be described by

EXPLANATION OF FIGURE 2

Fig. 2.—Types of leaves: *A*. Leaf of apple, and *B*, of Smilax showing blade (*b*), petiole (*p*), and stipules (*st*), those of Smilax modified into tendrils. *C*. Pinnately veined entire leaf of spice-bush. *D*. Palmately veined entire leaf of redbud. *E*. Pinnately lobed leaf of white oak. *F*. Palmately lobed leaf of sugar maple. *G*. Ternate or 3-foliate leaf of fragrant sumac. *H*. Pinnate or pinnately compound leaf of prickly-ash. *I*. Abruptly bipinnate leaf of honey locust. *J*. Palmately compound or digitate leaf of Virginia creeper.

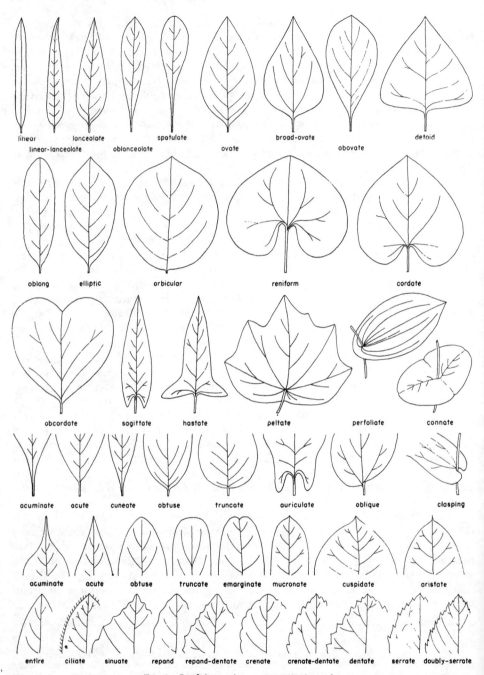

FIG. 3.—Leaf shapes, bases, apexes, and margins.

compounding the usual terms. Very small teeth may be described as denticulate, serrulate, crenulate—diminutive terms corresponding to dentate, serrate, and crenate.

The surfaces of leaves (and of stems also) differ greatly; this has resulted in the application of many terms, each of which has a special meaning. Leaves are **glabrous** when they are smooth and without hairs. A smooth leaf may be **glaucous,** that is, it may have a superficial, often removable, coating of a whitish, powdery, or waxy nature (like a Concord grape or blue plum). If the surface bears hairs, the leaf is **pubescent.** Many kinds of pubescence are seen and it is often convenient to use a separate term for each. These terms, which will be encountered in the keys and text, are defined in the Glossary; they cannot be illustrated by simple drawings. Among those most frequently used are canescent, hirsute, hispid, pilose, sericeous, tomentose, and villous. The Latinized form of these words will be noted among the specific names of plants—names which emphasize a character of the leaves: *Rhus glabra, Andromeda glaucophylla, Rubus pubescens, Populus canescens, Lonicera hirsuta, Robinia hispida, Salix sericea, Spiraea tomentosa.* Glandular or resinous dots, which usually glisten in sunlight, gland-tipped hairs, scalelike structures, and variously clustered and branched hairs are other leaf-surface features. Internal glands may, when a leaf containing them is held to the light, appear either as translucent dots because their cells are more transparent than chlorophyll-bearing cells, or as subsurface dots, because of tissue differences.

The **inflorescence** is the flowering part of a plant; it includes not only the flowers and subsequent fruits, but the stems and bracts as well; it is the entire flower-cluster. The stems of individual flowers of the inflorescence are **pedicels;** the stem which bears the inflorescence, or a solitary flower if there are no branches, is the **peduncle.** The simplest inflorescences are those in which a single flower is borne at the end of the stem or peduncle, as in *Liriodendron* (p. 146), or in which a single one or a few are borne from the axils of leaves of vegetative stems, as in *Nemopanthus* (p. 240), *Diospyros* (p. 304), and some species of *Lonicera* (p. 322).

Branched inflorescences are of two principal types, indeterminate and determinate. Mixed inflorescences also occur. In the **indeterminate inflorescence,** the flowers come from the axils of leaves or bracts, the lowest flowers on the stem opening first, the higher successively later; thus the axis of the inflorescence, often called the **rachis,** continues to elongate. This type includes the raceme (*Ribes,* p. 152; *Amelanchier,* p. 171), the corymb (*Physocarpus,* p. 163; *Kalmia,* p. 292), the umbel (*Smilax,* p. 70), the spike, the head (*Platanus,* p. 160; *Cephalanthus,* p. 316), the spadix (not represented by any of our woody species), and the catkin or ament (*Betula,* p. 108; *Alnus,* p. 111). In the **determinate inflorescence,** the terminal flower opens first; thus further elongation of the stem is not possible. Included here are the cyme, which may be simple (*Diervilla,* p. 316) or complex (*Hydrangea,* p. 152), the fascicle, the glomerule, and the oblique inflorescence or scorpioid cyme (characteristic of the Boraginaceae). In form, the cyme resembles the corymb, and cannot readily be distinguished from it except at flowering time. The fascicle is a crowded cyme; the glomerule resembles a head, and is best distinguished from it at flowering time. Development in the indeterminate type is centripetal; this is well illustrated by corymbs and umbels where the marginal flowers open first. In the determinate type development is centrifugal, as illustrated by the cyme (which is similar in shape to the corymb) where the central flowers open first. The **mixed inflorescence** is well illustrated by many members of the Compositae: *Liatris,* for example, in which the terminal head of flowers develops first—so that the inflorescence as a whole appears determinate—but the individual heads are centripetal. *Aesculus* also illustrates the mixed type, in which the primary branches are indeterminate, and the flowering of these branches is determinate. Compound flower clusters are often seen—compound umbels, compound racemes, compound corymbs, and so on. A **panicle** is a compound inflorescence of a racemose type; this term is usually applied to any loose much-branched flower cluster that is longer than wide (*Spiraea*, p. 163; *Rhus*, p. 234).

Although floral characters are not emphasized in this book, either in the keys or the text, it is well to have in mind the principal features. Observation of variation in shape and floral pattern and of features related to pollination add greatly to the interest of field studies.

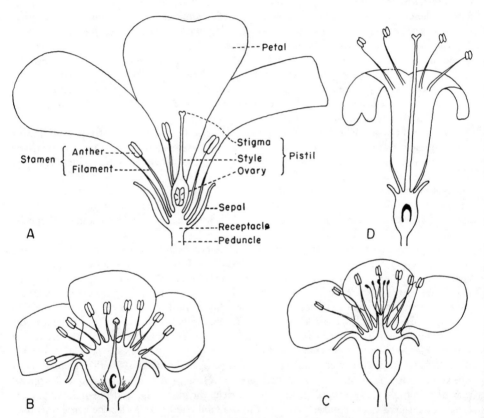

FIG. 4.—Flower-parts and their arrangement: *A*, hypogynous; *B*, *C*, perigynous; *D*, epigynous.

A **complete flower** consists of calyx, corolla, stamens, and pistil or pistils (Fig. 4, *A*). The calyx and corolla together constitute the perianth. An **incomplete flower** lacks one or more of these parts. If both stamens and pistils are present, the flower is **perfect**; if either is absent, the flower is imperfect and either **staminate** or **pistillate**. If both staminate and pistillate flowers are borne on the same plant, it is **monoecious** (*Carya*, *Fagus*); if on different plants, **dioecious** (*Salix*). The floral parts may be so arranged that the flower appears to be radially symmetric or regular, as is an apple flower, or bi-laterally symmetric and irregular, as is the flower of black locust. The calyx is composed of sepals, which may be separate or united, in which case we refer to calyx-lobes. The corolla is composed of petals which may be separate or united. The stamens, which may be few or numerous, are each made up of a filament and an anther, the latter being the pollen-bearing organ. The pistil, which is centrally placed, is composed of carpels. The hollow lower part is the ovary (ovulary);* the long or short slender column surmounting it is the style, with the stigma at the summit. A flower may contain more than one pistil; the pistils may be simple (of one carpel) or compound (of several carpels). The arrangement of the parts of the flower may be spiral (*Magnolia*) or cyclic.

The floral organs may be attached separately to the **receptacle**, which is the slightly enlarged top of the peduncle or pedicel (Fig. 4, *A*). In *Liriodendron*, the receptacle is greatly elongated, and may, after all floral organs have fallen, persist as a slender spire on which the scars of sepals, petals, stamens, and carpels are readily seen. In *Prunus* and comparable forms, a floral cup develops because of the union of the bases of sepals, petals, and stamens, and/or the upward growth of the receptacle. These outer cycles then appear to arise from the rim of the cup (Fig. 4, *B*); such a flower is **perigynous.** In the **epigynous** flower, the perianth and stamens appear to arise from the summit of an inferior ovary because the basal parts of the perianth are adnate to the ovary (Fig. 4, *D*); in fruit, the calyx-lobes may be persistent at the apex of the fruit, as in *Diervilla* (p. 316) and *Mitchella* (p. 316). If the receptacle is enlarged below the calyx, this enlarged structure is called a **hypanthium.** However, the difference between a hypanthium and a calyx-tube is not apparent externally, and the enlargements below the perianth in *B* and *C* are often referred to as hypanthium.

Botanically, a **fruit** is a ripened ovary, with or without accessory parts. It may be dry or fleshy; it may result from the growth of one, several, or many ovaries of a single flower (*Rubus*) or of many flowers (*Morus*); upon ripening, it may split (dehisce) or remain intact (indehiscent). In some instances, more than one fruit can develop from a single flower, as in *Asimina*: a papaw results from the ripening of one of the pistils of the papaw flower.

Many types of fruit are distinguished. Because the fruit, immature or mature, can be found on woody plants for a considerable period of time, recognition of fruit-types is desirable as a means of identifying genera.

Several types of fleshy fruits are recognized. An **aggregate fruit** is made up of the matured ovaries of a number of pistils of a single flower (*Rubus*, p. 208, *Rosa*, p. 212, *Calycanthus*, p. 146) with or without accessory parts. **Multiple fruits** are formed from the ovaries of many closely clustered flowers (*Morus*, *Maclura*). The **pome** is a fruit in which the part developing from the ovary is relatively small (such as the core of an apple) and the fleshy part is developed from the hypanthium or floral cup (Fig. 4, *C*). A **drupe** or stone fruit usually contains only one seed surrounded by a hard, stony layer or endocarp and an exterior fleshy layer or pericarp (*Prunus*, *Cornus*, *Viburnum*). Small drupes usually look like berries, and may be described as berry-like. A **berry** is a fleshy fruit developed from a single ovary, with from one to many seeds embedded in the pulp (*Ribes*, *Vitis*, *Vaccinium*, *Diospyros*). Some berries are simple fruits consisting only of developed

---

*All of our manuals refer to that part of the pistil that contains the ovules as the ovary. It has been suggested that since the ovary in animals is an egg-bearing organ, the ovule-bearing organ in plants should be called an ovulary (see Transeau, Sampson, and Tiffany, 1940, p. 354).

ovary (*Vitis*); others are complex fruits (*Vaccinium*) in which hypanthium and ovary make up the berry; in the latter (p. 300), the calyx-lobes may be seen at the apex of the berry.

Dry fruits, fruits which have no flesh or pulp, may be dehiscent or indehiscent. The **achene** (akene) is the commonest type of one-seeded, indehiscent dry fruit, but it is poorly represented among the woody plants of Ohio; the seedlike bodies in a rose hip are achenes, but in this case they are attached to a fleshy hypanthium. The **caryopsis** or **grain** is characteristic of the Gramineae. A **samara** is a winged fruit—dry, one-seeded, and indehiscent like an achene, but furnished with a wing or wings which are extensions of the outer wall of the ovary (*Ulmus*, p. 134, *Acer*, p. 246, *Fraxinus*, p. 308). The samaras of maple (p. 250) are commonly called keys. A **nut** is a dry, indehiscent fruit, usually one-celled and one-seeded, with a hard, bony wall, and is often enclosed, or partly enclosed, in a covering that is sometimes almost as resistant as the ovary wall or shell (*Juglans, Carya*, p. 99, *Quercus*, p. 119, *Fagus*, p. 114, *Castanea*, p. 114).

Dehiscent fruits may be formed from pistils with one carpel or with several carpels. The follicle and legume develop from one-carpel pistils; capsules, from pistils with two or more carpels. The **follicle** (of one carpel) splits along one side, exposing the few (rarely one) to many seeds; two or more follicles may result from a single flower (*Physocarpus*, p. 163, *Magnolia*, p. 144) in which case the one- or two-seeded follicles are grouped in an aggregate. The **legume** or true pod, formed from a simple pistil, splits along opposite sides (*Robinia*, p. 228, *Cercis*, p. 226, and the garden pea). This type of fruit (sometimes modified) is characteristic of the Leguminosae. The dry, dehiscent fruits of compound pistils (pistils with two or more carpels) are **capsules**. The splits may be through either the middle of the carpel (*Hamamelis*, p. 157, *Chimaphila*, p. 286, *Chamaedaphne, Oxydendrum*, p. 292, *Paulownia, Catalpa*) or along the partitions or septa (*Hypericum*, p. 272, *Rhododendron*, p. 288, *Kalmia*, p. 292, *Bignonia capreolata*, p. 314). If through the middle of the carpel, the split opens into a "cell" of the capsule, which then is a loculicidal capsule; if along the partitions, the capsule is divided into its constituent carpels and is septicidal. A few capsules dehisce by terminal pores. The silique and silicle (not represented by any of our woody species) are characteristic of the Cruciferae.

VEGETATION OF OHIO
AND
CORRELATION WITH ENVIRONMENT

"The State of Ohio, containing about 40,000 square miles, was once a magnificent hardwood forest. The forest types, thanks to the records of early surveyors, have been largely mapped. Yet it is almost impossible to form an adequate picture, from any surviving records, of the appearance of that forest. The state has its full share of memorials—statues, libraries, institutions; some useful, some not; some beautiful, many ugly. But somehow it never occurred to anyone to set aside a square mile, much less a township six miles square, of primeval vegetation for future generations to see and enjoy. Yet this could have been done for less than the cost of a single pile of stone of dubious artistic and cultural merit" (Sears, 1953).

More than thirty years ago, the map of the natural vegetation of Ohio referred to by Sears in the above paragraph was "prepared by transcribing the species of bearing trees recorded by the original surveyors who traversed the Territory and State of Ohio before the destruction of the virgin forest by the hands of white men" (Sears, 1925). In spite of the fact that some of the early land parcels were surveyed only in part or never surveyed systematically, the map clearly demonstrates the presence of more or less distinct vegetational areas (Fig. 5). Additional data obtained from notes in early geologic surveys and from local or county histories confirm the conclusions reached from the land-survey records, and give the location of and information about a number of more or less restricted types of habitats—wet and dry prairies, bogs, swamps, barrens, and oak openings. Under the title "treeless areas of natural vegetation in Ohio" these have been mapped, the locations based upon records made between 1750 and 1850 (Sears, 1926). In each, a few woody species occur. Herbarium specimens, some of them collected more than a hundred and twenty-five years ago, often supply records of plants that have long since become extinct in the places in which they were collected. They point unmistakably to the former existence of local habitats that have now been completely destroyed by man's activities— lumbering, farming, draining, and urban growth. Maps based on old records, and old herbarium specimens, emphasize the great changes which have taken place.

Old records and existing remnants of natural vegetation, together with the evidence of former vegetation that is given by disturbed and second-growth woodlands, make possible the preparation of fairly detailed maps of the natural vegetation of Ohio which preceded the white man's arrival. Such maps are now being prepared.

Ohio vegetation, originally mostly forest, is a part of the great Deciduous Forest of eastern North America (Fig. 6). Four of the major forest regions of the Deciduous Forest extend into the state. The Unglaciated Allegheny Plateau of eastern Ohio is within the limits of the **Mixed Mesophytic Forest** region, but since this area is in the northwestern angle of the region, its forests are not so complex as those farther south. Southern Ohio, west of the Appalachian escarpment and south of the Wisconsin glacial boundary, is a part of the **Western Mesophytic Forest** region, a region of diverse vegetation types. In extreme northeastern Ohio, a small lobe of the **Hemlock–White Pine–Northern Hard-**

19

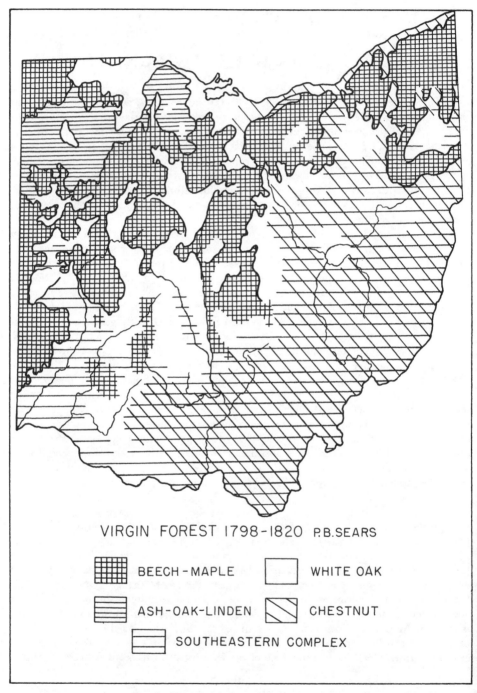

VIRGIN FOREST 1798-1820 P.B.SEARS

BEECH-MAPLE WHITE OAK

ASH-OAK-LINDEN CHESTNUT

SOUTHEASTERN COMPLEX

Fig. 5.—Map of original natural vegetation of Ohio.

wood **Forest** enters from the high plateau of western Pennsylvania. The rest of Ohio, more than half of the state, is in the **Beech–Maple Forest** region. Each of these regions has its own vegetational features; each contains a variety of unlike plant communities (Braun, 1950).

In the first two of these regions, rich mixed mesophytic forests may be seen, forests in which the canopy trees are beech, tuliptree, sweet buckeye, sugar maple, basswoods

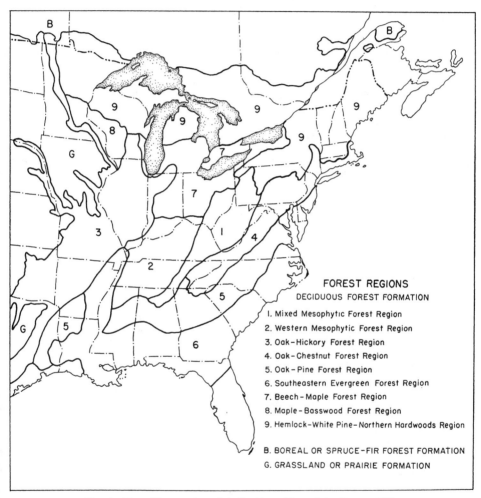

FOREST REGIONS
DECIDUOUS FOREST FORMATION

1. Mixed Mesophytic Forest Region
2. Western Mesophytic Forest Region
3. Oak–Hickory Forest Region
4. Oak–Chestnut Forest Region
5. Oak–Pine Forest Region
6. Southeastern Evergreen Forest Region
7. Beech–Maple Forest Region
8. Maple–Basswood Forest Region
9. Hemlock–White Pine–Northern Hardwoods Region

B. BOREAL OR SPRUCE-FIR FOREST FORMATION
G. GRASSLAND OR PRAIRIE FORMATION

FIG. 6.—Map of forest regions of the Deciduous Forest (after Braun, 1950).

(*Tilia heterophylla* and *T. neglecta*, but not *T. americana*), red oak, white oak, white ash, red elm, wild cherry, and many more, the first five usually the most abundant. In such a forest, understory trees and shrubs that require shade and rich soil occur. Because of the diverse topography of the Allegheny Plateau and its bordering hills on the west, many other forest types occur, from the most xeric pine forest of ridge tops to the mesic forests of ravines, and the flood-plain forests of alluvial flats; each has its own flora. In that part of the Western Mesophytic Forest region located in southwestern Ohio, forest types are fewer; however, the undissected Illinoian Till Plain, an area known as the "flats," is vegetationally unique in Ohio. It was originally occupied by extensive tracts of swamp forest, a forest composed of oaks (pin oak, swamp white oak, and white oak), red maple, sweet gum, and beech, with an admixture of hickories. Small semi-open and wetter spots afford favorable habitats for a variety of shrubs. The original forests are now largely replaced by farms and second-growth pin oak or pin oak and sweet gum stands (Braun, 1936).

The relatively large number of northern species which occur only in the northeastern part of Ohio (see distribution maps) are indicators of the Hemlock–White Pine–Northern Hardwood Forest. Its extent in Ohio is too small to illustrate well the features of the region. However, two of its more characteristic species, *Betula populifolia* and *Viburnum alnifolium*, occur here.

Within the area of the Beech–Maple Forest region, forest dominated by beech and sugar maple occupied the better sites. Extensive tracts of an elm–ash–maple type (black ash, white ash, elm, and red maple) occurred in depressions and intermorainal flats, reaching their best development in the Great Black Swamp area in northwestern Ohio. The morainal ridges and sandy beach-ridges of glacial Lake Maumee (the predecessor of Lake Erie) were favorable to drier types of forest that are dominated by oaks. The "treeless areas" of the old surveys, the bogs and prairies, increase the vegetational diversity of the Beech–Maple region.

Many of the species of our flora, both woody and herbaceous, range through much of the Deciduous Forest; some are confined (except for outlying stations) to one or two of the major forest regions of the East; their ranges can best be expressed in terms of forest regions, rather than by states. Others are more or less limited to particular habitats or are found in only one or two localities. Some species, such as *Quercus falcata* (p. 129) and *Bignonia capreolata* (p. 315), are at the margin of their more general range, which barely enters Ohio; others, such as *Magnolia macrophylla* (p. 145) and *Ledum groenlandicum* (p. 287), are widely separated from the area of general range (disjunct).

The range of species, whether wide or narrow, is a result of the influence of environmental factors, present and past.

All environmental factors—climatic or atmospheric, physiographic, soil or edaphic, and biotic, both present and past—affect the geographic distribution of species. Sometimes one factor, sometimes another, may be determinative; however, any factor acts in conjunction with all other environmental factors.

The influence of temperature and precipitation is clearly reflected in the distribution of some of our species. Temperature per se is less effective than length of frostless season, time of the last frost in spring and the first frost in autumn, and length and intensity of cold periods in winter; in a few instances an extremely low temperature is harmful. The accompanying maps (Fig. 7, *A*, *B*, *C*) show these temperature features, and may be correlated with certain of the distribution maps. The distribution patterns of many of our northern species, such as *Viburnum alnifolium* (p. 327), *Nemopanthus mucronata* (p. 239), *Chimaphila umbellata* var. *cisatlantica* (p. 285), and *Cornus canadensis* (p. 281), bear a resemblance to areas of low winter temperature, heavy snowfall, and a short growing season; those of certain of our southern species, such as *Phoradendron flavescens* (p. 139), *Ampelopsis cordata* (p. 259), *Chionanthus virginicus* (p. 312), and *Bignonia capreolata* (p. 315), to the area of higher winter temperature and a longer growing season.

Apparently, low winter temperature is harmful to only a few Ohio species whose northern limit of range may thus be determined. The flower-buds of cross-vine *Bignonia capreolata*, which reach considerable size the year before opening, are killed by intense cold; in Ohio this species is almost limited to the bluffs of the Ohio River, where it receives protection from cold dry north winds and where the close proximity of the river (and fogs) has an ameliorating effect. Here temperature is in part controlled by topography.

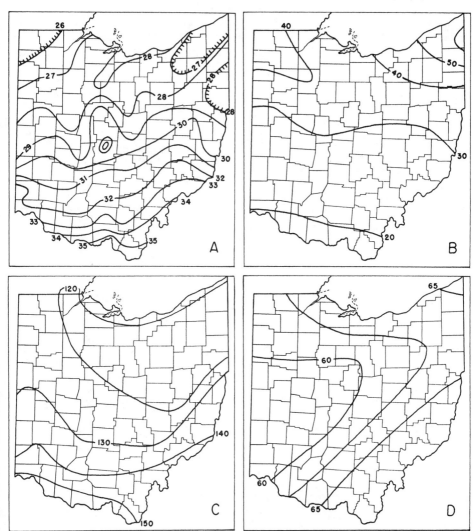

Fig. 7.—*A*. Average January temperature, ranging from below 26° F. in the northwest and below 27° F. in the northeast, to 33°–35° F. along the Ohio River. *B*. Average annual snowfall, in inches; note heavy snowfall of northeastern and northwestern Ohio. *C*. Length of available growing season, i.e. number of days without frost in 4 out of 5 years. Note shorter growing season of northern Ohio, except in the immediate vicinity of Lake Erie, and longer season along the Ohio River. *D*. Average relative humidity in July (8 P.M.) showing drier lobe over west-central Ohio. (*A*, from Ohio Department of Natural Resources, 1950; *B*, *C*, *D*, from Atlas of American Agriculture, 1918, 1922.)

The amount of annual precipitation and its seasonal distribution exert a controlling influence on major vegetation types.   Also important is the rate of evaporation in relation to precipitation.   Usable data for this factor are inadequate, but maps showing average relative humidity (Fig. 7, *D*) can be correlated with vegetation patterns.   The lower relative humidity in west-central Ohio is reflected in the abundance of prairie relics in that part of the state, and in the decreased frequency of the most mesophytic trees and shrubs (see map of *Liriodendron*, p. 147).   Unfortunately, distribution maps which merely show occurrence in a county do not give any idea of frequency.   The map of county distribution of beech (p. 113) shows it to be present in nearly every county, although beech occurred

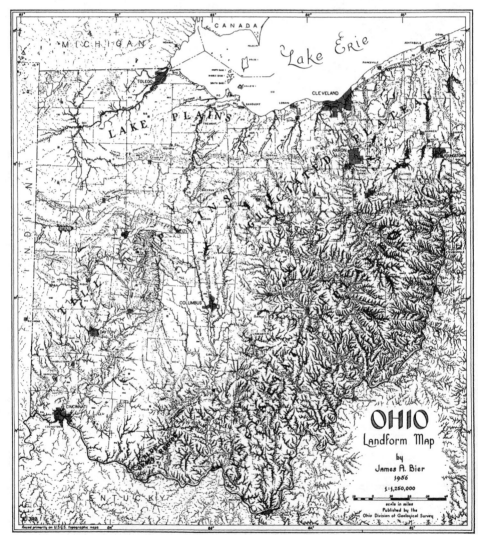

FIG. 8.—Relief map, showing principal physiographic areas of Ohio.

infrequently as a bearing tree in early surveys of the west-central counties (see Fig. 5). Such occurrences are made possible by the compensating influence of topography.

The physiography or geomorphology of the land surface, its topographic features and history, have a pronounced effect on the range of species. Within Ohio are a number of well-defined physiographic areas (Fig. 8), each of which has its distinctive topographic and soil features. Because land slope affects run-off and insolation, effective precipitation, temperature, and humidity are also affected. Areas of strong relief afford a greater variety of plant habitats than do areas of low relief. The number of native species in a hilly area is usually greater than in a flat area of equal size. Soils are related to physiographic areas, whether derived from underlying rock or from transported materials, for substratum in part determines topography.

The accompanying geologic map (Fig. 9) is designed to show the character of underlying rock—limestones, dolomites, and calcareous shales, sandstones, and non-calcareous shales. Within the glaciated areas, bedrock may be deeply covered by drift and exposed only in stream bluffs. In rugged areas of southeastern Ohio, strata of different ages, unlike in composition, may be exposed on a single slope. Narrow outcrops of calcareous shales or limestone may account for the presence in areas of prevailingly non-calcareous soils of species that are more widely distributed in calcareous soil areas (see map of *Crataegus mollis*, p. 197).

In Ohio, the Allegheny Plateau, a part of the Appalachian Plateau province of physiographers, is divided into the Glaciated Allegheny Plateau and the Unglaciated Allegheny Plateau. It is a hilly area—rugged in places; its western border, south of the glacial boundary, is higher and has more relief than the plateau to the east. Some of the most interesting and picturesque spots in Ohio are to be seen here; among them are the Hocking Valley parks in the Blackhand sandstone area, the Pottsville conglomerate area of Pike and Jackson counties, Fort Hill in southeastern Highland County, and the eastern half of Adams County. Each has its interesting species.

Soils in the Allegheny Plateau are prevailingly non-calcareous. In consequence, the ranges of species of non-calcareous soils often show a remarkable correlation with the Plateau (*Castanea dentata*, p. 113, found in every county of the Plateau; *Quercus montana*, p. 121; *Epigaea repens*, p. 295). However, local areas of calcareous soil do occur in the Allegheny Plateau; hence plants of calcareous soil may be local in the Allegheny Plateau section (*Crataegus mollis*, p. 197, and *Aesculus glabra*, p. 253). Some of the species of the Allegheny Plateau extend northward only as far as the glacial boundary, or fail to reach this limit (*Pinus rigida*, p. 63, *P. virginiana*, p. 63, *Oxydendrum*, p. 293). Others, of more or less general occurrence throughout the Allegheny Plateau, extend westward in acid soils of the Lake Plain (*Gaultheria procumbens*, p. 295, *Gaylussacia baccata*, p. 296); still others occur west of the Plateau in acid soils of the Illinoian Till Plain (*Smilax glauca*, p. 71, *Rhus copallina*, p. 235 and locally on leached hilltops (*Vaccinium stamineum*, p. 298).

Just west of the Appalachian Plateau escarpment in unglaciated southern Ohio is a very narrow northern extension from the "Knobs," a part of the Highland Rim section of the Interior Low Plateau; this is included in the Mixed Mesophytic Forest region (Knobs Border area). Adjacent to the Knobs on the west, is a triangular area projecting northward from the Bluegrass Section. Topography is more diverse in the Knobs than it is to the west; the soils on the knobs themselves are non-calcareous, and thus are favorable to the species of the Allegheny Plateau; the valleys and plateaus between the knobs are formed on limestones and dolomites and their soils are calcareous, as are those of the adjacent triangle of the Bluegrass Section.   Well-developed prairie communities containing a large number of prairie species and a number of southern or southwestern xerophytes, occur on

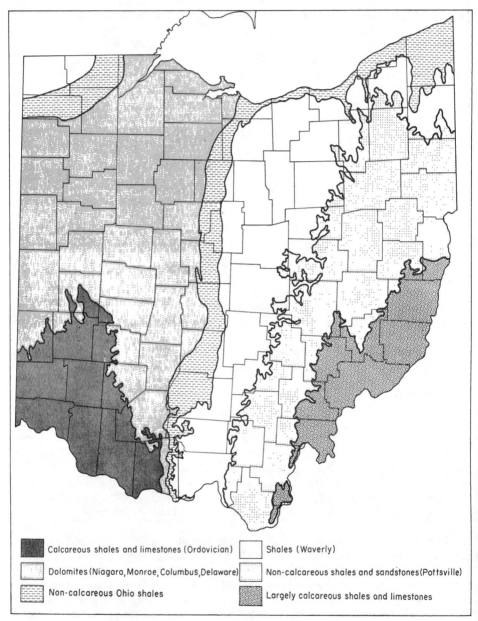

Calcareous shales and limestones (Ordovician)

Dolomites (Niagara, Monroe, Columbus, Delaware)

Non-calcareous Ohio shales

Shales (Waverly)

Non-calcareous shales and sandstones (Pottsville)

Largely calcareous shales and limestones

FIG. 9.—Map showing areas of calcareous and non-calcareous rock (after Stout, 1943).

the calcareous soils along this band (Braun, 1928).   In local spots, post oak and blackjack
oak are codominant and form an open stand—a woodland aspect not seen elsewhere; in
these situations a number of woody and herbaceous species of south-southwest range occur
(*Prunus americana* var. *lanata, Chionanthus virginicus,* and *Viburnum rufidulum*).

The Great Lakes Section of northern Ohio—which we often refer to as the Lake
Plains—is partly lacustrine, partly morainal, in origin.   Soils in some places are calcareous,
in others non-calcareous.   Some of the species common in the Allegheny Plateau extend

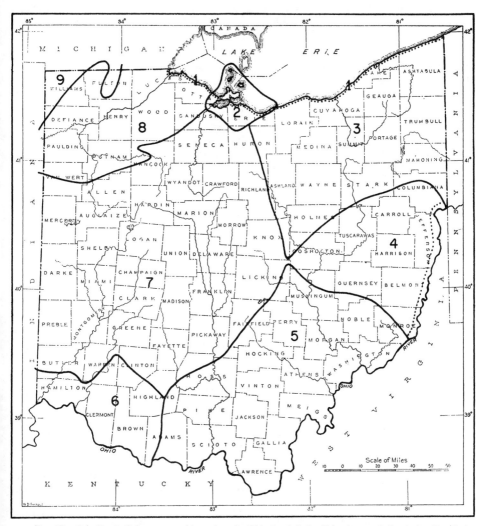

FIG. 10.—Schaffner's "phytogeographic regions in Ohio." *1.* Lake Erie shore. *2.* Sandusky Bay–Lake
Erie islands area. *3.* Northeastern or glaciated Allegheny Plateau region. *4.* Central nonglaciated area of the
Allegheny Plateau. *5.* Southeastern region or southern nonglaciated area of the Allegheny Plateau. *6.* The
Illinoian glaciated area, south of the Wisconsin glacier boundary; till plain. *7.* Miami region, mainly a calcareous,
glaciated region; till plain. *8.* Great Black Swamp area; lake plain. *9.* Northwestern or Williams County area;
till plain. (From Schaffner, 1932.)

westward in the Lake Plains. Marl prairies, sand dunes, the low limestone cliffs of the Lake Erie islands, and marshes, in addition to the flat areas and low morainal and beach ridges, afford a variety of habitats where some of our rarest species occur.

The Till Plains include most of Ohio west of the Allegheny Plateau escarpment and south of the Lake Plains. The underlying rock, usually deeply buried by drift, is mostly calcareous. The drift also is calcareous, although locally leached. Such local leached spots account for outliers of Allegheny Plateau species. In places streams have cut through the drift, exposing the bedrock beneath. Plants of calcareous banks here find suitable habitats (*Rhus aromatica*, p. 237, *Rhamnus lanceolata*, p. 257). In the main, the species of the area are calcareous-soil species.

Many species are indifferent to soil reaction and occur in all physiographic areas. Some are controlled by other soil factors, such as abundant water (*Cephalanthus*, many species of *Salix*) and are widely distributed, apparently without relation to physiographic areas.

In 1932, Schaffner, using ranges of groups of species, outlined nine phytogeographic regions in Ohio (Fig. 10). In part, these approximate some of the physiographic regions of the state. His "northeastern or glaciated Allegheny Plateau region" is extended far west of the Plateau boundary into the Till Plain so as to include outlying stations of characteristic trees of his region. His "central non-glaciated area of the Allegheny Plateau" is based on the absence of northern species and of certain species common farther south on the Plateau. The more complete records of distribution now available show that, although northern species and southern species are less frequent in this region than in those he outlined to the north and south, some are present; the area is transitional. The southwestern angle of his "southeastern region or southern non-glaciated area of the Allegheny Plateau" includes some of the area west of the Plateau which has a distinctive flora; it also includes a glaciated area north of Chillicothe. If used, the boundaries of the "regions" should be modified in accord with the more complete distribution records now available.

Glaciation has had a profound effect on the geographic range of many Ohio species. Its control is partly because of soil and topography, but some distributions cannot be explained by such present-day features. Glaciation in the Pleistocene, the "Ice Age," was not a single ice advance. Three or four great ice sheets (continental glaciers) advanced into Ohio (Fig. 11), separated by long intervals of mild climate. During periods of ice advance, southward migrations of plants and animals must have taken place; all life was eliminated in the glaciated area. With recession of the ice, northward migrations were possible.

The first ice advance, the Nebraskan, was about one million years ago. It is questionable whether or not it came far enough into Ohio to affect drainage. The second ice advance, the Kansan, about seven hundred thousand years ago, advanced far south. The deeply weathered drift south of the Illinoian glacial boundary in southwestern Ohio is referable to this age. The well-marked drainage changes in southern Ohio are ascribed to the damming of rivers by ice of Nebraskan age or, possibly, of Kansan age.

Relatively wide, now unoccupied valleys, are striking features of the landscape. Most prominent of these is the old Teays Valley and its tributaries. The Teays River was a large preglacial stream which included the present Kanawha River and its major tributary, the New River, whose headwaters are in the Blue Ridge of North Carolina; it also included the drainage basins of the Big and Little Sandy rivers and Tygarts Creek. In Ohio, it flowed north to Chillicothe in a course roughly parallel to, and just east of, the Scioto River, thence northwestward across Ohio, then across Indiana and Illinois to the Mississippi River (for details, see Fenneman, 1938, and Flint, 1947). Its flow was obstructed by an early ice advance. These events of the past would have no interest here were it not for the

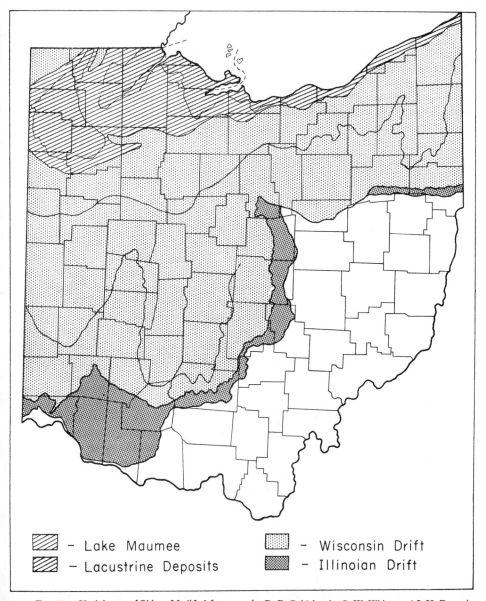

- Lake Maumee
- Lacustrine Deposits
- Wisconsin Drift
- Illinoian Drift

FIG. 11.—Glacial map of Ohio. Modified from map by R. P. Goldthwait, G. W. White, and J. H. Forsyth, to be published (in press) as separate map by U. S. Geological Survey. Lines within the area of Wisconsin drift indicate limits of readvance of ice and slight differences in clay-content of drift.

fact that the geographic distribution of a number of Ohio species seems intimately related to this drainage system (Transeau, 1941; Wolfe, 1942). The most interesting of these species is *Styrax grandifolia*, which must have reached its present site when the Teays River was in existence. *Pachistima canbyi*, found in two situations in the drainage of Ohio Brush Creek—in parts that were tributary to the Teays—occurs elsewhere (farther south) along this old drainage system; the nearest station is on the bluffs of Tygarts Creek in Kentucky (Massey, 1940; Braun, 1941).

The second ice advance, the Kansan, spread far south in Ohio, filling old valleys and perhaps further affecting drainage; or possibly, it was responsible for major drainage changes usually attributed to Nebraskan time.

After a long interglacial age, a third continental glacier, the Illinoian, advanced into Ohio, some three hundred thousand or more years ago. It reached farther south than the later advance, hence drift of Illinoian age is exposed south of the more recent drift. The ranges of some species in Ohio are closely correlated with the limits of this ice advance. Sweet buckeye (*Aesculus octandra*) is a common species of unglaciated southern Ohio; it extends only a mile or two into the area of Illinoian glaciation (map, p. 255); white basswood (*Tilia heterophylla*, p. 269) is comparable. Both are dominant and characteristic species in the Mixed Mesophytic Forest region. The Illinoian Till Plain of southwestern Ohio, which is flat, little dissected, and has a high water table and leached (acid) soils, affords suitable habitats for a variety of species, some of which are separated from the main area of their ranges in Ohio (*Spiraea tomentosa*, p. 162).

The last glacial age, the Wisconsin, was marked by a number of advances and partial recessions. Drift of this age covers a large area of Ohio. Its topography, still youthful, retains sites suitable for the persistence of northern species. Following the final ice retreat about ten thousand years ago, bogs developed in many depressions in the drift. Some of these have been obliterated by gradual filling, others have been destroyed by man. The distribution of all our bog species is dependent on the presence of bogs in the area of Wisconsin drift. Basswood (*Tilia americana*) is an abundant species in that part of Ohio covered by the Wisconsin drift; its range is essentially terminated on the south at the Wisconsin glacial boundary. Here, too, is the southern boundary of the Beech–Maple Forest region (see Fig. 6).

The physiographic areas of Ohio, which I have described briefly, have undergone long periods of erosional change. The Allegheny Plateau, with concordant levels of its hilltops, was in mid-Tertiary time a rolling plateau, the regional slopes of its southern half in general declining toward a former master stream, the Teays River. Its western border, because of more resistant rock, was rugged. Evidence indicates that a mixed forest covered the area, and that species for which ravine slopes and rugged topography were favorable became almost restricted to the western hilly border. Late in the Tertiary, uplift rejuvenated the streams and began the deepening of valleys. West of the Plateau, an extensive plain or low plateau developed; its features, except in the small triangular extension of the Bluegrass Section, are now everywhere modified by glaciation; its former valleys are to a great extent obliterated, but are known from well-borings.

All of the vegetation of the glaciated parts of Ohio is the result of migrations following recession of the glaciers. The vegetation of the unglaciated parts of Ohio has had a longer history, a history correlated with the events just outlined. It is believed that deciduous forest vegetation remained in southeastern Ohio throughout the Pleistocene, and that only near the ice border were vegetational changes pronounced. If this is true, certain mesophytes occurring locally in the rugged western border of the Allegheny Plateau may be called Tertiary relics. *Magnolia macrophylla* and *M. tripetala* are so interpreted: the former is limited to the very rugged Pottsville conglomerate area of Jackson County, the latter to local spots along the hilly border northward to Hocking County (p. 145). *Styrax* and *Pachistima*, whose ranges may be related to a former drainage pattern of comparable age, have already been mentioned. The termination of ranges of a number of species near the glacial boundary (even where favorable habitat conditions continue northward) may be explained by persistence during the Pleistocene (Ecol. Soc. symposium, 1951).

During warm interglacial ages, and in postglacial time, migrations have taken place. Certain species in the post oak area of the Bluegrass Section and in the Knobs Border area of Adams County, and the variability of redbud (*Cercis canadensis*) in this area caused by introgression from the more xeric *C. reniformis* of Texas and Oklahoma (Anderson, 1953; Braun, 1955), suggest an interglacial migration from a south-southwest direction.

Postglacial migrations are better known from information afforded by relic colonies and by buried plant fragments and fossil pollen in bogs. Relic boreal colonies bespeak northern species lagging behind, remaining in suitable habitats—north-facing cliffs and canyons, and bogs—during the great northward wave of migration following the ice retreat (*Acer spicatum* and *Sambucus pubens* in Clifton Gorge, *Taxus canadensis* in gorge of Baker Fork of Brush Creek, and so on; *Betula pumila* in Cedar Swamp and more northern bogs, *Andromeda glaucophylla*, *Ledum*, *Chamaedaphne*, *Vaccinium macrocarpon*). Relic colonies of prairie species are correlated with an eastward migration during the "Xerothermic Interval" or "climatic optimum" some three to five thousand years ago, at which time the "Prairie Peninsula" developed (Transeau, 1935). More recent increase in precipitation has been favorable to forest expansion, until now only small areas remain suitable to the prairie vegetation. The greater abundance of oaks and infrequency of the most mesophytic species in the area of the Prairie Peninsula are correlated with this dry interval, whose effects have not been obliterated in that part of Ohio which has a higher rate of evaporation in relation to rainfall. The xeric prairies and post oak woodlands of Adams County, if originating during an interglacial age, expanded again in the recent Xerothermic Interval, and are now being slowly curtailed in extent.

The very different distribution patterns illustrated by the woody plants of Ohio are correlated with environmental factors, with soil and soil moisture, with habitats related to the physiography and history of our area, and are everywhere under the influence of climate. Some knowledge of habitats in which the plants are growing is necessary for adequate interpretation of the many distribution patterns displayed.

REFERENCES

The following references describe in some detail certain areas of Ohio, or give more detailed information about topics included in this chapter. They are arranged chronologically to show the sequence of studies on the vegetation of Ohio.

1908. An ecological classification of the vegetation of Cedar Point. Jennings, O. E. Ohio Nat. 8: 291–340.

1912. An ecological study of Buckeye Lake. Detmers, F. Proc. Ohio Acad. Sci. Spec. Paper 19

1914. Botanical survey of the Sugar Grove region [the Hocking Valley parks area]. Griggs, R. F. Ohio Biol. Surv. Bull. 3.

1916. The physiographic ecology of the Cincinnati region. Braun, E. L. Ohio Biol. Surv. Bull. 7·

1922. The vegetational history of the Middle West. Gleason, H. A. Ann. Ass. Amer. Geogr. 12: 39–85; Contr. New York Bot. Garden 242.

1925. The natural vegetation of Ohio. Sears, P. B. Ohio Jour. Sci. **25**: 139–149; **26**: 128–146,. 213–231.

1928. Glacial and postglacial plant migrations indicated by relic colonies of southern Ohio. Braun, E. L. Ecol. **9**: 284–302.

1928. The vegetation of the Mineral Springs region of Adams County, Ohio. Braun, E. L. Ohio Biol. Surv. Bull. 30.

1928. Flora of the Oak Openings west of Toledo. Moseley, E. L. Proc. Ohio Acad. Sci. Spec. Paper 20.

1934. A history of Ohio's vegetation. Braun, E. L. Ohio Jour. Sci. **34**: 247–257.

1936. Forests of the Illinoian Till Plain of southwestern Ohio. Braun, E. L. Ecol. Monog. **6**: 89–149.

1941. Prehistoric factors in the development of the vegetation of Ohio. Transeau, E. N. Ohio Jour. Sci. **41**: 207–211.

1942. Species isolation and a proglacial lake in southern Ohio. Wolfe, J. N. Ohio Jour. Sci. **42**: 1–12.

1948. The flora of the Erie Islands. Core, E. L. Franz Theodore Stone Lab. Contr. 9.

1950. Deciduous Forests of Eastern North America. Braun, E. L. Blakiston, N. Y.

1951. The glacial border—climatic, soil, and biotic features. Symposium of Ecol. Soc. of Amer., Ohio Jour. Sci. **51**: 105–146, 153–167.

1959. The primeval forests of a periglacial area in the Allegheny Plateau (Vinton and Jackson counties). Beatley, J. C. Ohio Biol. Surv. Bull. N. S. 1.

KEYS TO GENERA

Synopsis of Keys

Key I

EVERGREEN AND HALF—EVERGREEN WOODY PLANTS

Leaves persistent, remaining green throughout the winter (evergreen), or, retaining green or bronzy leaves through mild winters and in protected situations (half-evergreen).

a. Evergreen, with persistent more or less coriaceous leaves.
  b. Leaves linear, needle-like or scale-like, not over 2 mm. broad.
    c. Leaves hoary pubescent, small and scale-like, appressed and overlapping; plant gray, low and bushy; fruit a small dry capsule.................HUDSONIA, p. 271
    cc. Leaves glabrous, green or glaucous, but not hoary; fruit a cone, or berry-like.
      d. Leaves elongate, 2, 3, or 5 in a cluster sheathed with brownish scales; cone-bearing trees..PINUS, p. 61
      dd. Leaves not in clusters.
        e. Leaves alternate.
          Leaves 4-sided, the angles apparent if leaves are twirled between the fingers
            PICEA, p. 67
          Leaves flat.
          Leaves green on both sides; ripe fruit red, berry-like, with pit at tip; shrubs..TAXUS, p. 58
          Leaves with white lines beneath; cone-bearing trees.
            Leaves decurrent on branchlets, on basal leaf-cushions; cones 1.5–2 cm. long..TSUGA, p. 59
            Leaves not decurrent; cones larger.
              Buds resinous; leaf-scars not raised; cone-scales deciduous, leaving persistent axis.................................ABIES, p. 59
              Buds not resinous; leaf-scars slightly raised; cones with exserted, apically 3-lobed bracts.................PSEUDOTSUGA, p. 67
        ee. Leaves opposite or whorled.
          Internodes flattened, sprays fan-like; leaves appressed, scale-like, in alternate pairs; fruit a small elongate cone....................THUJA, p. 65

33

Internodes not flattened and sprays not fan-like; leaves awl-like, stiff, sharp pointed and spreading, or scale-like and closely appressed; fruit blue, berry-like; trees or shrubs.............................JUNIPERUS, p. 66
bb. Leaves flat or somewhat rolled, not needle-like or scale-like.
  c. Leaves a foot or more in length, in large basal tufts; woody stem short and hidden by persistent leaves..YUCCA, p. 69
  cc. Leaves not as above.
    d. Leaves opposite.
      e. Plant parasitic on trees; leaves fleshy; fruit a white berry..................
          PHORADENDRON, p. 139
      ee. Plants not as above.
        Stems slender, creeping or trailing, scarcely woody; leaves small.
          Leaves entire, round-ovate; white flowers in short-peduncled pairs; fruit a red twinned berry............................MITCHELLA, p. 317
          Leaves crenate, round-oval; pink flowers on slender peduncles dividing into 2 divergent pedicels; fruit a small 1-seeded capsule..LINNAEA, p. 325
        Plants not creeping; woody shrubs.
          Large shrubs; leaves sometimes opposite, more often alternate, elliptical, 5–10 cm. long; capsules small, globose, in terminal cymes............
          KALMIA, p. 290
          Low shrubs, sometimes forming extensive patches; leaves linear to linear-oblong, 1–2 cm. long, serrulate; flower buds (as seen in winter) red, on short axillary pedicels.........................PACHISTIMA, p. 240
    dd. Leaves alternate.
      e. Stems scarcely woody, creeping or prostrate, or with short little-branched erect stems from subterranean creeping stems.
        Plants with few-leaved erect stems and horizontal subterranean stems.
          Leaves oval or obovate, 2–5 cm. long, mostly crowded near summit of stem; flowers axillary, short-pedicelled; fruit bright red, berry-like; leaves and fruit wintergreen flavored..GAULTHERIA procumbens p. 295
          Leaves lanceolate or oblanceolate, often sub-verticillate; flowers few, on an elongate terminal peduncle; fruit a 5-celled globose or depressed-globose capsule.............................CHIMAPHILA, p. 285
        Leafy stems creeping, prostrate, or some ascending.
          Leaves small, strongly to slightly revolute.
          Leaves glabrous, narrow (about 3 × as long as broad), pale and glaucous beneath; bog plants...........VACCINIUM (OXYCOCCUS), p. 303
          Leaves ovate or roundish, very little longer than broad, sparsely setose beneath; stems setose........GAULTHERIA (CHIOGENES), p. 296
          Leaves 2–7 cm. long, oval to suborbicular, more or less hirsute; flowers in short terminal racemes, the buds well developed in fall..EPIGAEA, p. 295
      ee. Stems definitely woody, trailing, ascending, or erect, much branched.
        Leaves spiny-toothed, mostly 5–6 cm. long; axillary fruit red, berry-like; tree
          ILEX opaca, p. 239
        Leaves not spiny-toothed; shrubs.
          Leaves conspicuously revolute; capsules in terminal clusters; bog shrubs.
          Leaves narrowly elliptic or oblong, densely covered beneath with matted brownish hairs.....................................LEDUM, p. 287
          Leaves narrowly oblong to linear, strongly revolute, lower surface whitened with minute puberulence...........ANDROMEDA, p. 291
         Leaves not revolute.
          Leaves bearing small peltate scales (widely spaced on upper surface, more or less contiguous on lower); capsules in leafy terminal racemes; bog shrub.....................CHAMAEDAPHNE, p. 293
         Leaves glabrous or nearly so; not bog shrubs.
          Not spiny; native.
            Leaves obovate, 1–3 cm. long; fruit red, berry-like; low trailing or ascending shrubs forming large patches.......................
            ARCTOSTAPHYLOS, p. 296
           Leaves larger, entire; tall erect shrubs with persisting 5-celled capsules in open terminal clusters.
            Leaves 1–2.5 dm. long, oblong to elongate-oblanceolate, pale beneath; the youngest slightly scurfy on prominent midrib; capsules 10–15 mm. long.........RHODODENDRON, p. 287

Leaves usually less than 1 dm. long, elliptic, green and glabrous
on both sides; capsules globose, 5–7 mm. in diam............
KALMIA, p. 290
Spiny; large densely branched exotic shrub; branchlets grayish
pubescent; leaves crenate-serrate; fruit orange-red, a small pome...
COTONEASTER, p. 199
aa. Half-evergreen, with thinner sometimes persistent leaves; leaves sometimes bronzy but not
dry and brown (the herbaceous *Cornus canadensis* may be looked for here).
  b. Bushy or climbing plants.
    c. Leaves alternate; stems armed with spines or prickles.
      Climbing, supported by paired tendrils born near base of petiole............
SMILAX glauca, S. hispida, p. 71
      Sprawling over the ground, or creeping. . RUBUS trivialis, RUBUS hispidus, p. 201
    cc. Leaves opposite; stems unarmed.
      Climbing.
      Leaves with 2 large leaflets with a terminal tendril between them............
BIGNONIA, p. 315
      Leaves simple.......................................LONICERA, p. 319
      Not climbing.
      Low, half-shrubby, the basal parts woody; stems 1–3 dm. long; larger leaves
1–2.5 cm. long, with axillary fascicles of small leaves; fruit a flattened 2-
valved capsule..ASCYRUM, p. 270
      Taller shrubs.
      Branchlets green, glabrous, somewhat 4-angled; leaves acute to acuminate,
almost sessile; fruit a broad lobed warty capsule, crimson when ripe......
EUONYMUS americanus, p. 241
      Branchlets gray, minutely puberulent; leaves obtuse to acute, on petioles
3–11 mm. long; fruit, black berries in panicles.......LIGUSTRUM, p. 312
bb. Plants with erect, hollow and conspicuously jointed canes and grass leaves............
ARUNDINARIA, p. 69

Key II

Summer Key

a. Leaves opposite.
  b. Leaves simple, entire..LONICERA, p. 319
  bb. Leaves compound.
    Leaves 3-foliate, leaflets toothed..........................CLEMATIS (excluded)
    Leaves with 2 large entire leaflets with a terminal tendril between them............
BIGNONIA, p. 315
    Leaves pinnate, with 9–11 ovate, acute, toothed leaflets; climbing by aerial roots.......
CAMPSIS, p. 315
aa. Leaves alternate.
  b. Leaves simple, entire, toothed, or lobed.
    c. Climbing by tendrils arising near base of petiole; leaves broad-oval to reniform,
rounded or cordate at base, the principal veins arcuate from base to apex; stems
armed with prickles...SMILAX, p. 71
    cc. Without this combination of characters.
      d. Stems scrambling or drooping, often with spinescent branchlets; leaves entire,
lanceolate to lance-ovate; flowers axillary, small, with 5-lobed corolla; fruit red;
introduced...LYCIUM, p. 313
      dd. Stems twining, or climbing by tendrils.
        e. Stems twining.
          Leaves ovate-oblong, finely serrate; flowers and fruit in terminal clusters;
flowers greenish, with 5 petals; fruit globular, orange, 3-valved, upon
splitting exposing the scarlet seeds.................CELASTRUS, p. 243
          Leaves ovate, acuminate, some with auricled or lobed base; flowers purple,
corolla 5-lobed; fruit a soft red berry.................SOLANUM, p. 313

Leaves more or less circular in outline, palmately veined, shallowly 5–7
lobed; petiole attached within the margin......MENISPERMUM, p. 141
ee. Stems climbing by tendrils which arise opposite the leaves; leaves palmately
veined..VITACEAE, p. 259
Bark shredding, without lenticels; pith brown; inflorescence paniculate;
leaves broad, coarsely toothed and often lobed, cordate......VITIS, p. 263
Bark close, lenticels abundant; pith white; inflorescence a forking cyme;
leaves thin, coarsely and sharply toothed, acuminate, not lobed..........
AMPELOPSIS, p. 259
bb. Leaves compound.
c. Stems armed with prickles, trailing or sprawling.
d. Leaves digitately 3- to 5-foliate....................RUBUS (dewberries), p. 205
dd. Leaves pinnately 3- to 5- (-7) foliate............................ROSA, p. 211
cc. Stems unarmed.
d. Leaves with 3 leaflets.
Stems twining (often widely trailing), young growth pilose; introduced.....
PUERARIA, p. 230
Stems not twining, climbing by aerial roots, or scrambling; in fall, with whitish
berries; *POISONOUS*.............................RHUS radicans, p. 238
dd. Leaves normally with more than 3 leaflets.
Leaves pinnate; stems twining..........................WISTERIA, p. 230
Leaves digitate, leaflets usually 5.
Climbing by means of tendrils, these often with adhesive disks; leaflets
coarsely toothed..........................PARTHENOCISSUS, p. 261
Twining; leaflets entire; introduced......................AKEBIA, p. 141

Key III

DECIDUOUS WOODY PLANTS WITH CLIMBING, TRAILING, OR
SPRAWLING BRANCHES

Winter Key

a. Leaf-scars opposite.
b. Stems soft and scarcely woody, freezing back in winter, finely ridged; bearing plumose
achenes in globular clusters.................................CLEMATIS (excluded)
bb. Stems woody.
Twigs coarse, light brown, often bearing patches of adventitious aerial roots below
the nodes; leaf-scars shield-shaped with 1 C-shaped bundle-scar; bushy to high-
climbing vines bearing elongate (1–2 dm.) 2-celled capsules.........CAMPSIS, p. 315
Twigs slender; leaf-scars small, crescent-shaped, on the ends of raised bases; bushy
or climbing berry-bearing shrubs with sprawling sometimes twining stems..........
LONICERA, p. 319
aa. Leaf-scars alternate.
b. Stems armed with spines or prickles, these sometimes few and widely spaced.
c. Stems trailing.
Leaf-scars torn and irregular, on raised persistent petiole-base......RUBUS, p. 199
Leaf-scars narrowly shallow U-shaped to almost linear, with 3 bundle-scars......
ROSA, p. 211
cc. Stems not trailing.
Climbing by means of paired tendrils born near base of petiole; stems terete or
angled...SMILAX, p. 71
Stems sprawling or scrambling, supported by other stems or brush.
Forming large fountain-like clumps; twigs with stout sharp detachable
prickles..ROSA setigera, p. 213
Scrambling or drooping; armed with spine-tipped twigs; introduced........
LYCIUM, p. 313
bb. Stems unarmed.
c. Stems twining or sprawling, without tendrils or aerial roots.*

---

*In addition to the genera given in this division of the key, *Akebia, Pueraria,* and *Wisteria* are twining vines
occasionally spreading from cultivation.

Scarcely woody twiners, often dying back in winter.
  Leaf-scars elliptic and slightly notched or sometimes V-shaped or narrowly
  U-shaped, then enclosing the lower of the superposed buds; bundle-scars 3
  or in 3 groups; fruit grape-like.................MENISPERMUM, p. 141
  Leaf-scars raised, half-round; bundle-scar 1, large; fruit in summer pulpy and
  bright red, shriveled remnants of panicles sometimes present.............
  SOLANUM dulcamara, p. 313
  Woody, often high-climbing vine bearing raceme-like clusters of orange capsules
  (fading in winter) splitting into 3 parts to expose the orange-red, fleshy-coated
  seeds; leaf-scars half-elliptic, flattened along upper edge; bundle-scar 1.........
  CELASTRUS, p. 243
  cc. Stems climbing by means of aerial roots or tendrils.
    Climbing by means of aerial roots; leaf-scars large, more or less triangular, bundle-
    scars numerous; terminal bud small, downy; loose axillary panicles of whitish
    berries; *POISONOUS*..............................RHUS radicans, p. 238
    Climbing by means of tendrils which are opposite the leaf-scars..VITACEAE, p. 259
      Tendrils often enlarged at tips into adhesive disks (these rarely present in the
      northern *P. inserta*); bark not shredding or flaking; pith white, continuous;
      high-climbing vines whose stems (at least the younger ones) usually adhere
      rather closely to supporting trees or other objects..PARTHENOCISSUS, p. 261
      Tendrils twining, and without terminal adhesive disks.
        Bark of older stems shredding into long strips; pith brown, interrupted at
        nodes; high-climbing often very large vines with hard wood...VITIS, p. 263
        Bark not shredding; branchlets with prominent lenticels; pith white, be-
        coming divided into thin plates.................AMPELOPSIS, p. 259

Key IV

DECIDUOUS TREES AND SHRUBS

Summer Key

KEY IV, A

Deciduous woody plants with compound opposite leaves.

a. Leaves digitate, with 5–7 leaflets; branchlets coarse....................AESCULUS, p. 251
aa. Leaves pinnate.
  b. Leaves with 3–5 leaflets.
    Tall shrub, often colonial, the branches dark with pale stripes; leaflets 3, serrulate, the
    terminal long-stalked.....................................STAPHYLEA, p. 244
    Tree, the branchlets green (or purple-red where exposed to sun); leaflets 3 or 5, with
    few coarse teeth or entire...............................ACER negundo, p. 251
  bb. Leaves with more than 5 leaflets (some leaves may have only 5).
    Tall trees; wood-cylinder of branchlets thick, pith moderate to small; fruit a more or
    less spatulate samara......................................FRAXINUS, p. 306
    Shrubs; wood-cylinder thin, pith moderate or large, stems weak.
      Flowers numerous in compound cymes, small, white, corolla deeply 5-lobed;
      small purple-black or red berries; pith large...............SAMBUCUS, p. 335
      Flowers few, trumpet-shaped, orange, in terminal clusters; fruit a large long 2-
      valved capsule; usually climbing by aerial roots...............CAMPSIS, p. 315

KEY IV, B

Deciduous woody plants with compound alternate leaves.

a. Leaves digitately 3-7-foliate.*
  b. Stems usually prickly; erect, arching (or trailing) shrubs.................RUBUS, p. 199
  bb. Commonly climbing by tendrils, but climbing habit may not be discerned in young plants
  PARTHENOCISSUS, p. 261

---

*Pinnate and digitate 3-foliate leaves may be difficult to distinguish, hence plants with 3-foliate leaves should
be looked for in more than one division of key.

aa. Leaves pinnately 3-many-foliate.
  b. Stems armed with spines or prickles.
    c. At least some of the leaves twice or 3 times compound.
      d. Tree, leaves once or twice abruptly pinnate (without terminal leaflet), 1–2 dm.
        long; leaflets numerous, very small; thorns branched, on trunk as well as branch-
        lets (rarely without thorns); fruit a large flat but curly pod..GLEDITSIA, p. 223
      dd. Large shrub or small tree; leaves very large, up to 1.5 m. long, twice pinnate;
        leaflets large (5–8 cm. long); leaf-rachis, branchlets, and trunk armed with stout
        prickles; small berries numerous in compound terminal panicle....ARALIA, p. 275
    cc. Leaves once pinnate.
      d. Leaves with foliaceous stipules adnate to petiole (tips free), leaflets 3–9 (–11),
        toothed; stems commonly prickly, irregularly so and/or with stout infrastipular
        prickles; shrubs...ROSA, p. 211
      dd. Without this combination of characters.
        Leaves mostly 3-foliate, occasionally some simple or 5-foliate, coarsely
        toothed, the linear-attenuate stipules inconspicuous; stems commonly
        prickly; erect, sprawling or depressed shrubs................RUBUS, p. 199
        Leaflets more numerous, entire or crenulate.
        Stems with stout detachable prickles sometimes paired at nodes; petiole
        and leaf-rachis often prickly; leaflets 5–11, entire or crenulate; inflores-
        cence axillary on 2-year old stems, pedicels short, flowers small, greenish,
        the 1-seeded globose capsules berry-like, with pitted surface, aromatic;
        shrub.....................................XANTHOXYLUM, p. 230
        Stems with stout or bristle-like stipular spines; leaflets 7–21, stipellate;
        petiole with pulvinus, i.e., basal part distinctly swollen; flowers large and
        showy in drooping racemes: fruit a flat pod; trees or shrubs............
                                    ROBINIA, p. 229
  bb. Stems not armed with spines or prickles (spineless individuals of *Gleditsia*, *Robinia*, and
    Rosa occur; see under *b* above).
    c. Leaves twice pinnate, 3–9 dm. long; leaflets mostly 3–4 cm. long, entire; fruit a heavy
      pod 1–1.5 dm. long, the seeds large; tree with coarse branchlets................
                                  GYMNOCLADUS, p. 223
    cc. Leaves once-pinnate.
      d. Leaves 3-foliate.
        Terminal leaflet long-stalked, margin entire or with few coarse teeth; axillary
        panicles of greenish flowers and later of whitish berries; low shrub or high-
        climbing by aerial roots; *POISONOUS*..............RHUS radicans, p. 238
        Terminal leaflet sessile or short-stalked.
        Leaflets coarsely crenate or crenate-dentate, more or less pubescent; flowers
        yellow, in short axillary spikes appearing with or before the leaves; fruit
        berry-like, red, densely hairy; diffuse shrub........RHUS aromatica, p. 237
        Leaflets entire, undulate, or serrulate, glabrous, pellucid-dotted; flowers in
        early summer, greenish, in compound terminal cymes; fruit a circular
        samara; large shrub..................................PTELEA, p. 231
      dd. Leaves with more than 3 leaflets.
        e. Leaflets entire.
          Tree, with smooth gray bark; leaflets 7–11, large, oval or ovate; petiole
          hollow at base; flowers white, large and showy, in panicled racemes......
                                    CLADRASTIS, p. 227
          Without this combination of characters.
          Trees or large often tree-like shrubs with coarse branchlets.
          Petiole with pluvinus; stems usually with stipular spines............
                                    ROBINIA, p. 229
          Petiole without pulvinus.
           Leaf-rachis winged; flowers and red fruit in compact terminal
           panicles.....................................RHUS copallina, p. 235
           Leaf-rachis not winged; flowers and white fruit in loose axillary
           panicles; *POISONOUS*......................RHUS vernix, p. 238
          Many-stemmed shrubs, with slender branchlets.
          Low shrub (3–10 dm.) with narrow silky-pubescent, revolute-margined
          leaflets; leaflets usually 5; flowers yellow, rose-type, 2 cm. across;
          fruit a head of small hairy achenes............POTENTILLA, p. 199

Loosely branched shrub, 2–4 m. tall, with locust-like glabrous to sparingly pubescent leaves; leaflets 11–27; flowers dark purple, in crowded peduncled spike-like racemes; fruit a small warty pod..............
                                                        AMORPHA, p. 227
ee. Leaflets toothed.
Tree with coarse twigs with large continuous colored pith; leaves large, 4–6 dm. long; leaflets 11–41, acuminate, coarsely few-toothed near base or some entire; veins running to teeth ending in gland just within margin; fruit a 2-winged twisted samara.........................AILANTHUS, p. 231
Without this combination of characters.
Pith chambered, with thin cross plates; leaflets (9–) 11–23; trees........
                                                        JUGLANS, p. 93
  Pith not chambered.
  Branchlets very tough; leaf-scars more or less shield-shaped; leaflets 5–11 (–15 in one sp.), the 3 terminal often the largest; nut-bearing trees...CARYA, p. 94
  Branchlets thick but weak (easily broken); leaf-scars U- or C-shaped, partly surrounding bud; leaflets 11–25 and sharply serrate, or fewer and with few teeth; inflorescence a compact terminal panicle; large or tree-like shrubs.....................RHUS (the sumacs), p. 233
  Branchlets moderate; leaf-scars linear to crescent-shaped; leaflets 11–17, serrate; inflorescence a compound cyme; small trees or shrubs; northern...............................PYRUS (SORBUS), p. 169
  Scarcely woody little-branched shrub, 1–2 m. tall; leaf-scars half-round; leaflets 13–21, sharply serrate, long acuminate; inflorescence a terminal panicle; introduced shrub.................SORBARIA, p. 163

KEY IV, C

Deciduous woody plants with simple (entire, toothed, or lobed) opposite or whorled leaves.

  1. Leaves entire (margin sometimes irregularly undulate)

a. Leaves very large and broad (to 2–4 dm. long), broad-ovate to rotund, ⅔ or more as wide as long; branchlets stout; flowers large and showy, in terminal panicles; introduced trees.
  b. Leaves usually 3 at a node, sometimes slightly lobed; flowers white spotted with yellow and purple; fruit a long cylindric 2-valved capsule..................CATALPA, p. 317
  bb. Leaves 2 at a node, sometimes angled but scarcely lobed; flowers violet, fragrant; fruit a large ovoid 2-celled capsule.................................PAULOWNIA, p. 313
aa. Leaves smaller, rarely more than 1 dm. long, not broad-ovate; branchlets slender to moderate.
  b. Leaves dotted with translucent internal glands, sessile or short (1–5 mm.) petioled, only the midrib readily visible without lens.
    Low diffuse shrub, 1–3 dm. tall; leaves small (1–2.5 cm. long), usually oblanceolate to spatulate, the dots visible on both surfaces.....................ASCYRUM, p. 270
    Much branched shrubs, 3–15 dm. tall; leaves linear-oblong to oblanceolate, 2–10 cm. long, the dots visible only on upper surface..................HYPERICUM, p. 271
  bb. Leaves not dotted with internal glands.
    Leaves densely silvery-downy and scurfy with rusty scales on under surface........
                                                        SHEPHERDIA, p. 274
    Under surface of leaves not silvery and scurfy.
    Branchlets somewhat 4-angled, the ridges forming the angles extending downward from petiole-margin; leaves mostly elliptic or oval, acute at both ends, 8–20 cm. long...CHIONANTHUS, p. 312
    Branchlets not 4-angled.
    Leaves usually 3 at a node, sometimes opposite, lustrous above, oval to lanceolate, acute or acuminate; flowers (white) and fruit in long-peduncled compact globose terminal or terminal and axillary heads..................
                                                        CEPHALANTHUS, p. 318
    Without this combination of characters.
    Leaves and branchlets aromatic when crushed.....CALYCANTHUS, p. 147
    Not aromatic.
    Leaves small, 1.5–4 (–5) cm. long, broad (½ to as wide as long), pubescent beneath; flowers and berries in short-peduncled axillary or axillary and terminal clusters.............SYMPHORICARPOS, p. 324

Leaves small, 3–6 cm. long, mostly less than ½ as wide as long, glabrous beneath; 4-parted white flowers and black berry-like fruit in terminal panicles.....................................LIGUSTRUM, p. 312

Leaves mostly larger, with slender, usually long petioles; leaves more or less pubescent beneath, with some 2-armed usually appressed hairs; inflorescence a terminal cyme, the 4-parted flowers white, the small drupes white or blue; *or*, in 2 species, a dense cluster, the greenish-yellow flowers surrounded by 4 large white bracts, the drupes red...... CORNUS, p. 277

Leaves various, some small, some 10 cm. long, glabrous or pubescent, sometimes ciliate; inflorescence terminal, often with a pair of connate leaves below, or axillary, flowers then paired on long or short peduncle, corolla 2-lipped or 5-lobed; berries red (or black).....LONICERA, p. 319

2. *Leaves toothed* or *lobed*

(Plants with leaf-margins irregularly slightly undulate, or with some leaves somewhat angled rather than lobed, should be looked for under *1*.)

a. Leaves 3–5-lobed (lobes occasionally almost obsolete in some 3-lobed species); palmately veined.
  b. Trees or large shrubs; flowers red or reddish in compact sessile clusters, yellow with elongate pendent pedicels, or in peduncled corymbs, panicles, or racemes; fruit a paired samara.. ACER, p. 244
  bb. Shrubs, the leaves dotted beneath or with petiolar glands; flowers white, in erect terminal compound cymes; fruit a red or purple-black drupe.................................. VIBURNUM acerifolium, V. trilobum, p. 325
aa. Leaves finely or coarsely toothed, not lobed; shrubs, sometimes tree-like.
  b. Leaves with 1–3 principal lateral veins on each side from basal half of midrib.
    Leaves glabrous, finely serrate with gland-tipped teeth, the 2–3 principal lateral veins prominent, curving strongly almost to leaf apex; some stems end in spine-like tips; flowers small, greenish, in axillary clusters; fruit berry-like, black................. RHAMNUS catharticus, p. 257
    Leaves pubescent beneath, at least on veins or in basal vein-angles, coarsely few-toothed, usually only 1 pair of lateral veins prominent; flowers large, with 4 white petals; fruit a 4-valved capsule........................PHILADELPHUS, p. 151
  bb. Leaves with lateral veins more equally spaced.
    Leaves sharply doubly serrate, ciliate; petiole short, 2–7 mm.; flowers solitary, terminal, white, with 4 sepals (and 4 alternate bracts), 4 petals, numerous stamens, and 4 carpels which develop into black drupes..................RHODOTYPOS, p. 211
    Without this combination of characters.
      Branchlets green (at least first year) or reddish on sunny side; leaves finely toothed.
        Branchlets quadrangular; leaves ovate-lanceolate, elliptic, to obovate, serrulate or crenulate, sessile or with short petiole (or to 2 cm., but less than ⅙ blade length); flowers small, flat, axillary, greenish to red-purple, with 4 or 5 petals; fruit a red, smooth or tuberculate lobed capsule...EUONYMUS, p. 241
        Branchlets not quadrangular; leaves ovate to lanceolate, crenate-serrate and ciliate; petioles 3–8 mm. long, ciliate; flowers tubular, yellow turning red, in few-flowered terminal and axillary cymes; fruit a slender pointed capsule.... DIERVILLA, p. 319
      Branchlets not green; inflorescence a terminal compound cyme.
        Soft-wooded, many-stemmed shrub usually about 1 m. tall; leaves large ovate or oblong to nearly orbicular, with short-acuminate tip, serrate to dentate, long petioled; cyme often with 2 kinds of flowers—numerous small fertile and few large sterile; fruit a small 2-celled capsule........HYDRANGEA, p. 151
        Hard-wooded, large or tree-like shrubs; leaves various; flowers white with 5-lobed corolla, in 1 sp. with large marginal sterile flowers; fruit a red to dark purple drupe.....................................VIBURNUM, p. 325

KEY IV, D

Deciduous woody plants with simple alternate leaves.

1. Leaves entire (occasionally a few leaves slightly toothed or angled)

a. Leaves narrowly linear or filiform, needle-like but soft; fruit a cone.
  b. Leaves spirally arranged on shoots of the season, in bunches of 20–50 on very short lateral branchlets of older wood; cones ovoid, 1–2 cm. long, the scales thin.LARIX, p. 61

bb. Leaves widely spreading in one plane except when young, the branchlets resembling finely pinnate leaves; leaves and part of branchlets deciduous; cones globular, about 2.5 cm. in diam., the scales thick, angular, somewhat peltate.................TAXODIUM, p. 67

aa. Leaves not as above; fruit not a cone.

   b. Leaves broad-ovate to suborbicular, palmately veined, open cordate at base, the slender petiole swollen just below blade; flowers pink, in lateral clusters; fruit a flat pod........
                              CERCIS, p. 225

   bb. Leaves not broad-cordate; pinnately veined.

      c. Leaves large, more than 1.5 dm. long.

         d. Stipule-scars encircling twig; leaves 1.5–10 dm. long; flowers terminal, large, white or greenish-yellow, with 6–9 petals; fruit cone-like...........MAGNOLIA, p. 143

        dd. Stipule-scars absent; leaves 1.5–3 dm. long; flowers lateral, maroon-red, before the leaves, with 6 petals; fruit pulpy, large.....................ASIMINA, p. 149

     cc. Leaves averaging less than 1.5 dm. long.

        d. Leaves abundantly resinous-dotted beneath, less so above; shrubs.

          e. Leaves elliptic to oblanceolate or obovate, entire or few toothed, aromatic; fruit lateral, berry-like, waxy-coated, pale gray, remaining in winter........
                           MYRICA, p. 91

          ee. Leaves oval, oblong-ovate, or oblong, obscurely short-ciliate; berries black, in summer...GAYLUSSACIA, p. 296

        dd. Leaves not resinous dotted.

          e. Leaves and branchlets aromatic, fragrant; flowers yellow, opening before or with the leaves.

            f. Tree; leaves ovate or oval to obovate, entire or with 1–3 lobes; branchlets green; flowers in clustered peduncled racemes which appear terminal until new shoot elongates............................SASSAFRAS, p. 149

           ff. Shrubs; larger leaves oblong-obovate; branchlets dark; flowers in almost sessile lateral clusters................................LINDERA, p. 150

          ee. Not aromatic fragrant, some strong-smelling.

            f. Midrib excurrent, leaves bristle-tipped; trees; fruit an acorn............
                QUERCUS imbricaria, Q. phellos, pp. 129

           ff. Leaves not bristle-tipped.

              g. Leaves silvery-scurfy beneath; flowers and berry-like fruit axillary; introduced shrubs.........................ELAEAGNUS, p. 274

              gg. Leaves not silvery-scurfy beneath.

                h. Branchlets armed with spines.

                  i. Tree, or large hedge-shrub (if pruned), armed with stout axillary spines; leaves glossy, ovate to oblong-lanceolate, long-acuminate; fruit globular, multiple, about 1 dm. in diam.....
                        MACLURA, p. 138

                  ii. Much branched shrub, 1–2 m. tall, armed with slender sharp usually simple spines subtending very short spur-branchlets with fascicled oblanceolate cuneate leaves.................
                      BERBERIS thunbergii, p. 141

              hh. Not armed with spines or prickles.

                i. Trees with deeply furrowed bark, the ridges broken into short lengths; leaves 8–15 cm. long; flowers axillary, appearing when leaves about half grown.

                  j. Leaves oval to obovate, mostly broadest above middle, sometimes with a few coarse teeth; small greenish-white flowers and blue ellipsoid drupes long-pedunculate........
                       NYSSA, p. 274

                  jj. Leaves ovate to oval, mostly broadest at or below middle; flowers yellow, short-stalked; fruit pinkish-orange, depressed globose, 2–3 cm. in diam., subtended by enlarged calyx...........................DIOSPYROS, p. 303

                ii. Shrubs, some tall, some low; leaves mostly smaller.

                  j. Young branchlets pubescent; leaves ciliate, pubescent beneath at least on midrib; flowers large, showy, orange or pink, in terminal clusters; capsules 5-celled, more or less cylindric.........RHODODENDRON (AZALEA), p. 287

                jj. Without this combination of characters.

k. Leaves broad, half or more as wide as long.
    Leaves crowded near ends of branchlets, long-petioled, usually arranged in a horizontal mosaic; lateral veins curving strongly toward apex; 4-petaled flowers and blue-black berry-like fruit in flat cymes
    CORNUS alternifolia, p. 281
    Leaves not crowded near ends of branchlets; inflorescence not a cyme.
        Petioles long, slender; leaves oval or oblong to obovate.
            Lateral veins prominent, not branched except near margin, there curving strongly and joining next vein; berry-like drupe black.............
            RHAMNUS frangula, p. 258
            Without this combination of characters.
                Leaves blunt at apex, mucronate or sometimes acute; red berry-like fruit axillary on long pedicels; northern bog shrub...............
                NEMOPANTHUS, p. 239
                Leaves rounded or emarginate at apex; inflorescence a much-branched panicle becoming plumy; introduced...... COTINUS, p. 233
        Petioles short, or leaves sessile.
            Shrub, tree-like in form; branchlets very flexible, each year's growth appearing to come from a socket; leaves broad-oval, oval-obovate, to rhombic; flowers yellow, tubular, before the leaves........................ DIRCA, p. 273
            Much-branched shrubs; branchlets not flexible; flowers small, white, greenish or reddish, open campanulate, cylindric, or broad urn-shaped...
            VACCINIUM, p. 297
            Very rare shrub; young branchlets densely pale stellate pubescent; leaves obovate, acute, densely pale stellate pubescent beneath.......
            STYRAX, p. 305
kk. Leaves narrow, less than $\frac{1}{2}$ as wide as long; shrubs (narrow-leaved forms of *Nemopanthus* might be looked for here; see above).
    Branchlets and leaves glabrous; leaves oblong to oblanceolate; calyx corolla-like, lilac- or rosy-purple, in sessile lateral clusters of 3; fruit red; intr., rarely escaped....................... DAPHNE, p. 274
    Branchlets glabrous or pubescent; buds with one exposed scale; leaves from narrow obovate or elliptic to almost linear; flowers imperfect, in erect aments; fruit a small 2-valved capsule......... SALIX, p. 72
    Branchlets glabrous or pubescent; corolla 5-lobed, white, greenish or reddish, open campanulate, cylindric, or broad urn-shaped...................
    VACCINIUM, p. 297

2. Leaves toothed.

Teeth sometimes inconspicuous, or confined to part of margin; plants with only occasional leaves toothed should be looked for under *1.*

a. Leaves (or most of them) inequilateral, asymmetric or oblique at base.
    b. Leaves wavy-toothed, 5–14 cm. long and almost as wide; flowers in autumn, axillary, yellow, with 4 narrow curly petals; capsules obovoid, maturing 1 year after flowering; large shrub... HAMAMELIS, p. 156
    bb. Leaves serrate or dentate.
        c. Leaves long-petioled, broadly ovate to rounded, abruptly acuminate, with 3–5 veins from base but little more prominent than laterals; margin sharply serrate, the attenuate teeth gland-tipped; trees............................... TILIA, p. 266

cc. Leaves short-petioled, elliptic, oblong, or obovate.
   Lateral veins prominent, straight, only occasionally forking, ending in marginal teeth; trees..ULMUS, p. 133
   Lateral veins slightly curved (the basal pair more prominent), forking and curving within margin, only fine branch-veins reaching teeth; trees or shrubs............
   CELTIS, p. 135
aa. Leaves, or most of them, bilaterally symmetrical at base.
   b. Lateral veins extending to leaf-margin.
      c. Lateral veins and teeth equal in number, a vein extending to each tooth; lateral veins not forking; trees.
         Leaf-margin more or less deeply scalloped with rounded to angular teeth; leaves varying from narrow-elliptic or lanceolate (high canopy leaves) to broad-oval or obovate in outline; trees.................................QUERCUS, p. 115
         Leaf-margin not scalloped, straight or nearly straight between the small teeth; leaves tapering or subcordate at base, ovate to broad- or rhombic-elliptic; trees
         FAGUS, p. 113
         Leaf-margin coarsely toothed, teeth ending in very slender almost bristle-like tips; leaves elliptic to lanceolate or oblanceolate; trees or large shrubs.........
         CASTANEA, p. 113
      cc. Teeth more numerous than lateral veins, unequal in size, margin sometimes lobulate, a principal lateral vein extending to each large tooth; lateral veins forking or branched, a branch extending to each smaller tooth.
         Shrubs of wet soil (or introduced tree), with fruiting catkins cone-like, 1–2 cm. long.
            Cones solitary, short-stalked, the deciduous bracts 3-lobed; leaves broad-oval to orbicular, 2–3 cm. long, coarsely dentate..........BETULA pumila, p. 106
            Cones in irregular open clusters, the persistent bracts woody; leaves elliptic or oval, 5–10 cm. long, finely toothed (sometimes lobulate), teeth callous-tipped
            ALNUS, p. 110
         Without this combination of characters.
            Trees, with slender branchlets and enlarged fruiting catkins.
               Small tree with smooth dark gray bark, fluted or muscular in appearance; leaves glabrous beneath except on veins; fruiting catkins with few-toothed bracts subtending nutlets.........................CARPINUS, p. 105
               Small tree with brown-gray bark longitudinally shredding; leaves soft-pubescent beneath; fruiting catkins of bladdery ellipsoid bracts enclosing nutlets...OSTRYA, p. 104
               Trees with bark exfoliating in thin curling sheets, chalky-white and not exofliating, or cherry-like with wintergreen-flavored branchlets; fruiting catkins cylindric, the bracts 3-lobed....................BETULA, p. 105
            Trees or shrubs with coarser flexuous usually thorny branchlets............
            CRATAEGUS, p. 172
            Shrubs.
               Stoloniferous shrubs to 3 m. tall; leaves large, 6–12 cm. long and almost as wide; fruit a nut enclosed in toothed leafy involucre....CORYLUS, p. 103
               Shrubs with slender little-branched stems, compound terminal panicles; leaves narrow-oblong to lanceolate or oblanceolate.......SPIRAEA, p. 161
   bb. Lateral veins curving before reaching leaf-margin.
      c. Trees or shrubs armed with spines or spine-tipped branchlets.
         d. Low to medium much-branched shrubs; spines simple or forking, subtending fascicles of leaves; flowers yellow, berries red...............BERBERIS, p. 139
         dd. Trees or large coarse shrubs usually with spine-tipped branchlets.............
            e. Fruit a pome...................................PYRUS, p. 164
            ee. Fruit a drupe.........................PRUNUS (Plums), p. 215
      cc. Not armed with spines or spine-tipped branchlets.
         d. Lateral veins curving strongly toward apex, seldom branched (except for fine veinlets), thus the only prominent veins.
            e. Shrubs; lowest (basal) pair of veins prominent, leaves appearing 3-veined; fruit dry, 3-lobed, umbellate in compound panicles.....CEANOTHUS, p. 258
            ee. Shrubs or small trees; lateral veins about equal; fruit axillary, berry-like, black when ripe.......................................RHAMNUS, p. 255
         dd. Principal lateral veins forking or branched (thus losing prominence toward margin), or outer branches curving strongly, joining branches of next vein (thus enclosing areas of leaf-blade).

e. Lowest pair of laterals longer and more prominent than others, strongly branched on basal side.
  f. Trees; leaves coarsely and sharply toothed, or some of them lobed; flowers in short spikes, the whole producing a juicy multiple fruit.......
                                        MORUS, p. 138
  ff. Trees, leaves finely to coarsely toothed or lobulate; teeth mostly rounded or blunt; staminate and pistillate flowers in pendant catkins............
                                   POPULUS, p. 85
ee. Lateral veins more or less equal.
  f. Shrubs or small trees with stellate pubescence; leaves large, ½ or more as wide as long.
    Leaves densely stellate pubescent beneath, sparingly toothed or entire; inflorescence and young branchlets densely pubescent..STYRAX, p. 305
    Petioles and veins beneath stellate pubescent; leaves serrulate; inflorescence puberulous..........................HALESIA, p. 305
  ff. Trees or shrubs, not stellate pubescent (some hairs may be in bunches).
    g. Teeth inconspicuous, confined to apical half of margin, or some leaves entire, or with scattered teeth basally (some species of *Salix* might be looked for here; see gg, below).
      Leaves resinous-dotted, aromatic, elliptic to oblanceolate or obovate, few-toothed or entire, often apiculate; fruit gray, wax-covered; shrub.....................................MYRICA, p. 91
      Leaves not resinous-dotted.
        Leaves large, averaging about 12 cm. long and ⅓–½ as wide; margin finely saw-toothed to touch; inflorescence a drooping panicle; fruit an oblong-ovoid 5-celled capsule; tree...........
                                 OXYDENDRUM, p. 293
        Leaves small, averaging 5–6 cm. long, elliptic to obovate, pubescent beneath at least on veins; inflorescence a series of stiff ascending racemes; fruit a subglobose 5-celled capsule; shrub....
                                 LYONIA, p. 291
        Leaves small, 3–5 cm. long, oblanceolate to obovate, cuneate at base, glabrous and paler beneath; fruit a small drupe; shrubs...
              PRUNUS pumila, P. susquehanae, p. 221
  gg. Teeth not confined to apical half or margin, minute or larger.
    Trees or shrubs, with one bud-scale visible; leaves various, linear, oval, obovate, mostly 3 or more times as long as wide; inflorescence an upright catkin................................SALIX, p. 72
    Trees or shrubs with 2 or more bud-scales visible.
      Leaves with a row of slender glands on midrib on upper side; fruit a small pome; shrubs..............PYRUS (ARONIA), p. 167
      Leaves without such glands; fruit fleshy, a pome, drupe, or berry.
        Leaves doubly serrate or serrulate, sometimes irregularly so, or if simply serrate then with petiolar glands.
          Fruit a drupe..........................PRUNUS, p. 215
          Fruit a small pome................AMELANCHIER, p. 169
        Leaves not doubly serrate, often irregularly toothed, and without petiolar glands.
          Leaf-margin finely and obscurely toothed; shrubs with many-seeded berries capped by persistent calyx-lobes; shrubs.....
                              VACCINIUM, p. 297
          Leaf-margin evidently toothed.
            Fruit a large pome; trees........PYRUS (MALUS), p. 165
            Fruit a small berry-like pome; small trees or shrubs......
                            AMELANCHIER, p. 169
            Fruit axillary, a red berry-like drupe.........ILEX, p. 238

3. Leaves (at least some of them) lobed.

a. Leaves palmately 3–5 (–7) veined and lobed.
  b. Leaves strongly whitened beneath with dense white tomentum, lobed on elongate shoots, dentate on shorter shoots.....:........................POPULUS alba, p. 88
  bb. Leaves not strongly whitened beneath.
    c. Sinuses between lobes definitely rounded; leaves symmetrically lobed, or lobed only on one side, or not lobed, coarsely toothed.

Leaves grayish and velvety pubescent beneath, scabrous above, occasionally opposite; branchlets pubescent; intr. tree.........................BROUSSONÉTIA, p. 138
Leaves glabrous or thinly pubescent beneath, scabrous to nearly smooth above; branchlets pubescent at first, soon glabrous......................MORUS, p. 138
cc. Sinuses between lobes not or but slightly rounded, shallow or deep; leaves more or less symmetrically lobed.
Tree, with exfoliating bark and white branches; leaves with shallow sinuses and broad triangular lobes; margin irregularly few-toothed; petiole hollow at base, encasing bud; fruit a light brown long-peduncled spherical head..PLATANUS, p. 159
Without this combination of characters.
Leaves large, mostly more than 10 cm. long and wide.
Tree, with deeply furrowed bark and dark- or silvery-gray branches; leaves glabrous, with deep acute sinuses, and long pointed lobes, star-shaped, commonly with 5 lobes; margins serrate; fruit a spiny ball..LIQUIDAMBAR, p. 158
Shrub; petioles and young shoots villous and glandular; leaves pubescent, with 3-5 triangular lobes, the sinuses extending ⅓ to ½ way into blade, margins irregularly serrate; flowers rose-purple, 3-5 cm. across; fruit a flattish red raspberry......................................RUBUS odoratus, p. 203
Leaves smaller.
Tall shrub, commonly with but 1 stem at base; leaves glabrous, ovate to rhombic-ovate, usually 3-lobed, lobes acute, margins cut-toothed; flowers large, white, pink, or violet; capsule 5-celled, angular-ovoid; commonly planted, sometimes escaped.......................................HIBISCUS syriacus, p. 270
Low to medium-size much-branched shrubs.
Bark of older stems separating in thin layers; leaves heart-shaped or ovate in outline, obscurely to clearly 3-lobed, crenate-dentate; flowers white, in many-flowered corymbs; inflated follicles, commonly 3 from a flower, glabrous and lustrous, often reddish........................PHYSOCÁRPUS, p. 161
Bark not separating; stems often bristly with nodal and/or internodal prickles; leaves alternate on young wood and in fascicles on older wood, deeply or shallowly lobed; flowers solitary, in small clusters, or in racemes; fruit a many-seeded berry......................................RIBES, p. 151
aa. Leaves pinnately veined and more or less lobed.
  b. Low much-branched shrub; leaves 5-12 cm. long, linear-lanceolate to linear-oblong, with 15-25 lobes..COMPTONIA, p. 91
  bb. Trees or shrubs.
    c. Thorn-bearing small trees or shrubs; flowers white or pink, pedicellate in umbel-like clusters; fruit a pome (apple).
Spine-tipped branchlets usually present; pome large, green to yellowish, 2.5 cm. or more in diam...PYRUS, p. 164
Thorns usually present; pome small, red, dull red, or rarely yellowish, less than 2 cm. in diam...CRATAEGUS, p. 172
    cc. Without thorns or spine-tipped branchlets; large trees.
Branchlets green, aromatic fragrant; leaves bilaterally symmetric, 3-lobed, or asymmetric, with a lobe on one side only, or not lobed; flowers before the leaves, yellow, in clustered peduncled racemes which appear terminal until growth of new shoot..
SASSAFRAS, p. 149
Branchlets not green, not aromatic fragrant; leaves and flowers not as above.
Leaves as broad or broader than long, 4-lobed (rarely with a pair of additional lobes toward base), truncate or retuse at apex; flowers large, solitary, terminal, yellow and orange; fruit cone-like..................LIRIODENDRON, p. 145
Leaves longer than broad, shallowly or deeply 3-9 (-11) lobed, the lobes rounded at apex or acute, with or without bristle-tips; flowers inconspicuous, the staminate in pendant aments; fruit an acorn.......................QUERCUS, p. 115

KEY V

DECIDUOUS TREES AND SHRUBS

Winter Key

a. Leaf-scars opposite or whorled, i.e. 2 or 3 at a node.
  b. Terminal bud absent.

 c. Leaf-scars meeting, fan-shaped; bundle-scars 5 or 7, separate and distinct...........
                      SAMBUCUS, p. 235
 cc. Leaf-scars not meeting.
  d. Leaf-scars connected by stipule-scars, lines, or transverse ridges.
   e. Bundle-scars 3; leaf-scars half-round, or later crescent-shaped by rupture; buds with 2 almost valvate scales................PHILADELPHUS, p. 151
   ee. Bundle-scar 1, C-shaped.
    f. Twigs slender, dark reddish; leaf-scars usually 3 at a node; heads of fruit globose......................................CEPHALANTHUS, p. 318
    ff. Twigs medium, light brown; leaf-scars opposite; fruit a large long pendant capsule splitting into two valves......................CAMPSIS, p. 315
  dd. Leaf-scars not connected by stipule-scars or line; stipule-scars, if present, distinctly separate.
   e. Leaf-scars horseshoe-shape, partially surrounding the bud (which is an aggregate); buds roundish, brown-hairy.................CALYCANTHUS, p. 147
   ee. Leaf-scar distinctly beneath bud, not encircling it.
    f. Leaf-scars large, roundish; buds small; twigs stout.
     Leaf-scars dark; bundle-scars arranged in C-shape.................
                   PAULOWNIA, p. 313
     Leaf-scars not dark, usually 3 at a node; bundle-scars in an ellipse....
                   CATALPA, p. 317
    ff. Leaf-scars small, half-round to oval; twigs slender or medium.
     Stipule-scars minute; twigs often spine-tipped; visible bud-scales 4 or more.............................RHAMNUS cathartica, p. 257
     Stipule-scars distinct; twigs never spine-tipped; visible bud-scales 3 (rarely 2 or 4).............................STAPHYLEA, p. 244
bb. Terminal bud present (some twigs may have a fruit-scar terminal).
 c. Terminal bud with one pair of scales visible (sometimes parted at apex), or naked.
  d. Buds naked or with scarcely specialized scales.
   e. Buds small, opening partially or developing into very small leafy shoots in fall or winter; scales and leaves punctate; leaf-scars rhombic to angular lens-shaped; shrubs with erect ovoid capsules.............HYPERICUM, p. 271
   ee. Buds large, composed of miniature scurfy leaves, these clasping the terminal flower buds.............................VIBURNUM alnifolium, p. 327
  dd. Buds with well-developed bud-scales.
   e. Terminal bud smaller than the lateral; leaf-scars half-round or elliptic; bundle-scar 1, crescent-shaped.........................BROUSSONETIA, p. 138
   ee. Terminal bud larger than the lateral or about the same size; or, if a flower-bud, much enlarged.
    f. Leaf-scars somewhat crescent-shaped; bundle-scars 3.
     Buds scurfy-pubescent, flower buds swollen at base, with long pointed apex formed of 2 outer scales.................VIBURNUM, p. 325
     Buds not scurfy-pubescent.
      Lateral buds sometimes hidden; leaf-scars commonly raised; flower buds oblate, with 2 outer scales meeting or nearly meeting over flattened summit.......................CORNUS, p. 277
      Lateral and tip buds similar, large and prominent..............
                  VIBURNUM, p. 325
    ff. Leaf-scars small, half-round; bundle-scar 1; twigs and buds scurfy, red-brown.......................................SHEPHERDIA, p. 274
 cc. Terminal bud with 2 or more pairs of scales visible.
  d. Bundle-scars numerous, almost contiguous in a U-shaped or C-shaped line.
   e. Buds with 2–4 pairs of scales visible...................FRAXINUS, p. 306
   ee. Buds with more than 4 pairs of scales..............CHIONANTHUS, p. 312
  dd. Bundle-scars 1, or 1-several, not contiguous.
   e. Bundle-scar 1, transverse, or indistinct.
    f. Leaf-scars raised, small; twigs terete.
     Shrubs, often retaining leaves in winter; fruit black, in terminal panicles...................................LIGUSTRUM, p. 312
     Low colonial shrubs with axillary or axillary and terminal clusters of coral-red or white berry-like fruits.......SYMPHORICARPOS, p. 324
    ff. Leaf-scars not raised, half-elliptic or crescent-shaped; twigs more or less 4-angled, green (or reddish on side exposed to sun)...EUONYMUS, p. 241
   ee. Bundle-scars 3 or more.

f. Leaf-scars on the ends of raised bases; buds often superposed............
LONICERA, p. 319
ff. Leaf-scars not raised.
Leaf-scars narrow, V- to U-shaped....................ACER, p. 244
Leaf-scars broader, or at least not V- or U-shaped.
Leaf-scars semicircular to oval, widely separated, with distinct
stipule-scars; twigs green or green-brown, later striped..........
STAPHYLEA, p. 244
Stipule-scars absent; leaf-scars may be connected; twigs straw-
color to brown.
Buds large, twigs stout; leaf-scars large and triangular; bundle-
scars in 3 groups.......................AESCULUS, p. 251
Buds small; twigs weak with moderate to large pith.
Lenticels prominent; leaf-scars large, fan-shaped, usually
meeting; bundle-scars 5 or 7, distinct...SAMBUCUS, p. 335
Lenticels not prominent; leaf-scars crescent-shaped, con-
nected by a line; bundle-scars 3 (rarely 5).
Twigs coarse; many small capsules in flat-topped cymes
HYDRANGEA, p. 151
Twigs slender; capsules cylindric, 1 cm. long, 2-valved. .
DIERVILLA, p. 319
aa. Leaf-scars alternate, i.e. one at a node.
b. Leaf-scars minute, on raised bases, or indistinct.
c. Leaf-scars crowded on short thick lateral alternate branchlets, and spirally arranged
on first-year growth; twigs yellow-brown, longitudinally ridged.......LÁRIX, p. 61
cc. Leaf-scars distinctly alternate.
d. Scars of 2 kinds, minute flat crescent-shaped leaf-scars, and round or oval branch-
let-scars; twigs very slender, blackish....................TAXODIUM, p. 67
dd. Leaf-scars on the ends of definitely elevated structures or petiole remnants.
Base clasping, 3-nerved, bearing stipules at its tip; bundle-scar 1; shrub with
shreddy bark..........................POTENTILLA fruticosa, p. 199
Leaf-scars half-round or crescent-shaped; bundle-scar 1; remains of paniculate
inflorescence on some stems............................SPIRAEA, p. 161
Leaf-scars torn and shriveled, on petiole-base; bundle-scars not distinguishable
(3 seen in cross-section of petiole); stipules often persistent on petiole-remnant;
shrubs with hairy or prickly twigs......................RUBUS, p. 199
bb. Leaf-scars, or most of them, on ordinary twigs, and not on greatly elevated bases.
c. Twigs armed with definite spines or prickles, not spine-tipped branchlets.
d. Spines at nodes, sometimes internodal prickles also present.
e. Spines 2 at each node.
Spines (prickles) infrastipular, broad-based in an up-and-down direction,
detachable, often curved or hooked; internodal prickles or bristles often
present; leaf-scars narrow, buds small, with 3–4 exposed scales..........
ROSA, p. 211
Spines slightly below and to the side of the semi-circular leaf-scar which
subtends red-tomentose bud; terminal bud present...................
XANTHOXYLUM, p. 230
Spines placed slightly above and to the side of the irregular leaf-scar in
which are sunken concealed buds; terminal bud absent...ROBINIA, p. 229
ee. Spines 1 from a node or from some nodes only, sometimes branched.
Spines simple.
Spines of leaf-origin, subtending short leaf-bearing spurs, slender, sim-
ple or branched....................................BERBERIS, p. 139
Spines with twig-like base; terminal bud present. .CRATEAGUS, p. 172
Spines axillary; terminal bud absent..............MACLURA, p. 138
Spines branched.
Spines supra-axillary, branched, shining reddish-brown...............
GLEDITSIA, p. 223
Spines (prickles) below leaf-scars, often divided at base into 3 parts,
sometimes simple.................................RIBES, p. 151
Spines subtending short leaf-bearing spurs, simple or branched........
BERBERIS, p. 139

   dd. Prickles of unequal size scattered over surface of twig.
         Twigs very stout; leaf-scars U-shaped, nearly or at least half encircling twig;
         bundle-scars about 12, in one series......................ARALIA, p. 275
         Twigs slender; leaf-scars linear to crescent-shaped; bundle-scars 3.
         Prickles detachable..ROSA, p. 211
         Spines not detachable; bark shredding; bud-scales loose......RIBES, p. 151
         Twigs moderate; leaf-scars irregular and torn, on persistent petiole-bases;
         bundle-scars not distinguishable except in section............RUBUS, p. 199
  cc. Twigs not armed with spines or prickles; some branchlets may be spur-like or thorn-like.
    d. Stipule-scars encircling or nearly encircling twig.
      e. Stipule-scars encircling twig; buds with 1–2 scales.
        f. Leaf-scar encircling bud............................PLATANUS, p. 159
        ff. Leaf-scar not encircling bud.
          g. Buds somewhat flattened, oblong; leaf-scars nearly round; a small leaf-
            scar at base of terminal bud..............LIRIODENDRON, p. 145
          gg. Buds ovate to long conical; leaf-scars not round; a small leaf-scar on tip
            of ridge at side of and ¼–½ as long as terminal bud...........
                                         MAGNOLIA, p. 143
      ee. Stipule-scars nearly encircling twig; buds long and slender, with many over-
        lapping bud-scales..FAGUS, p. 113
  dd. Stipule-scars, if present, not nearly encircling twig.
      e. Terminal bud present. (If terminal fruit-cluster of preceding year is present,
        genus will be found in this section.)
        f. Buds naked, or the bud-scales leaf-like, or valvate, or buds so covered with
        hair that scales obscured.
          g. Terminal bud small, sharp-conical or slender ovoid, its scales indistinct
            (obscured by hair); leaf-scars large, more or less triangular and in-
            dented above; bundle-scars prominent; axillary panicles of whitish
            berries; *POISONOUS*.......RHUS radicans and RHUS vernix, p. 238
         gg. Not as above.
           h. Pith chambered............................JUGLANS, p. 93
          hh. Pith not chambered.
            i. Buds pubescent; lateral buds not valvate.
            j. Stipule-scars present.
              k. Buds olive-brown tomentose, stalked...................
                          HAMAMELIS, p. 156
             kk. Buds and twigs grayish.............RHAMNUS, p. 225
            jj. Stipule-scars absent; buds dark brown, silky pubescent,
             more or less stalked....................ASIMINA, p. 149
          ii. Buds yellow-glandular, or brown-glandular and pubescent,
            lateral buds valvate, somewhat stalked....................
               CARYA illinoensis, C. cordiformis, p. 95
      ff. Buds covered with definite overlapping bud-scales.
        g. Bundle-scar 1, sometimes compound.
          h. Twigs green, with spicy odor; terminal flower buds much larger
            than lateral buds; trees, suckering freely and forming patches.....
                         SASSAFRAS, p. 149
          hh. Twigs not green *and* spicy; shrubs, mostly low, or up to 3–4 m.
            i. Terminal flower buds much larger than lateral buds; higher
             lateral crowded toward tip of stem, larger than lower; leaf-
             scars half-round to triangular or shield-shape..............
                 RHODODENDRON (Azalea), p. 287
            ii. Buds about equal in size; leaf-scars crescent-shape to somewhat
             triangular or notched-elliptic.
             j. Low shrubs, easily recognized by the persistent cup-like bases
              of capsules in axillary or terminal peduncled clusters......
                         CEANOTHUS, p. 258
             jj. Not as in j.
              k. Pith chambered; young twigs slightly stellate-scurfy;
               outer bud-scale spreading-pubescent; leaf-scars half-
               round to notched elliptic...........HALESIA, p. 305
             kk. Pith not chambered; young twigs not stellate-scurfy,
               sometimes pubescent.

l. Slender-stemmed shrubs with small pubescent buds and somewhat elevated leaf-scars; panicles terminating some stems..............SPIRAEA, p. 161

ll. Fruit, if present, berry-like, in axillary clusters or terminating short lateral branches.

    m. Terminal bud deciduous, twigs green or red; leaf-scars narrow; much branched shrubs forming large patches in acid soil..............
                VACCINIUM, p. 297

    mm. Terminal bud persistent; twigs gray or brownish; leaf-scars more or less triangular.

        n. Buds with 2 ciliate exposed scales, sometimes parted at tip; twigs ashy or glaucous
              NEMOPANTHUS, p. 239

        nn. Buds with 4–6 exposed scales; some twigs with superposed buds; red fruits persistent into winter.................ILEX, p. 238

gg. Bundle-scars 3 or more.

  h. Bundle-scars 3 or in 3 groups.

    i. Pith chambered........................JUGLANS, p. 93

    ii. Pith not chambered.

      j. First scale of lateral bud directly above leaf-scar; buds not stalked or constricted toward base; trees...............
              POPULUS, p. 85

      jj. First scale of lateral bud not directly above leaf-scar.

        k. Buds covered with shiny resinous coating (varnished); stipule scars present, indistinct; twigs often corky ridged....................LIQUIDAMBAR, p. 158

        kk. Buds not covered with resinous coating, but may be shiny.

          l. Buds distinctly long stalked.........ALNUS, p. 110

          ll. Buds not stalked, those of some shrubs may be constricted at base.

            m. Bundle-scars 3, sometimes more by division.

              n. Trees (or shrubs with 3-toothed bud-scales) with brown, red-brown, or grayish twigs; leaf-scars not crescent-shaped or very narrowly so (*Cornus alternifolia*, in nn, might be looked for here).

                o. Lateral buds widely divergent from twigs; leaf-scars with 3 prominent bundle-scars, or bundles divided.
Stipule-scars absent...NYSSA, p. 274
Stipule-scars present.
Exposed bud-scales 2 or 3.......
                  CASTANEA, p. 113
Exposed bud-scales about 6......
                  CRATEAGUS, p. 172

                oo. Lateral buds not widely divergent; leaf-scars narrow, often inconspicuous.
Terminal bud 7–12 mm. long, 3–4 times as long as broad, greenish or yellowish- to reddish-brown and sometimes tinged with purple, somewhat silky hairy; some of the scales 3-toothed; stipule-scars absent....AMELANCHIER, p. 169
Terminal bud not as above.
At least some of the lateral twigs short, pointed or thornlike; buds ovoid or conical...PYRUS, p. 164
None of the twigs pointed or thorn-like; some may be short and stubby, crowded with scars.

Lenticels very prominent, raised; bark otherwise smooth and shiny.

Catkins usually present on some twigs; buds about 3 times as long as broad, twigs very slender. BETULA, p. 105

No catkins present; some lateral branches often short and stubby..PRUNUS, p. 215

Lenticels not very prominent.

Buds tomentose............ PRUNUS persica, p. 221

Buds glabrous, usually shining, nearly spherical....... CRATAEGUS, p. 172

nn. Shrubs, tall or low, with several to many stems from ground, not tree-like.

Flower buds numerous, spherical, axillary superposed and accessary; leaf-scars broad crescent-shaped; twigs greenish or greenish-brown, with spicy odor...............LINDERA, p. 150

Flower buds terminal, oblate; leaf-scars broad crescent-shaped; twigs greenish or brownish, without spicy odor..... CORNUS alternifolia, p. 281

Flower buds not present, or at least not greatly different from leaf buds.

Buds superposed; fruits are small pustulate pods in terminal clusters of spike-like racemes............ AMORPHA, p. 227

Buds not superposed.

Low shrubs with persistent cup-like bases of capsules in axillary or terminal peduncled clusters.. CEANOTHUS, p. 258

Shrubs with terminal cymes bearing berry-like pomes; buds elongate and appressed............ PYRUS (Aronia), p. 167

Shrubs with twigs decurrently ridged below nodes; bud-scales loose; leaf-scars slightly raised; bark shredding.

Stipule-scars absent.......... RIBES, p. 151

Stipule-scars present; clusters of follicles in terminal corymbs. PHYSOCARPUS, p. 161

mm. Bundle scars more than 3, in 3 groups.

Lateral buds divergent, valvate or imbricate; bundle-scars numerous; fruit a nut; trees...............CARYA, p. 94

Buds small, with loose scales; bundle-scars 5 (rarely 3); shrubs with shredding bark and follicles in corymbs................. PHYSOCARPUS, p. 161

Lateral buds appressed, minute, the two outer scales meeting in line above center of leaf-scar.......RHUS (Cotinus), p. 233

hh. Bundle-scars more than 3 and not arranged in 3 groups.
  i. Stipule-scars present.
    j. Tall shrubs, sometimes tree-like; fruit persistent, a large 5-parted capsule......................HIBISCUS, p. 270
    jj. Trees.
      k. Buds usually clustered at tips of branches; bud-scales imbricate.........................QUERCUS, p. 115
      kk. Buds not clustered at tips of branches.
        l. Outer bud-scale striate; bundle-scars about 5, often divided, in an ellipse......BROUSSONETIA, p. 138
        ll. Outer bud-scales not striate; bundle-scars usually 3, more by division...............CASTANEA, p. 113
  ii. Stipule-scars absent.
    j. Leaf-scars large, more or less triangular.
      k. Bundle-scars numerous, arranged in a closed ring or curved line; terminal bud small, downy; axillary panicles with white waxy berries often present; *POISONOUS*....RHUS radicans, RHUS vernix, p. 238
      kk. Bundle-scars numerous, irregularly scattered or collected in 3 groups; fruit a nut.........CARYA, p. 94
    jj. Leaf-scars crescent-shaped to U-shaped; bundle-scars 5, in curved line; terminal bud large, 10–15 mm. long, curved at tip, more or less hairy or gummy..PYRUS (Sorbus), p. 169
ee. Terminal bud absent.
  f. Buds buried or partially buried or hidden in the bark.
    g. Buds in the leaf-scar and surrounded or nearly surrounded by it and rupturing or distorting it on swelling.
      h. Twigs dark brown, usually with 2 stout stipular spines..........ROBINIA, p. 229
      hh. Twigs greenish brown, zigzag; several small superposed buds; some twigs usually with branched thorns.........GLEDITSIA, p. 223
    gg. Buds in axil of leaf-scar or above it; leaf-scar not disturbed by growth of bud.
      h. Twigs very stout; 2–3 round superposed buds in tomentose depressions above axil; pith large, salmon pink.....................GYMNOCLADUS, p. 223
      hh. Twigs slender.
        i. Buds partially hidden in bark, depressed globular, brownish...MACLURA, p. 138
        ii. Buds entirely hidden in transverse fold in axil of leaf-scar; clusters of short aments usually present...................RHUS aromatica, p. 237
ff. Buds not buried or hidden in the bark.
  g. Leaf-scar surrounding or nearly surrounding bud.
    h. Pith whitish.
      i. Buds brown, silky tomentose or glabrous, forming a high conical group; twigs red-brown...............CLADRASTIS, p. 227
      ii. Buds pale, silky tomentose.
        j. Buds angular pyramidal; leaf-scar pale with numerous dark bundle-scars; twigs light brown, very flexible..DIRCA, p. 273
        jj. Buds flat, little raised above surface of twig; leaf-scar brownish; twigs brown, stiff..............PTELEA, p. 231
    hh. Pith brownish; twigs stout; buds tomentose, conical............RHUS (sumac), p. 233
  gg. Leaf-scar not nearly surrounding bud.
    h. Buds depressed, little raised above surface of bark.
      i. Twigs very stout; leaf-scars large, heart-shaped; pith large.
        j. Pith salmon pink; buds superposed, in tomentose depressions GYMNOCLADUS, p. 223
        jj. Pith brownish, sometimes pale; buds light brown, pubescent AILANTHUS, p. 231
      ii. Twigs not very stout; pith light.

  j. Leaf-scars very broad, V- to U-shaped; some twigs usually with branched thorns above axil......GLEDITSIA, p. 223

  jj. Leaf-scars oval or narrow elliptic.

   k. Twigs yellowish or greenish brown; some twigs usually with simple thorns from axils; bark of roots orange...
.   MACLURA, p. 138

   kk. Twigs generally red; bundle-scar horseshoe-shaped or crescent-shaped.............OXYDENDRUM, p. 293

 hh. Buds not depressed, some may be closely appressed.

  i. Twigs brown- or silvery-scurfy, or somewhat stellate-scurfy, at least near buds; twigs slender.

  j. Pith chambered........................HALESIA, p. 305

  jj. Pith not chambered.

   k. Buds naked, scurfy, superposed; very rare native shrub
              STYRAX, p. 305

   kk. Buds with about 4 exposed bud-scales; twigs often with glistening peltate scales; planted and sometimes escaped.................ELAEAGNUS, p. 274

  ii. Twigs not brown- or silvery-scurfy.

  j. One bud-scale visible; buds not round.....SALIX, p. 72

  jj. Two or more bud-scales visible; some round buds show but a small part of 2nd bud-scale.

   k. Bundle-scar 1 or appearing as 1.

    l. Twigs zigzag; leaves 2-ranked; upper edge of leaf-scar convex......................CELTIS, p. 135

    ll. Twigs not zigzag, leaves more than 2-ranked.

     m. Bundle-scar crescent-shaped; trees or large shrubs.

      Twigs slender, lenticels inconspicuous; buds with about 4 exposed scales; pith finally chambered.............HALESIA, p. 305
      Twigs moderate, lenticels prominent; buds with 2 overlapping scales; pith often spongy or chambered.................
            DIOSPYROS, p. 303
      Twigs moderate, usually red; buds small, with about 6 bud-scales; pith not chambered.....OXYDENDRUM, p. 293

     mm. Bundle-scars not crescent-shaped; shrubs, occasionally tree-like.

      Buds ovoid to rotund, acute or obtuse; a fruit-scar to one side of some buds may be mistaken for a stipule-scar.............
            RHAMNUS, p. 255
      Buds red, elongate, appressed; small globose capsules in racemose clusters.......
            LYONIA, p. 291
      Buds small, 1–2 mm.; flower-buds larger, spherical; much-branched shrubs........
      GAYLUSSACIA & VACCINIUM,
            pp. 267–68

  kk. Bundle-scars 3 or more.

   l. Pith large, usually dark.

    m. Leaf-scars broadly U-shaped, raised on salient projections of twig; lenticels prominent and raised...............RHUS copallina, p. 235

    mm. Leaf-scars large, heart-shaped, little raised; lenticels inconspicuous....AILANTHUS, p. 231

   ll. Pith moderate or small.

    m. Leaf-scars 2-ranked.

     n. Exposed bud-scales 2, sometimes tip of third showing.

Buds very small, closely appressed;
flower-buds ovoid or spherical, conspicuous on older wood..............
     CERCIS, p. 225
Buds larger.
 Buds asymmetric, glabrous, green to
 red.................TILIA, p. 266
 Buds symmetric, brown and rusty-
 pubescent......CASTANEA, p. 113
nn. Exposed bud-scales 3 or more.
 Buds oblique, directed to one side of
 twig, i.e., axis of bud not perpendicular to leaf-scar.
  Bundle-scars 3 or in 3 groups,
  prominent.........ULMUS, p. 133
  Bundle-scars numerous, in an ellipse
      MORUS, p. 138
 Buds not directed to one side. tips may
 be curved.
  Tip of bud closely appressed; buds
  broadly triangular; pith closely
  chambered.........CELTIS, p. 135
  Tip of bud not closely appressed:
  pith continuous.
   Catkins usually present in winter.
   Buds round or broad-ovoid, obtuse; shrubs.
    Twigs pubescent with perpen-
    dicular hairs; catkins becom-
    ing long...CORYLUS, p. 103
    Twigs densely soft-pubescent;
    catkins short; bog shrub....
     BETULA pumila, p. 106
   Buds ovoid to lance-ovoid,
   acute; trees.
    Buds ovoid, bud-scales striate;
    bark on trunk scaly, splitting
    into narrow shaggy plates..
      OSTRYA, p. 104
    Buds lance-ovoid; bark pa-
    pery and flaking or white, or
    twigs with strong winter-
    green taste..BETULA, p. 105
   Catkins not present in winter;
   buds may be of two kinds, smaller
   leaf-buds and larger flower-buds;
   bark dark gray, smooth, trunk
   and limbs fluted..............
     CARPINUS, p. 105
mm. Leaf-scars more than 2-ranked, at least on erect
 branches.
 n. Buds with more than 3 exposed scales,
  round-ovoid to conical, collateral often
  present; short spur-like fruit-bearing
  branchlets usually present..............
    PRUNUS (plums), p. 215
 nn. Buds with 2 or 3 exposed scales.
  Much-branched shrubs; buds small;
  leaf-scars somewhat raised.
   Buds single or superposed; small
   pustulate pods in clustered spike-
   like racemes.....AMORPHA, p. 227

Buds solitary; twigs resinous-dotted when young, aromatic; leaf-scars half elliptic to triangular.
Stipule-scars present............
COMPTONIA, p. 91
Stipule-scars absent; fruit encrusted in wax..MYRICA, p. 91
Trees, native, now represented only by saplings or coppice-shoots; buds 3–5 mm. long.........CASTANEA, p. 113

INDEX TO GENERA IN KEYS

55

Fig. 12.—Reference map for location of counties.

SYSTEMATIC TEXT

In the following treatment, species native to Ohio are indicated in the text by boldface type (see p. 1). The text does not include detailed descriptions of genera or species; for the most part, characters given in the keys are not repeated (see p. 8). Range of species is usually expressed in general terms and, where possible, is correlated with the forest regions of eastern North America (map, p. 21). More detailed range, by states, can be found in standard manuals. Ohio distribution of all native species is shown by maps; a dot is placed in each county in which the species is known to occur. These records are based on specimens that have been examined and are deposited in herbaria (see p. 2). Dots are usually placed in the center of the county. Departures from this plan are desirable in a few cases: (1) where a species is restricted to one or a few localities (and that fact is known); (2) where a species is at the margin of range, if range is correlated with particular physical features. Such placing of dots facilitates comparison with maps of physical features.

The student desiring more complete descriptions of species, further information on uses, or other general information, may refer to books by Berry (1923), Fernald (1950), Fernald & Kinsey (1943), Gleason (1952), Rehder (1940), Sargent (1922) and, for winter characters, Trelease (1918). Specific names may differ from book to book; correlations must be made carefully.

GYMNOSPERMAE (Gymnosperms)

Only two families, both of the order Coniferae (or Coniferales) are represented in our flora. The flowers are unisexual; most of our species are monoecious. The class is characterized in part by the naked ovules, i.e., ovules not enclosed in an ovary (ovulary).

TAXACEAE

A family of three genera, only one of which is represented in Ohio. Another, *Torreya*, includes the rare Florida endemic, *T. taxifolia* Arn. (with Taxus-like foliage), one Californian, and three Asiatic species. Flowers in the leaf-axils; staminate globular or ellipsoid, composed of a few naked stamens under a peltate scale; pistillate only a single ovule (subtended by a pair of scales) which develops into a bony-coated seed more or less surrounded by a fleshy disk or aril.

TAXUS L. Yew

Only one of the several species, all of which are native to the north temperate zone, occurs in the Northeast. Our horticultural forms are derived from *T. baccata* L. of Europe and *T. cuspidata* Sieb. and Zucc. of eastern Asia. In all, the aril-covered seed is scarlet and berry-like; the flesh is edible, the stone poisonous. The wilted foliage is also poisonous due to the development of a toxic alkaloid. The fresh foliage of American yew is heavily browsed by deer in the North.

1. **Taxus canadensis** Marsh. YEW.

GROUND-HEMLOCK

A straggling shrub of northern range, widely scattered in Ohio, but local except in the northeast. Various habitats: wooded bluffs, cliffs, boggy woods. At its southernmost Ohio locality (Highland County) it festoons dolomite cliffs; in west-central Ohio (Champaign County) it grows under arbor vitae in a bog forest. Young individuals of hemlock are sometimes mistaken for yew, but can be distinguished readily by the two white lines on the lower side of the leaves; leaves of yew are green beneath.

Taxus canadensis

PINACEAE

A large family in which distinct subfamilies, sometimes elevated to family rank, may be recognized; about 33 genera of which five are represented in the native flora of Ohio. The two kinds of "flowers" or cones of pines differ greatly in appearance. The staminate, clustered near base of new growth, live but a short time and are seen only in spring. The staminate cone is composed of cone-scales spirally arranged; each bears two stamens or pollen sacs. The pistillate cones, green at first (or reddish), are borne singly (or two or three close together) toward the upper part of the new growth. The pistillate cone, also, is composed of spirally arranged cone-scales, each of which bears two ovules, which later develop into seeds; in some of the pines, cones remain on the trees two or more years. The cone-scales are woody or papery in most members of the family; fleshy in *Juniperus*, resulting in a berry-like fruit.

ABIES Mill. FIR

A genus of some 40 species of the north temperate zone; one occurs in Ohio as an escape. Three species have been planted in small quantities on certain state forests: *A. balsamea* (Waterloo Forest), *A. concolor* (Hocking, Scioto Trail forests), and *A. fraseri* (Dean, Waterloo forests). Others are seen in ornamental plantings. The cones of all species are erect on the upper side of branches, and the cone-scales are deciduous at maturity of cone, so that fir cones (unless cut off by squirrels) are not found on the ground beneath the trees.

1. ABIES BALSAMEA (L.) Mill. BALSAM FIR

A tree of the Northern Coniferous Forest, whose natural range is to the north of Ohio. "Escaped into Mitchells Mills bog," Chardon Twp., Geauga County.

TSUGA (Endl.) Carr. HEMLOCK

Tsuga is the Japanese name for the native hemlocks of Japan. Ten of the 14 species of the genus are Asiatic in distribution, two western American, and two eastern American; only one of these occurs naturally in Ohio, the other is local in the Southern Appalachians.

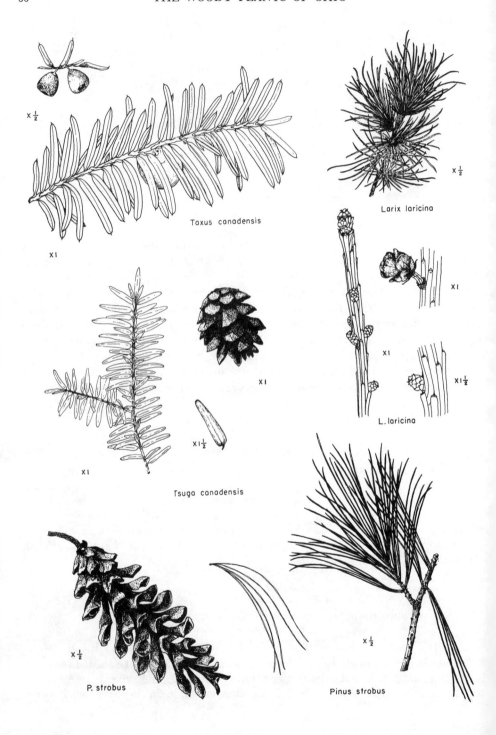

x ½

x 1

Taxus canadensis

Larix laricina

x ½

x 1

x 1

x 1½

L. laricina

x 1

x 1½

Tsuga canadensis

P. strobus

x ½

Pinus strobus

x ½

1. Tsuga canadensis (L.) Carr. HEMLOCK

A large forest tree of ravines, stream bluffs and steep, usually northerly, slopes, frequent and often abundant but local in the Allegheny Plateau section of Ohio, very restricted elsewhere. Its general range, exclusive of outlying stations, corresponds approximately with the combined area of the Hemlock-White Pine-Northern Hardwoods, Mixed Mesophytic, and Oak-Chestnut forest regions. The largest trees of this species in Ohio are in the Hocking Valley forest parks, where they exceed 100 ft. in height (the tallest is said to be 149 ft.). Leaves linear, 8–13 mm. long, obtuse, short-petioled, whitened beneath, spreading laterally, thus forming flattened sprays. Cones small, 1.5–2.5 cm. long.

LARIX Mill. LARCH

Deciduous conifers of the cooler parts of the north temperate zone and arctic zone; the soft but needle-like leaves crowded on very short stubby lateral branchlets. Of the ten species, three are American: two western, one eastern.

1. Larix laricina (DuRoi) K. Koch, TAMARACK. LARCH

A tree of the North, extending across the continent along the northern limit of trees, and southward through much of New England and the northern Great Lakes area. In winter, readily distinguished from all other of our deciduous trees by its unbranched mast-like trunk extending up through the crown, and its horizontal branches. Local in wet soil of bogs and borders of glacial lakes in northeastern and northwestern Ohio. The European larch, *L. decidua* Mill., is often planted; it may be distinguished by its more numerous and pubescent cone-scales. The Japanese larch, *L. leptolepis* (Sieb. & Zucc.) Gord. (*L. kaempferi* Sarg.), has been planted on several of our state forests—Scioto Trail, Dean, Hocking, Waterloo; it may be distinguished by the conspicuous white stripes on the lower leaf-surface and recurving cone-scales.

Tsuga canadensis   Larix laricina

PINUS L. PINE

A genus of some 80 species of the northern hemisphere. In addition to the four species indigenous to Ohio, several others are used in reforestation projects and still others in ornamental plantings. Of these, red pine (*P. resinosa* Ait.), which is a tree of the Hemlock-White Pine-Northern Hardwoods forest region to the north, is most extensively planted for reforestation purposes; jack pine (*P. banksiana* Lamb.), Scotch pine (*P. sylvestris* L.), and Austrian pine (*P. nigra* Arnold) are also used extensively, both in reforestation and in ornamental plantings.

The needle-leaves of our pines are in clusters or fascicles of 2, 3, or 5, each cluster surrounded at the base by a sheath composed of numerous brownish bud-scales, the whole a dwarf branchlet in the axil of an inconspicuous scarious bract. This bract is a primary leaf produced on the elongating stem; the needle-leaves are secondary leaves borne on branchlets which never elongate. These needle-leaves are commonly referred to as leaves and the bractlike leaves are disregarded.

a.   Leaves 5 in a fascicle, slender, bluish-green; cones cylindric and often curved, brown, with thin
       scales...1.   *P. strobus*
aa. Leaves 2 or 3 in a fascicle; cones conic-ovoid, with thickened scales.
       b.   Leaves in 2's, short, less than 8 cm. long.
              Leaves twisted, very short and thick, 2–4 cm. long, obtusish; cones 3–5 cm. long, asym-
                metric; scales without prickles; cones remaining closed for many years..2.   *P. banksiana*
              Leaves longer, 3–8 cm., acute, usually twisted.
              Leaves 4–8 cm. long; cones 4–6 cm. long, persisting on tree for several years; umbo
                of scales tipped with prickle.....................................3.   *P. virginiana*
              Leaves 3–7 cm. long, grayish- or bluish-green; cones 3–6 cm. long, slender conic, the
                umbo of the thickened scales ending in a tubercle rather than prickle. .4.   *P. sylvestris*
       bb. Leaves in 2's or 3's, more than 7 cm. long.
              Leaves flexible, but breaking when bent, in 2's, 12–17 cm. long; cones 4–6 cm. long,
                scales without prickles......................................5.   *P. resinosa*
              Leaves not breaking when bent, in 2's or 3's.
              Twigs brown; leaves dark green.
              Leaves usually in 3's, 7–14 cm. long; cones conic-ovoid, 3–8 cm. long, light brown;
                umbo of cone-scale prominent, with a sharp prickle..................6.   *P. rigida*
              Leaves in 2's, 9–16 cm. long; cones ovoid, 5–8 cm. long, yellowish brown, umbo
                with a short prickle...7.   *P. nigra*
              Twigs pale and glaucous; leaves in 2's or 3's, slender, 7–12 cm. long, dark bluish-green;
                cones 4–6 cm. long, dull brown, scales with a weak prickle; resin-ducts large and
                easily seen on smooth plates of bark...........................8.   *P. echinata*

In addition to the species in the key, four of which are indigenous to Ohio, two indigenous to the north, and the others extensively planted and sometimes spontaneous, a dozen or more species are used in ornamental plantings and might be seen in arboretums and state forests.

1. **Pinus strobus** L.   WHITE PINE
     Our only soft pine, and the largest northeastern conifer, commonly reaching a height of 100 ft. and rarely, 200 ft. One of the dominant trees of the Hemlock-White Pine-Northern Hardwoods region, and frequent in the Appalachian Mountains south to northern Georgia and along the western border of the Appalachian Plateau to northern Alabama; very local to the westward. In Ohio, most frequent in the northeastern quarter of the state. Readily distinguished from other pines at a distance by its blue-green color and softer contour. It is more mesophytic than our other native pines, usually occupying moister and less exposed sites. In addition to the distribution shown on the map, it has been recorded from Geauga County (Dean & Chadwick) and as an escape in Adams and Athens counties.

2. PINUS BANKSIANA Lamb.   JACK PINE
     A small tree of the far North, extending south about half-way down the Lower Penin-sula of Michigan and locally to northeastern Illinois and northwestern Indiana. Planted on several of our state forests (Shawnee, Tar Hollow, Scioto Trail, Dean, Hocking, Waterloo, Maumee), and frequent in the Christmas-tree market. Its short, stout, twisted needles should serve to distinguish it.

3. **Pinus virginiana** Mill. VIRGINIA PINE. SCRUB PINE
Usually a small and more or less irregularly-branched tree, 30–40 ft. in height or rarely taller; whose range roughly coincides with the unglaciated area of the Appalachian Highland (farther eastward extending to New Jersey and eastern Virginia, and locally westward into southern Indiana, western Kentucky, and northeast Mississippi). In Ohio, common and widespread on the non-calcareous soils of the unglaciated Allegheny Plateau; an aggressive invader of eroded slopes and worn-out soil, where it frequently forms pure stands. The species is intolerant of shade; in closed older stands the lower branches die but persist on the tree for many years. The short needles (shorter than those of any other native Ohio species), the persisting but open cones (progressively older cones seen from near tips of branches backward), and (in old stands) the persisting dead branches, are good characters for field identification. Frequently seen in the Christmas-tree market in southern Ohio.

4. PINUS SYLVESTRIS L. SCOTCH PINE
Native of Europe; reported as an escape in Ashtabula, Erie, Franklin, and Lorain counties. Bark of large branches of old trees orange-brown; crown of old trees round-topped and irregular. Used in plantation plantings on most of our state forests.

5. PINUS RESINOSA Ait. RED PINE. NORWAY PINE
A large tree of the Hemlock-White Pine-Northern Hardwoods region, its natural range not reaching as far south as Ohio. Very extensively planted on state forests. The long, slender, dark green leaves which break when bent, and reddish bark will distinguish this species from others in Ohio.

6. **Pinus rigida** Mill. PITCH PINE
A medium size tree (50–75 ft.) of southern range, extending into southeastern Ohio where it is widely distributed on the Allegheny Plateau. It grows on dry ridges and slopes, usually in non-calcareous soils. Often associated with Virginia pine and yellow pine; can be distinguished by its bright green foliage, fairly dense crown, and dark rough bark. Reported also from Guernsey county (Dean & Chadwick).

7. PINUS NIGRA Arnold AUSTRIAN PINE
A European tree, commonly planted in parks and on some of our state forests. Variable in form; old trees sometimes flat-topped.

Pinus strobus

Pinus virginiana

Pinus rigida

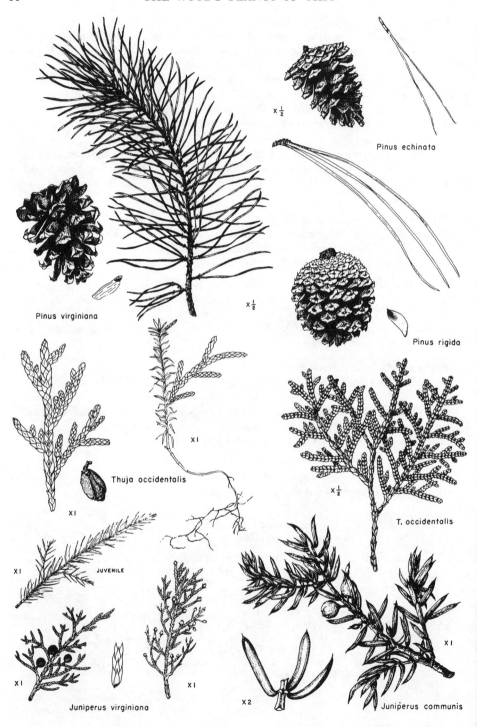

$\times\frac{1}{2}$

Pinus echinata

Pinus virginiana

$\times\frac{1}{2}$

$\times\frac{1}{2}$

Pinus rigida

X1

Thuja occidentalis

X1

$\times\frac{1}{2}$

T. occidentalis

X1

JUVENILE

X1

X1

X1

Juniperus virginiana

X2

X1

Juniperus communis

8. Pinus echinata Mill. Yellow Pine. Shortleaf Pine

A southern tree of wider range than Virginia or pitch pine; extends from New Jersey westward to southern Missouri, Arkansas, and Oklahoma, skipping the calcareous regions of middle and western Kentucky and Tennessee. Somewhat more circumscribed in range in Ohio than pitch pine, but growing in similar habitats. Readily distinguished from that species by thinner crown, dark bluish-green foliage, and conspicuous resin ducts showing on surface of plates of bark. Reported also from Adams and Lawrence counties (Dean & Chadwick).

THUJA L. Arbor Vitae

A genus of six species, two in America (one eastern, one western) and four in eastern Asia. Our eastern species, *T. occidentalis* L., and one of the Asiatic species, *T. orientalis* L. (distinguished by the orientation of the branchlets in vertical planes), are much used in ornamental plantings where they are represented by a large number of horticultural forms. The generic name is from the Greek, *thuia*, an aromatic wood highly prized in ancient times.

1. **Thuja occidentalis** L. Arbor Vitae. Northern White-cedar

A medium-size tree, most abundant in the Northern Coniferous forest and Hemlock-White Pine-Northern Hardwoods region; local south of these regions to northern Illinois, northwestern Indiana, southern Ohio, and in the Appalachian Valley to Tennessee. A tree of bogs and boggy lake borders (usually with calcareous substratum) and thin soil on limestone ledges and cliff-faces. In Ohio, local west of the Appalachian Plateau, in marl bogs and on calcareous outcrops; in Scioto County, on the McDermott "sandstone," a silt-stone with calcareous cement. Some of the largest trees of this species formerly grew in Adams County (before the last period of intensive logging in the early 40's) where trees with trunks 5–7 feet in circumference and age around 300 years could be seen on the dolomite outcrops, and even larger individuals in the deeper soil of adjacent forest (Braun, 1928, pp. 392, 402, 404–5, and Fr.). Cedar Swamp, a state reservation in Champaign County, is so named from the trees of the bog forest surrounding the marl bog.

Leaves closely appressed, imbricated, in 4 rows, the branchlets appearing flattened and arranged in fan-like sprays; leaves of seedlings awl-shaped, spreading. Cones drooping, about 1 cm. long, narrowly ellipsoid (broader when scales spread).

Pinus echinata

Thuja occidentalis

JUNIPERUS L. Juniper

A genus of some 40–50 species of trees or shrubs widely distributed through the north temperate zone and represented in arctic regions and the mountains of the tropics; many inhabit arid or semi-arid regions. A number of the species are in cultivation; a few of these (including *J. communis* and *J. virginiana*) have many color and form variants distinguished as horticultural varieties. Usually dioecious, hence only some of the individuals produce the blue berry-like fruit characteristic of the genus, and which is composed of 3–6 coalescent fleshy scales, each bearing one ovule.

Leaves in whorls of 3, all needle-like or awl-shaped, sharp-pointed, spreading, green and convex below, concave above, with a central white band one-third as wide as leaf; flowers axillary . .
1. *J. communis*
Leaves mostly opposite, scale-like and appressed (needle-like or awl-shaped only on young trees and then in 3's only on leading shoots); flowers terminal on short lateral branches.
Prostrate or creeping shrub, the appressed leaves acutely cuspidate; fruit 6–10 mm. in diam . . . . .2. *J. horizontalis*
Tree, the appressed leaves broadly deltoid, subacute or obtuse; fruit 5–6 mm. in diam . . . .
3. *J. virginiana*

1. **Juniperus communis** L. Common Juniper
A circumpolar species with several geographic varieties and a number of garden forms.

Arborescent, sometimes attaining 35 ft. in ht., with the main trunk vertical or nearly so; leaves about 15 mm. long and 1 mm. wide; fruit 5–8 mm. in diam . . . . .var. *communis*
Decumbent, forming mats up to 5 ft. in diam.; leaves generally somewhat sharter and broader; fruits 6–10 mm. in diam . . . . .var. *depressa*

Var. *communis* is rare and local throughout its range; it is often confused with juvenile forms of *J. virginiana*. Var. *depressa* Pursh, recognized by its characteristic growth-form, occurs in several northern Ohio counties.

2. Juniperus horizontalis Moench. Creeping-cedar
A northern species which cannot be considered as indigenous to Ohio. One collection from Williams County bears the note "from seeds dropped by birds along a fence."

3. **Juniperus virginiana** L. Red-cedar
Our most frequent evergreen, ranging almost throughout the Deciduous Forest formation of the eastern half of the United States. Widely distributed in Ohio; most abundant on calcareous outcrops and limestone soils of southwestern Ohio; and, in second-growth stands, on eroding slopes and worn-out soils. Young trees (and sometimes vigorous

Juniperus communis

Juniperus virginiana

shoots on older trees) bear little resemblance to older trees, because of their sharp-pointed, spreading, acicular leaves. Wood fragrant, heart-wood dull red; much used for fence posts, lead pencils, and cedar chests.

Two varieties have been distinguished, var. *virginiana* of more southern range, and var. *crebra* Fern. & Grisc., more northern.

> Trees with ovoid or pyramidal, broad crown; leaves (on adult branches) tightly appressed, broadly deltoid, and obtuse to subacute; seeds deeply pitted..................var. *virginiana*
> Trees with columnar, narrow-conic to spire-shaped crown; leaves less closely appressed, narrowly ovate and acute; seeds shallowly pitted.................................var. *crebra*

The two varieties intergrade; in southern Ohio, where both occur together, specimens cannot always be distinguished. Variation in *J. virginiana* is further complicated by introgression from *J. horizontalis* in northern part of range, and from *J. mexicana* in southwestern quarter of range (Anderson, 1953; Hall, 1952). The varieties are not distinguished on the map.

OTHER GYMNOSPERMS

A number of other gymnosperms have been used in state forest plantations and arboretum plantings, of which three, *Ginkgo biloba*, *Picea abies*, and *Taxodium distichum*, occur in the larger plantings.

GINKGO BILOBA L. GINKGO. Readily recognized by the fan-shaped parallel-veined leaves. Fruit fleshy, drupe-like. From eastern China, and long cultivated in Japan.

PICEA ABIES (L.) Karst. (*P. excelsa* Link.). NORWAY SPRUCE. This (and other species of spruce) recognized by the 4-angled leaves (most easily discerned by twirling a leaf between thumb and forefinger). Twigs subglabrous; leaves dark green and glossy, slender and sharp-pointed; cones 1–1.5 dm. long. From Europe. Several other species of *Picea* are used sparingly.

TAXODIUM DISTICHUM (L.) Richard. BALD-CYPRESS. A deciduous conifer with tapering trunk buttressed at base; produces "knees" when growing in areas subject to flooding. Native in swamps and on alluvial bottoms of the Atlantic and Gulf Coastal Plain and Mississippi Embayment, north to southwestern Indiana. Where planted in suitable environment (for example, swampy valley-flats of Zaleski State Forest), the trees might erroneously be regarded as indigenous, especially when accompanied by other planted species (as sweet gum) common to such habitats. Leaves soft, linear, 2-ranked, deciduous, as are some of the branchlets; cone globular.

PSEUDOTSUGA TAXIFOLIA (Poir.) Britt. (*P. mucronata* Sudw.). DOUGLAS-FIR. When fruiting, easily recognized by the conspicuous exserted bracts in the cones. Leaves flattened, spreading into two opposite rows; buds not resinous (as in *Abies*).

ANGIOSPERMAE (Angiosperms)

This class includes all our woody plants except the conifers. It is characterized in part by the closed ovary (ovulary) in which the ovules are borne.

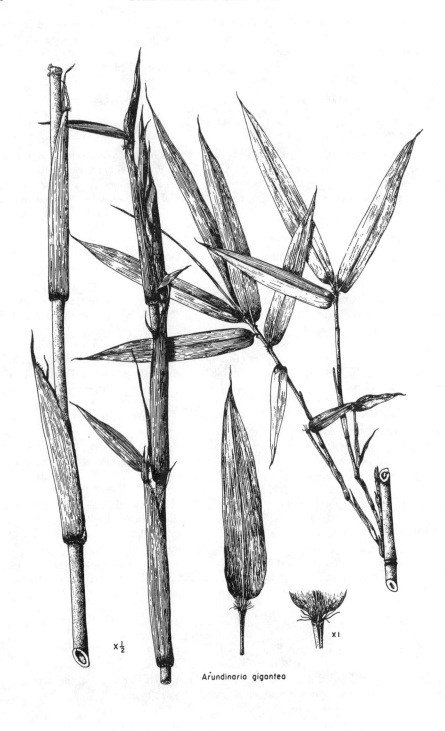

X ½

X I

Arundinaria gigantea

MONOCOTYLEDONEAE *(Monocotyledons)*

Leaves usually parallel-veined; stems (in transverse section) with wood-bundles separate and scattered; parts of flowers usually in threes or sixes, rarely in twos or fours, never in fives; embryo with one cotyledon. A subclass containing some 40,000 species (about one-fifth of the flowering plants); ours mostly herbaceous.

GRAMINEAE

Cane *(Arundinaria)* is the only woody member of the grass family native to the United States, and the only woody representative in our area of the large tribe of grasses known as bamboo.

ARUNDINARIA Michx. BAMBOO. CANE

Easily recognized by the hollow woody stems, solid at the nodes. A genus of about 25 species of America and Asia.

1. **Arundinaria gigantea** (Walt.) Chapm. CANE. LARGE CANE.
   A. macrosperma Michx.

Stems 2–3 m. in height, 1–2 cm. in diam., but in the South attaining a height of 8–10 m. and stem diameters 2 or 3 times as great, unbranched the first season, later developing almost erect branches from many of the nodes; flowering infrequently, at intervals of many years, the flowering branchlets borne toward the ends of branches. A southern species extending north to southern Ohio, Indiana, and Illinois. Although it seems now to occur in Ohio only in disturbed areas or in patches where it may have spread from cultivation, the distribution of these patches is such as to suggest that the species is indigenous to a few counties of west-southern Ohio; it is known to have been abundant 15–20 miles southward in

Arundinaria gigantea

Kentucky where cane-land is mentioned in early literature (Jillson, 1930). Heavy grazing of the very palatable young shoots soon eradicates the species. It grows on ravine slopes and flats, and in moist or swampy spots.

LILIACEAE

A large and cosmopolitan family of about 240 genera and 3600 species, ours mostly herbaceous. Only *Yucca* (with one introduced species) and *Smilax* are represented among Ohio's woody plants. Flowers 6-parted, regular and symmetrical, large and showy in *Yucca* (and many other genera), small and green or greenish in *Smilax* (and a few other genera).

YUCCA L.

Stems short and woody, or in some species of the Southwest, tall and branched, arborescent. Pollination of the large 6-parted creamy-white flowers is brought about by the activity of the pronuba moth (*Tegeticula*) which visits the flowers in the evening, collecting pollen into a ball, then pushing it into the angle between the stigmas, thus insuring fertilization of the ovules, which are numerous in each locule of the 3-valved capsule. Larvae hatching from eggs laid in the ovary of the pollinated flower consume some of the developing seeds. The small round holes which are always seen in the walls of ripened capsules are bored by the larvae when they finish feeding and emerge from the capsule, to drop to the ground and pupate. A genus of tropical and warm-temperate (mostly arid) North America; one species is occasionally found as an escape in Ohio. Yucca is the Haitian name for a species of this genus.

1. YUCCA FILAMENTOSA L. COMMON YUCCA.

Woody caudex very short, the elongate stiff narrow leaves ascending to spreading; sometimes along roadsides and near old cemeteries. Reported as an escape in Adams, Coshocton, Franklin, Hamilton, Highland, and Meigs counties.

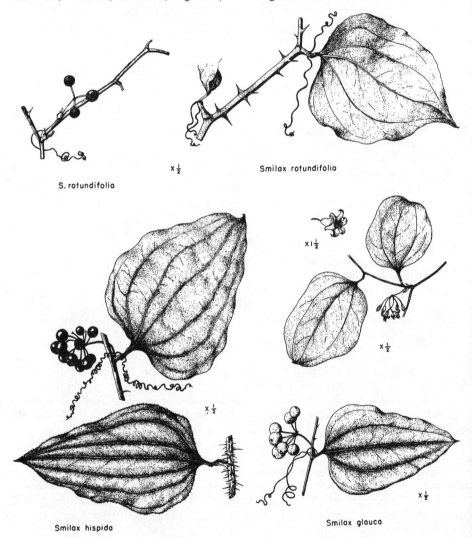

S. rotundifolia

x ½

Smilax rotundifolia

x 1½

x ½

x ½

Smilax hispida

Smilax glauca

x ½

SMILAX L. Greenbrier

Woody and herbaceous plants of the tropics, extending into temperate North America and eastern Asia; also in the Mediterranean region. A genus of about 300 species; four woody species in the Ohio flora. Because of the broad leaves of most species of *Smilax* and the prominent cross or net-veined feature, the parallel venation is not as obvious as in most monocots. Dioecious, the perianth segments greenish. The berries are blue-black or black, in umbels, and often persist through the winter. All of our woody species have tendrils which are stipular in origin.

Leaves whitened beneath, ovate or rounded; stems slender, terete, irregularly armed with slender prickles; berries blue-black, glaucous...............................1. *S. glauca*
Leaves green on both sides.
Stems terete, usually more or less densely covered, except on young branchlets, with sharp but slender terete black bristles; peduncles longer than petioles; berries black, without bloom...2. *S. hispida*
Stems angled (or sometimes terete), armed with stout green spine-like prickles somewhat flattened toward base.
Stems 4-angled or terete, not striate, green, rigid, with stout flattened prickles; leaves broadly ovate or suborbicular, sometimes narrowly ovate; leaf-margins more or less denticulate or erose (under a lens), not conspicuously thickened; peduncles about as long as petiole or shorter.............................3. *S. rotundifolia*
Stems rounded or with one prominent angle, scurfy at base; leaves from ovate to ovate-hastate, with much constricted sides; margins colorless but noticeably thickened, entire, or on young shoots denticulate to erose; peduncles longer than petioles....4. *S. bona-nox*

1. **Smilax glauca** Walt. Greenbrier. Sawbrier

A vine often forming dense tangles in clearings and second-growth thickets where the tough wiry and spiny stems make walking difficult. A plant of acid soil, hence found in the flats of southwestern Ohio and through much of the Allegheny Plateau, but more frequent southward. Usually deciduous, but in mild winters, retaining its leaves. Two varieties are distinguished:

Leaves densely papillose or hirtellous-puberulent beneath........................var. *glauca*
Leaves smooth beneath...var. *leurophylla* Blake

The typical variety is much more frequent in Ohio than the smooth variety. Although the extremes are distinct, gradation in puberulence or papillosity connects the two varieties.

2. **Smilax hispida** Muhl. Bristly Greenbrier

S. tamnoides L., var. *hispida* (Muhl.) Fern.

Usually in circumneutral soil of rich woodlands; often high-climbing but not forming

Smilax glauca

Smilax hispida

extensive thickets. Ranging throughout the Deciduous Forest; widespread in Ohio. The thin dark-green leaves (darker than those of our other species of *Smilax*) frequently retained through the winter, at least in sheltered situations. Occasionally, plants are found with leaf-blades constricted in the middle. (For synonymy, see Clausen, 1951.)

3. **Smilax rotundifolia** L. Greenbrier. Sawbrier

Smilax rotundifolia

One of the most viciously spiny of our plants; the thick rigid but sharp-pointed prickles readily penetrate the skin. A common and wide-ranging species, but not extending as far westward as the two preceding. In Ohio, common in the acid soils of the Allegheny Plateau, more local westward; often forming dense tangles particularly in thinned woodlands and where the dying of chestnut opened up the forest canopy. Vines with leaves somewhat constricted in the middle have been mistaken for *S. bona-nox*.

4. **Smilax bona-nox** L. Fringed Greenbrier

A southern species, extending north to Virginia, Kentucky, southern Indiana, and Missouri; not definitely known from Ohio. The scurfy base of the stem (rarely shown in herbarium specimens) and thickened leaf-margin, entire or (on young shoots) with spinulose-denticulate fringe, will distinguish this species from the much more common *S. rotundifolia*.

DICOTYLEDONEAE   (*Dicotyledons*)

Leaves usually netted-veined; stems (in transverse section) with wood-bundles arranged in a ring, and, in the woody plants, increasing in size by annual growth toward the stem circumference, becoming hard and woody, and united into a ring around the pith; parts of the flowers usually in fours or fives; embryo with two cotyledons. A large subclass containing some 160,000 species (about four-fifths of the flowering plants).

SALICACEAE

A family of only two genera, all of whose species are woody plants. Dioecious (very rarely monoecious), with the flowers (without perianth) arranged in aments, each flower subtended by a small bract. Staminate catkins soon deciduous, pistillate increasing in size and persisting until after maturing of capsules and shedding of seeds, each of which is furnished with a tuft of long silky hairs.

SALIX L. Willow

A large genus of about 300 species, mostly in cooler latitudes of the northern hemisphere, but represented on all continents except Australia. Dioecious trees or shrubs, mostly in wet ground; those of arctic and alpine situations very small, sometimes but a few centimeters in height. Winter-buds covered by a single scale; leaves alternate (subopposite in the introduced *S. purpurea*), relatively narrow in most of our species; aments expanding before the leaves (as in the familiar pussy willow), with the leaves (as in the common black willow), or, in a few species, after the leaves (as in the sandbar willow). Over half of our species belong to the second group; only two (nos. 3 and 10) to the third; the remainder (nos. 12, 14, 17, 21, 22, 23, 24, 25) flower in early spring, before the leaves develop. Winter-buds of early-flowering species are of two sizes, the ament buds the larger. The goat willow, *S. caprea* (the pussy willow of florists), has the largest buds of any of our species. Fruit a small one-celled pointed capsule, sessile or pedicelled, dividing at maturity into two valves; seeds retain their viability for only a few days.

In addition to our native species, a number of introduced ones occur more or less frequently; of these, *S. alba* and *S. fragilis* are widely naturalized. The flexible slender branches of several species are used in basket-making.

Species vary as to leaf shape, pubescence, and glaucescence, characters by which varieties are sometimes distinguished. Shape and persistence of stipules, especially those of vigorous shoots, may aid in identification. Willows are largely insect-pollinated. Hybrids, both between Old World species and between native species, and occasionally between species from the two hemispheres, occur and add to the difficulty of identification. Botanists sometimes disagree as to the interpretation of certain species. All county records shown on our maps are based on specimens identified by the late Dr. Carleton R. Ball. Our interpretation of species follows Ball (see Gleason, 1952) with synonomy or notes correlating these with Fernald's treatment (1950). Our key is based principally on mature vegetative characters; species with glabrous mature leaves often have pubescent young leaves. Because willows are dioecious, and because specimens rarely are complete, certain identification is often difficult. The student will need to refer to more complete descriptions in Gleason (1952) and Fernald (1950).

a. Leaves and buds distinctly alternate.
  b. Branchlets glabrous or but thinly pubescent, often lustrous; leaves glabrous at maturity.
    c. Leaves relatively broad, sometimes lanceolate, broadest below or near middle, glandular toothed; petioles with coarse glands near blade.
      d. Leaves green beneath, or gray or glaucescent; aments developing with the leaves.
        e. Leaves lanceolate to ovate-lanceolate, long-acuminate to attenuate; stipules (on sprouts) oblong to semi-circular . . . . .1. *S. lucida*
        ee. Leaves ovate, short-acuminate, gummy when expanding; stipules small, early deciduous; introduced shrub . . . . .2. *S. pentandra*
      dd. Leaves pale or glaucescent beneath.
        e. Aments expanding after the leaves, capsules maturing in late summer; leaves short-acuminate, finely glandular-serrate; stipules absent; native shrub . . . . .
                                                3. *S. serissima*
        ee. Aments expanding with the leaves; leaves acuminate; introduced trees.
          Leaves coarsely glandular-serrate (5–6 teeth per cm.); stipules small, semi-cordate or wanting . . . . .4. *S. fragilis*
          Leaves with 9–10 glandular serrations per cm., often more or less silvery-silky; stipules minute, deciduous, or wanting . . . . .5. *S. alba*
    cc. Leaves relatively narrow, or if broad, without petiolar glands.
      d. Vein-eyelets minute, 4–5 per mm.
        e. Leaves green on both sides, narrowly lanceolate, sometimes falcate; stipules (on sprouts) acute, serrate . . . . .6. *S. nigra*
        ee. Leaves glaucous beneath.
          Branchlets not pendulous.
            Leaves linear-lanceolate to lanceolate; stipules on sprouts conspicuous, obtuse; branchlets yellowish to dark brown, brittle-based; southern . . . .
                                             7. *S. caroliniana*
            Leaves lanceolate to ovate-lanceolate; stipules wanting or minute; petioles 6–20 mm. long, slender, often twisted; branchlets not brittle-based; northern . . . . .8. *S. amygdaloides*
            Branchlets elongate, pendulous or "weeping" . . . . .9. *S. babylonica*
      dd. Vein-eyelets larger, often 1 mm. across.
        e. Leaves green beneath, sometimes paler but scarcely glaucous.
          Leaves linear or linear-oblong, the margins more or less parallel except near base or apex; petioles very short.

Leaves linear to linear-oblanceolate, spinulose-denticulate, younger leaves
  silky on one or both sides; aments after the leaves........10.   *S. interior*
Leaves linear-oblong, glaucescent and purple-tinged, entire or irregularly
  toothed toward apex, often subopposite; aments before the leaves......
                                                            25.   *S. purpurea*
  Leaves not as above.
    Leaves broadest above middle.
      Leaves oblong to oblanceolate, short-acute to obtuse, revolute margined;
        lateral veins subequal and parallel.................11.   *S. pedicellaris*
      Leaves oblanceolate to narrowly obovate, acute to short-acuminate;
        margin somewhat revolute; lateral veins unequal, curving...........
                                                            23.   *S. humilis*
    Leaves broadest below or near middle, oblong-oblanceolate to ovate,
      acuminate; margin not revolute.
      Margin closely glandular-serrate; stipules on vigorous shoots large, 10–15
        mm. long, semi-ovate to subreniform...................12.   *S. rigida*
      Margin coarsely undulate-serrate with 5–6 glandular teeth per cm.;
        stipules small or absent, early deciduous; branchlets brittle at base;
        intr. tree.......................................4.   *S. fragilis*
  ee. Leaves whitened or glaucous beneath, rarely somewhat silky.
    Stipules, at least on vigorous sprouts, conspicuous and more or less persistent.
      Leaves closely crenate or crenate-serrate with gland-tipped teeth, rounded
        to cordate at base or somewhat tapering, white-glaucous beneath;
        northern shrub..............................13.   *S. glaucophylloides*
      Leaves entire, undulate, or crenate-serrate, generally tapering toward
        base; buds large; widespread shrub....................14.   *S. discolor*
      Leaves closely glandular-serrate, rounded to subcordate at base, glauces-
        cent beneath; widespread shrub.......................12.   *S. rigida*
    Stipules absent, or very small and early deciduous.
      Introduced trees, widely naturalized; branchlets brittle at base; leaves
        narrowly lanceolate to lanceolate.
        Branchlets yellow to greenish; leaf-margin with 9–10 gland-tipped
          teeth per cm...5.   *S. alba*
        Branchlets greenish to dark red, very brittle at base; leaf-margins with
          5–6 gland-tipped teeth per cm........................4.   *S. fragilis*
      Native shrubs, northern.
        Branchlets yellowish to dark brown or reddish; leaves narrow-lanceolate
          or -oblanceolate, subentire to serrulate; capsules lanceolate-rostrate, on
          pedicels 2.5–5 mm. long............................15.   *S. petiolaris*
        Branchlets olive-green to dark brown; leaves narrowly to broadly ellip-
          tic, oblanceolate to obovate, entire, often revolute; capsules narrowly
          lanceolate, on slender pedicels 2–4 mm. long....................
                                              11.   *S. pedicellaris* var. *hypoglauca*
bb. Branchlets silky, puberulent, or pubescent; leaves more or less pubescent, often densely so
    beneath.
    c.   Introduced tree; leaves silky beneath, narrowly lanceolate to lanceolate; branchlets
        yellow to greenish..5.   *S. alba*
    cc. Shrubs, sometimes tree-like.
      d.   Leaves pubescent on both sides, often densely so beneath.
        e.   Leaves relatively narrow, less than ⅓ as wide as long.
          Leaves glaucous and densely silky to glabrate beneath, pubescent above, at
            least on midrib, linear-lanceolate to oblanceolate; stipules absent or minute
                                                            16.   *S. subsericea*
          Leaves densely white-tomentose beneath, thinly tomentose above with
            deeply impressed midrib, linear-oblong to linear-lanceolate; stipules
            lanceolate, glandular................................18.   *S. candida*
          Leaves green on both sides, usually more or less silky pubescent, linear to
            linear-oblanceolate; stipules absent....................10.   *S. interior*
        ee. Leaves broader, ⅓–½ or more as wide as long.
          Leaves densely lustrous pubescent on both sides, oblong-ovate, glandular-
            serrate with divergent teeth; stipules large, cordate-ovate, toothed, longer
            than the short thick petiole...........................19.   *S. syrticola*

Leaves more or less gray-pubescent on both sides.
Leaves rugose-veiny beneath and glaucous, elliptic to broadly rhombic-oblanceolate or obovate-oval, toothed or subentire; stipules small; native shrub..20. *S. bebbiana*
Leaves densely pubescent beneath, obovate to elliptic-lanceolate, crenate or remotely serrate; intr. shrub........................21. *S. cinerea*
dd. Leaves glabrous or nearly so above, except on midrib.
e. Leaf margin finely and regularly serrulate, pubescence of lower leaf surface silvery-silky, leaves lanceolate; branchlets brittle at base; foliage blackening in drying...17. *S. sericea*
ee. Leaf margin entire, undulate, remotely or irregularly dentate or serrate, often slightly revolute; veins prominent beneath; leaves not silvery-silky.
Leaves relatively broad, ⅓ or more as wide as long, margins usually toothed, teeth large; stipules large, half-ovate; large shrubs.
Native shrub; leaves elliptic to elliptic-oblanceolate...................
14. *S. discolor* var. *latifolia*
Intr. shrub, often tree-like; leaves broadly ovate to elliptic, suborbicular or obovate, soft-pubescent beneath......................22. *S. caprea*
Leaves relatively narrow, less than ⅓ as wide as long; margins subentire, teeth, if present, small and far apart; stipules lanceolate, or wanting; low shrubs.
Shrub, 1-3 m.; leaves oblanceolate to narrowly obovate, often glabrate and pale beneath; aments 1.5-3 cm. long................23. *S. humilis*
Shrub, 0.5-1 m.; leaves narrowly oblanceolate; aments 1-1.5 (-2) cm. long.
24. *S. tristis*
aa. Leaves and buds subopposite; leaves linear-oblanceolate, linear-oblong, or spatulate, irregularly serrate above middle...25. *S. purpurea*

1. **Salix lucida** Muhl. Shining Willow

Large shrub or small tree, 1-6 m. tall. The large leaf-blades (up to 15-20 cm. long and 5-8 cm. wide), green and shining on both sides but slightly paler beneath, the long-acuminate or attenuate apex, gland on outer end of petiole, and shining chestnut-brown twigs and buds characterize this species. A northern species of low ground. Three varieties are distinguished; our specimens (with one exception) belong to the typical variety. One specimen from Catawba Island, Ottawa County, is referred to var. *intonsa* Fern., which has young branchlets and lower leaf-surface more or less hairy. A hybrid between *S. alba* and *S. lucida* is recorded from Logan County.

2. Salix pentandra L. Bay-leaf Willow

Small Eurasian tree, introduced into the United States and Canada for basket making, or as an ornamental; an occasional escape. Resembles *S. lucida*. Our records, for Athens, Auglaize, Belmont, Franklin, Greene, Hamilton, and Lake counties, doubtless include planted specimens.

Salix lucida

Salix serissima

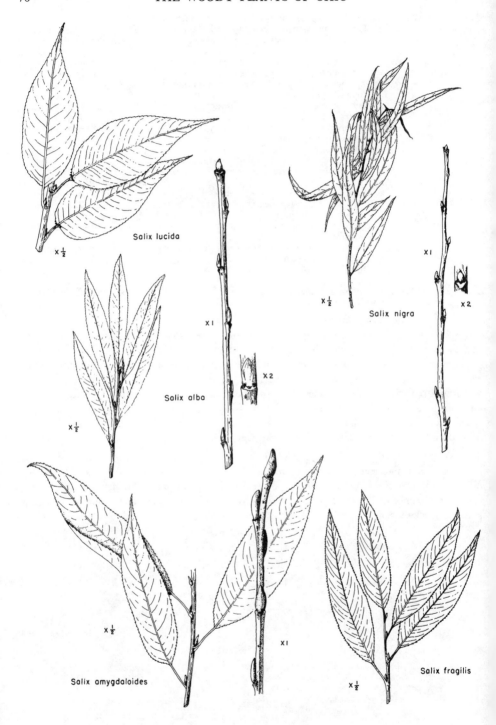

Salix lucida
x ½

Salix nigra
x ½
x 1
x 2

x 1

x 2

Salix alba
x ½

Salix amygdaloides
x ½
x 1

Salix fragilis
x ½

3. **Salix serissima** (Bailey) Fern. Autumn Willow

A northern shrub of swamps and bogs. Similar to *S. lucida*, from which it differs in its smaller size and its short-acuminate very firm, more elliptic leaves whitened beneath.

4. Salix fragilis L. Crack Willow; Brittle Willow

Large tree, early introduced from Europe; similar to *S. alba*, but the branchlets not as yellow, and very brittle at base, the leaves glabrous (glaucous) beneath, and the serrations coarser. Aments in both species developing with the leaves. Recorded from over one-third of Ohio's counties.

5. Salix alba L. White Willow

Large tree, early introduced from the Old World. Ohio's largest white willow, in Ashtabula County, is 26 ft. 2 inches in circumference at 4¼ feet, 60 ft. in height, and has a crown spread of 110 ft. (Ohio Forestry Assn.). The yellowish branchlets, somewhat brittle at base, and bright green leaves silky and glaucous beneath, identify this common tree in the field. Trees with more yellow branchlets and less pubescent leaves are common; these are referred to var. *vitellina* (L.) Stokes by Fernald (1950) and by Deam (1953), but are considered by Ball (Gleason, 1952) to be hybrids of *S. alba* and *S. fragilis*, both European species. Hybrids with native species also occur; *S. alba* × *S. lucida* is recorded from Logan County. White willow, together with its forms and hybrids, is widely distributed in Ohio.

6. **Salix nigra** Marsh. Black Willow

Tree to 20 m. tall, sometimes shrub-like; occasionally taller, and with massive trunk. A Wayne County tree 70 ft. in height has a trunk 21 ft. 5 inches in circumference, and crown spread of 84 ft. (Ohio Forestry Assn.). Our commonest species, growing in ravines and along almost every stream, if there is sufficient light. The yellow-brown to dark brown branchlets, brittle at base, and narrowly lanceolate leaves green on both sides, with tiny vein-eyelets, characterize black willow. Ranges almost throughout the Deciduous Forest and its outliers. Hybrids between *S. nigra* and *S. amygdaloides* are recorded from Ashtabula, Erie, Franklin, and Lorain counties.

7. **Salix caroliniana** Michx. Carolina Willow; Ward's Willow

S. longipes Shuttlew., *S. longipes* var. *wardii* (Bebb) Schneider, *S. wardii* Bebb.

A southern shrub or small tree, 1–3 m. tall, barely entering our range; usually in rocky soil along streams. Resembles shrubby plants of *S. nigra*, from which it is distinguished by the glaucous under surface of leaves.

Salix nigra

Salix caroliniana

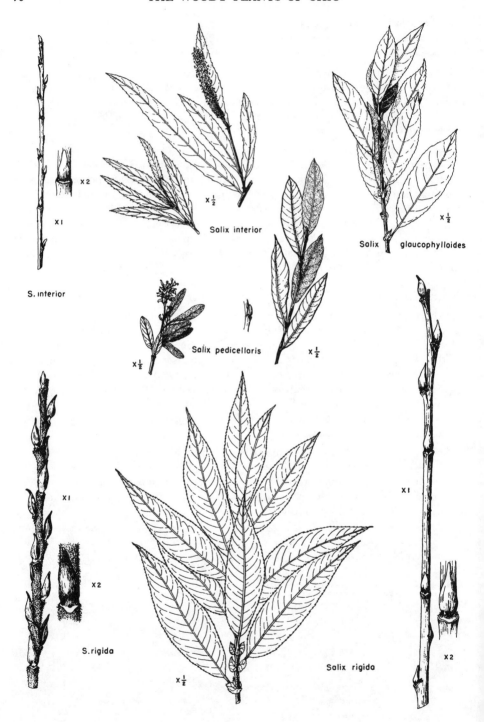

Salix interior

S. interior

Salix glaucophylloides

Salix pedicellaris

S. rigida

Salix rigida

8. **Salix amygdaloides** Anderss. PEACH-LEAF WILLOW
Small tree or shrub, to 15 m. tall; "branchlets longer and less brittle than those of *S. nigra*, yellowish to reddish brown, usually somewhat drooping, giving a 'weeping' appearance, which, with its color, makes the species easy to recognize at a distance" (Deam, 1953). Along streams and in swamps in the northern part of Ohio; wide-ranging, extending almost across the continent (sw. Quebec and Vermont west into Washington and Oregon). Hybrids between this species and *S. nigra* are recorded from Ashtabula, Erie, Franklin, and Lorain counties.

9. SALIX BABYLONICA L. WEEPING WILLOW
A well known and frequently planted tree, easily recognized by the slender pendulous branchlets; rarely occurring as an escape propagating from broken branchlets. The specific name is inappropriate, for this species is a native of northern China, where it is scarcely weeping. It was early introduced into Europe, and later given its specific name by Linnaeus, apparently under the impression that it was the "willow" mentioned in Psalm 137. Hybrids between *S. babylonica* and *S. fragilis* are recorded from Ashtabula and Erie counties.

10. **Salix interior** Rowlee SANDBAR WILLOW
S. longifolia Muhl.
A very common willow, 1.5–4 or 5 m. tall, forming large colonies along streams and shores of ponds and lakes; often a pioneer on depositing shores. The slender leaves with small outwardly-pointing teeth, usually silky pubescent when young and later glabrous, the late and prolonged period of flowering, and habit of growth readily distinguish this species. Commonly infested with pine-cone willow-galls, caused by the larvae of a gall-gnat. A widespread species, extending from New Brunswick to Alaska and south to Virginia, Louisiana, and southern Texas; common throughout Ohio. A variable species in which several varieties and forms have been distinguished. Most of our specimens are referable to the typical variety. *S. interior* var. *wheeleri* Rowlee (forma *wheeleri* (Rowlee) Rouleau), differing in its relatively shorter and broader leaves, more densely and permanently clothed with long appressed hairs, is recorded from Erie, Ottawa, and Warren counties; var. *pedicellata* (Anderss.) Ball, a western variation with leaves only 3–6 mm. wide, is recorded from Ottawa County.

Salix amygdaloides

Salix interior

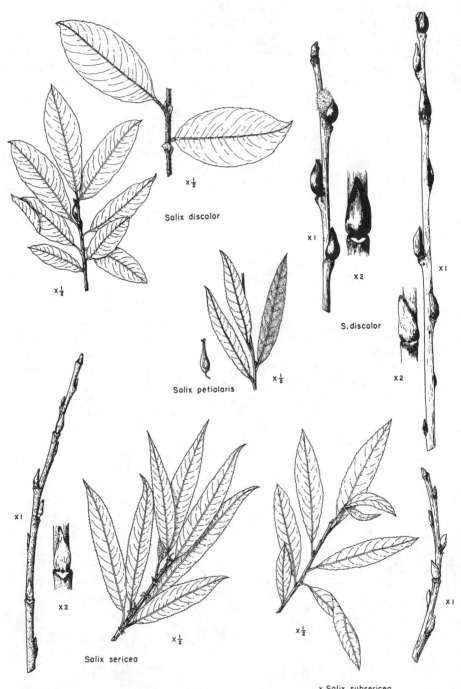

Salix discolor

$x\frac{1}{2}$

$x\frac{1}{2}$

Salix petiolaris $x\frac{1}{2}$

X I

X 2

S. discolor

X I

X 2

X I

X 2

$x\frac{1}{2}$

Salix sericea

$x\frac{1}{2}$

x Salix subsericea

X I

11. **Salix pedicellaris** Pursh   Bog Willow

Slender creeping and stoloniferous bog shrub 3–10 dm. tall, glabrous throughout. As now understood, the typical variety is less common and more circumscribed in range than var. *hypoglauca* Fern., to which all our records are referred. This has the thick leaves whitened and distinctly glaucous beneath; this character, together with the slightly revolute margin and usually obtuse apex distinguish this from our other willows. Northern in range; in our area, local in bogs.

12. **Salix rigida** Muhl.   Heart-leaf Willow

　　S. cordata Muhl., not Michx.

Shrub, 2–6 m. tall, of streambanks, ditches, and swamps over a wide range, from the maritime provinces of Canada westward to Saskatchewan and Montana, and southward, in outlying stations, to the Gulf States. Represented in our area by two intergrading varieties, extremes of which are quite distinct.

Leaves broad, 3–3.5 cm. wide...var. *rigida*
Leaves narrower, 1–2 cm. wide................................var. *angustata* (Pursh) Fern.

The large stipules of vigorous shoots are conspicuous on both varieties, which are similarly distributed over Ohio. *S. rigida* hybridizes with a number of other species; Ohio records include a hybrid of the typical variety and *S. sericea*, and a hybrid of var. *angustata* and *S. petiolaris*. Pine-cone willow-galls are common on this species.

13. **Salix glaucophylloides** Fern.   Dune Willow; Blueleaf Willow

　　S. glaucophylla Bebb, of Ed. 7 and Schaffner (1932), *S. glaucophylloides* var. *glaucophylla* (Ball) Schneid.

Low spreading shrub, 1–2.5 m. tall, of sandy shores, dunes, and sometimes wet soil. Represented in Ohio by two varieties, the typical, and var. *albovestita* (Ball) Fern., differing by having branchlets densely white-pilose and somewhat larger leaves more or less densely white-pubescent at least when young. The latter recorded only from Erie County (Cedar Point) where it grows with the typical variety. The general range of the species is northern, from Quebec eastward through the Great Lakes area to Wisconsin, with occasional inland stations.

14. **Salix discolor** Muhl.   Pussy Willow

Large shrub, sometimes tree-like, 2–6 m. tall, flowering very early, before leaf-buds start growth. Winter-buds of two sizes, the larger ament-buds conspicuous in late summer

Salix pedicellaris
var. hypoglauca

Salix rigida
↓var. rigida ←var. angustata

Salix glaucophylloides

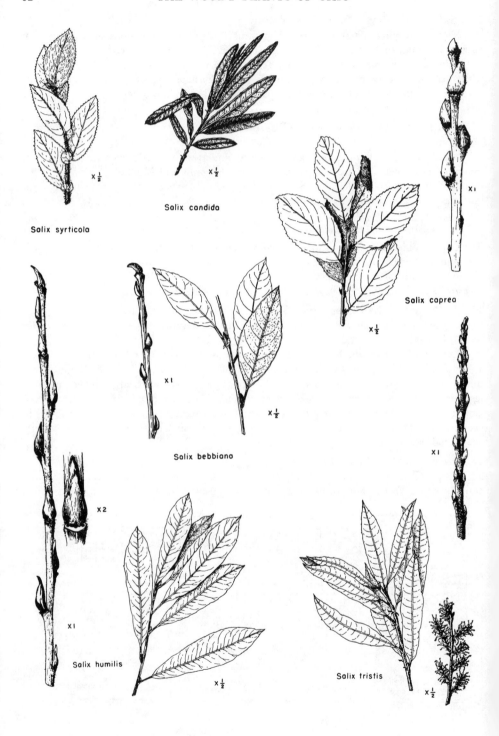

Salix syrticola

Salix candida

$\times\frac{1}{2}$

$\times\frac{1}{2}$

$\times 1$

Salix caprea

$\times\frac{1}{2}$

$\times 1$

Salix bebbiana

$\times 1$

$\times 2$

$\times 1$

Salix humilis

$\times\frac{1}{2}$

Salix tristis

$\times\frac{1}{2}$

and all through the winter. A common species of swamps and wet ground, transcontinental in the North, and ranging south to Maryland, Kentucky, and Missouri. Represented in Ohio by two varieties:

Twigs, buds and leaves glabrous...var. *discolor*
Twigs, buds, and often lower leaf-surface pubescent...................................
var. *latifolia* Anderss. (var. *eriocephala* of Ed. 7)

The latter variety usually has narrower leaves than the typical variety, and is less frequent in Ohio.

15. **Salix petiolaris** J. E. Smith
S. gracilis Anderss. var. *textoris* FERN.
Shrub or small tree, 1–3 m. tall, with gray bark. Swamps, wet prairies, and stream-banks; northern in range, sporadic southward. The linear-lanceolate, closely glandular-serrate leaves with long slender petioles give this species a distinctive appearance.

16. × **Salix subsericea** (Anderss.) Schneid.
Diffuse shrub of meadows and moist ground, mostly near streams. Considered to be a hybrid: *S. petiolaris* × *S. sericea*. Northern in range, as is *S. petiolaris*.

17. **Salix sericea** Marsh. SILKY WILLOW
Shrub with clustered stems, 1–4 m. tall; in moist soil and on the banks of streams through much of the Deciduous Forest. The appressed silvery or glistening silky pubescence of the lower leaf-surface and capsules is distinctive.

Salix discolor
↓ var discolor   ←var. latifolia

Salix petiolaris

Salix subsericea

Salix sericea

18. **Salix candida** Fluegge SAGE-LEAF WILLOW; HOARY WILLOW

Low shrub 2–10 dm. tall, the branches widely spreading, thè branchlets densely white-tomentose. The dark, dull green leaves with sunken veins (leaves densely white-tomentose beneath) suggest sage leaves, hence the common name. A transcontinental northern species extending southward into northern Ohio, where it occurs sparingly in wet prairies and bogs.

19. **Salix syrticola** Fern. SAND-DUNE WILLOW

S. adenophylla of Amer. auth., not Hook. *S. cordata* Michx. (as in Gleason) in part.

A large straggling shrub of fore-dunes along the Great Lakes. In Ohio, known only from Cedar Point, and probably extinct. The crowded broadly ovate leaves (⅔ as wide as long) with cordate base, short petioles, and large stipules, and the dense lustrous pubescence give this willow a distinct aspect.

20. **Salix bebbiana** Sarg. LONG-BEAKED WILLOW

S. rostrata Richards

Shrub, or occasionally a small tree, 2–5 m. tall, with one or a few stems. The common name refers to capsule shape—lanceolate or flask-shaped, 6–10 mm. long, tapering to a long beak. A northern species of moist soil; several varieties are recognized; ours are referable to the typical variety.

21. SALIX CINEREA L. GRAY WILLOW

Similar to *S. caprea*, and used in the same way, but less frequently planted. The hybrid, *S. cinerea* × *S. caprea*, is doubtfully recorded from Lake and Lorain counties.

22. SALIX CAPREA L. GOAT WILLOW

The pussy willow of florists; native of Europe, widely planted, and rarely escaped. Tree-like shrub, most conspicuous when flowering in early spring. Recorded for Brown, Franklin, Hamilton, Highland, and Lake counties.

23. **Salix humilis** Marsh. UPLAND WILLOW

Shrub, 1–2 m. tall, with many stems. Aments in early spring, before the leaves, resemble those of *S. discolor* but are much smaller, hence the common name, small pussy willow, sometimes applied to this species. Usually found in dry situations, but also in swamps

Salix candida

Salix syrticola

Salix bebbiana

and swampy woodlands; ranges throughout the area of the Deciduous Forest. A highly variable species, represented in Ohio by three intergrading varieties (four, if Fernald's interpretation, 1950, is followed) differing in leaf-shape and pubescence:

Leaves 1.5–2 (–3) cm. wide, oblanceolate to narrowly obovate, loosely pubescent to glabrate...
var. *humilis*
Leaves 0.7–1.5 (–2) cm. wide, narrowly oblanceolate, glabrate...........................
var. *hyporbiza* Fern. (var. *rigidiuscula* Anderss.)
Leaves 1.5–5 cm. wide, obovate to broadly obovate, densely pubescent....................
var. *keweenawensis* Farw.

The typical variety is most abundant; the second is recorded in seven counties, often with the typical; the third, a far-northern form, is known from one specimen from Margaretta Ridge, Erie County (fide Ball). The var. *microphylla* (Anderss.) Fern., included in *S. humilis* by Fernald (1950), is by Ball here interpreted as a distinct species, *S. tristis*.

24. Salix tristis Ait. Dwarf Upland Willow

S. humilis Marsh. var. *microphylla* (Anderss.) Fern., *S. humilis* var. *tristis* (Ait.) Griggs

A low shrub, usually in dry habitats. Similar to *S. humilis* but smaller in every way; leaves smaller and narrower than in that species, densely gray-tomentose beneath. Range similar to that of *S. humilis*, but more local.

25. Salix purpurea L. Purple Osier; Basket Willow

Introduced from Europe in colonial days for basket-making, and now sparingly escaped. Our records, for 16 widely scattered counties, do not distinguish between planted and escaped specimens: Ashtabula, Auglaize, Belmont, Delaware, Erie, Fairfield, Franklin, Hamilton, Jefferson, Lake, Lorain, Montgomery, Ottawa, Ross, Scioto, Stark. The sub-opposite position of leaves and buds, usually seen on at least some of the branchlets, distinguishes this from other willows.

Salix humilis
↓ var. humilis ←var. hyporhysa

Salix tristis

POPULUS L. Poplar

A genus of about 30 species; trees of the north temperate zone. All of the poplars are intolerant to shade and reproduce only in open situations. Five species indigenous to Ohio. In addition, several Eurasian species and cultivated forms occur as escapes. These cultivated forms have been multiplied by vegetative propagation, in some cases of a single distinctive individual, perhaps of hybrid origin; i.e., they are clones. As all

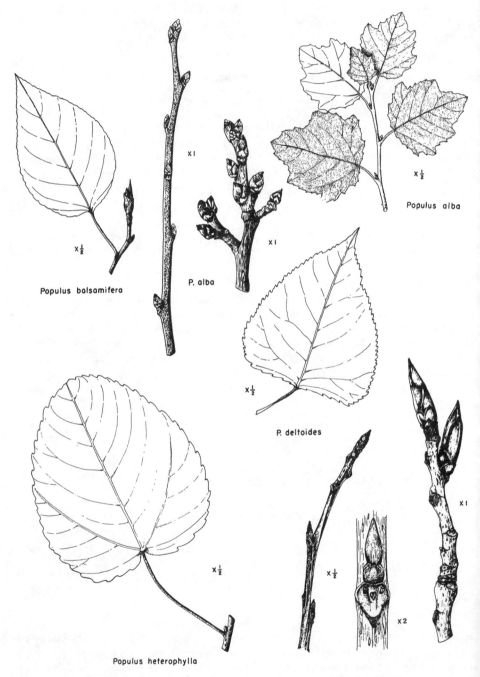

Populus balsamifera

P. alba

Populus alba

P. deltoides

Populus heterophylla

Populus deltoides

poplars are unisexual, such species are incapable of producing seed, hence of becoming naturalized. They may spread by root sprouts which persist long after the tree from which the sprouts came has been removed. Although these "species" cannot become a part of our flora, they are mentioned below because of their frequent occurrence.

a. Petioles terete or nearly so, sometimes channeled above.
  b. Leaves permanently white- or whitish-tomentose or felted beneath, bright green above, with coarse angular teeth; escaped species spreading by root-suckers.
    Leaves heavily white-felted beneath, those of long shoots somewhat lobed; bark of branches smooth and white, of old trunks black and cracked...............1. *P. alba*
    Leaves whitish to grayish beneath, less heavily felted except on vigorous young shoots, not lobed; bark of branches and young trunks green-white; tree with aspect of silver poplar...2. ×*P. canescens*
  bb. Leaves not white-felted beneath at maturity (woolly on both surfaces when young in no. 3), finely serrate or crenate.
    Leaves obtuse or bluntly acute at apex, rounded or cordate at base, usually somewhat pubescent beneath at least on veins, margins crenate-serrate; young twigs and lower part of buds downy...3. *P. heterophylla*
    Leaves acute to acuminate at apex, more or less lustrous beneath, fragrant when crushed, finely serrate; buds viscid and fragrant.
      Twigs glabrous, leaves glabrous, with metallic luster beneath........4. *P. balsamifera*
      Twigs and leaves beneath somewhat pubescent................5. ×*P. gileadensis*
aa. Petioles flattened laterally.
  b. Leaves deltoid to deltoid-rhombic, with a narrow colorless border, the teeth with hard incurved tip; trees with dark furrowed bark, smoother and yellowish on branches.
    Leaves with basal glands.
      Basal glands 2 or more................................6. *P. deltoides*
      Basal glands 1–2, or absent................................7. ×*P. canadensis*
    Leaves without basal glands......................................8. *P. nigra*
  bb. Leaves round-ovate to elliptic, without colorless border, the teeth without callous tip; trees with pale smooth bark.
    Leaves coarsely toothed, usually with fewer than 12 teeth to a side; expanding leaves white-pubescent...9. *P. grandidentata*
    Leaves finely toothed, usually with more than 12 (to 40) teeth to a side.............
          10. *P. tremuloides*

WINTER KEY

a. Buds glutinous or resinous, at least when swelling; glabrous.
  b. Buds small, less than 1 cm. long, slender; lateral buds closely appressed; tree with erect branches and narrow crown................................8. *P. nigra italica*
  bb. Buds larger, usually 1 cm. long or more; tips, at least of lateral buds, divergent; tree with spreading crown.
    Terminal bud very large, about 2 cm. long, very resinous and fragrant; twigs reddish-brown.................................4. *P. balsamifera* and 5. ×*P. gileadensis*
    Terminal bud smaller, about 1 cm. long; twigs often strongly angled, yellowish or greenish-brown..............................6. *P. deltoides* and 7. ×*P. canadensis*
aa. Buds not resinous or only slightly so.
  b. Twigs light yellowish; pith orange; buds reddish-brown, scales pubescent toward base....
       3 *P. heterophylla*
  bb. Twigs not light yellowish; pith light.
    Twigs greenish or greenish-brown and conspicuously cottony pubescent.
      Twigs green and sparsely cottony pubescent; buds brown, appressed; tree with erect branches and narrow crown................................*P. alba pyramidalis*
      Twigs greenish-brown and cottony pubescent; buds white-tomentose, not appressed; tree with spreading crown...1. *P. alba*
    Twigs brownish and glabrous or sparsely pubescent.
      Twigs dull brownish and often grayish-tomentose; bark of branches and young trunks green-white; trees strongly resembling silver poplar...............2. ×*P. canescens*
      Twigs reddish-brown.
        Twigs slender; buds glabrous, lateral buds appressed; bark often whitish, then resembling a birch...10. *P. tremuloides*
        Twigs not slender; buds finely tomentose, lateral buds not appressed; bark of branches and young trunks light yellowish or buff............9. *P. grandidentata*

1. POPULUS ALBA L. SILVER POPLAR. WHITE POPLAR

From Europe; formerly commonly planted; patches of root-sprouts often seen along roadsides or along the borders of fields. Widespread in Ohio; specimens from more than three-fourths of the counties. A number of horticultural varieties, of which var. *pyramidalis* Bge. (*P. alba boleana* Lauche), a columnar tree, is sometimes planted.

2. × POPULUS CANESCENS (Ait.) Sm. GRAY POPLAR

P. alba × tremula L.

A hybrid of white poplar and European aspen; sometimes planted, and escaping in the same manner as does *P. alba*. Southwestern Ohio; specimens from Brown and Hamilton counties.

3. **Populus heterophylla** L. SWAMP COTTONWOOD

A tree of inundated swamps and bottomlands of the Atlantic and Gulf Coastal Plain and Mississippi Valley north to the lower Wabash Valley in southwestern Indiana; local northward of this area in Indiana and Ohio. Its range in Indiana and Ohio suggests that it may have reached northern Ohio by migration along the Wabash-Maumee valleys.

4. **Populus balsamifera** L. BALSAM POPLAR

A tree of the North, where it is the largest of the sub-arctic trees, ranging from the Atlantic Coast of New England, Newfoundland, and northern Quebec westward and northwestward across Canada to the Yukon; in Ohio, local in a few northern counties. Sometimes confused with Balm-of-Gilead. The species includes two varieties:

P. BALSAMIFERA var. BALSAMIFERA

P. tacamahaca Mill. (name used by Sargent, 1919)

P. balsamifera var. **subcordata** Hylander

P. balsamifera var. *fernaldiana* Rouleau, and known until recent years as:

P. balsamifera var. *candicans* (Ait.) A. Gray, and as *P. balsamifera* var. *michauxii* (Dode) Henry

(However, the original specimens on which the two latter names are based belong to the typical variety.) Our specimens can be referred to var. *subcordata*.

5. × POPULUS GILEADENSIS Rouleau BALM-OF-GILEAD

P. candicans Ait. (name used by Schaffner, 1932).

Widely planted, and spreading by sprouts. Origin not definitely known; perhaps a sterile hybrid of *P. deltoides* and *P. balsamifera*, or a sterile clone of *P. balsamifera* var. *subcordata*, with which it is often confused. Reported from a score of counties.

Populus heterophylla

Populus balsamifera

6. **Populus deltoides** Bartr. Cottonwood

P. balsamifera L. in part (sensu Sargent, 1933), *P. monolifera* Ait.

A common and widespread tree, ranging over the eastern half of the United States, except in the higher mountains. A large tree of floodplains and bottomlands; can endure long periods of partial submergence, hence is frequent on the alluvial bottoms of large streams subject to overflow. Fruiting catkins 15–20 cm. long, from very large outwardly curving buds, the 3–4 valved capsules opening in May, releasing the seeds, each furnished with a tuft of long cottony hairs; seeds retain their viability but a short time, germinate readily on mud left by receding flood-water, on moist depressions between dunes, etc. Generally distributed in Ohio in suitable habitats.

Geographic varieties can be recognized, of which two occur in Ohio:

Leaves of short branches triangular-ovate, 6–12 cm. long and broad, with 2–3 basal glands...
var. *deltoides*

Leaves definitely longer than broad, sometimes 16 cm. long, with 3–4 basal glands..........
var. *missouriensis* (Henry) Rehd.

7. × Populus canadensis Moench Carolina Poplar

× *P. eugenei* Simon-Louis, *P. deltoides* × *nigra* L., *P. canadensis* var. *eugenei* (Simon-Louis) Schelle.

The commonly planted Carolina poplar, which resembles cottonwood (*P. deltoides*), one of the parent species, is vegetatively reproduced. It includes hybrid clones which originated in a number of different places. The var. *eugenei*, included in × *P. canadensis*, has a narrow pyramidal crown; the name "Carolina poplar" is more correctly applied to this variety. Leaves with or without 1 or 2 basal glands.

8. Populus nigra L. Black Poplar

Populus nigra var. italica Muench. Lombardy Poplar

The narrowly columnar Lombardy poplar is interpreted as a clone derived from a single freakish individual (Fernald, 1950) or as a hybrid clone (Little, 1953). Reported from a dozen or more widely scattered counties.

9. **Populus grandidentata** Michx. Bigtooth Aspen

A tree of northern United States and adjacent Canada, local south of the Ohio and Potomac rivers. Widely distributed in Ohio, but more common in the northern part of the state where it is usually in moist soil near streams, ponds, or lakes; in the southern part, generally found in clearings, on eroded slopes, or in poor second-growth woodland. Young leaves often densely white-pubescent.

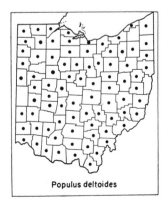

Populus deltoides

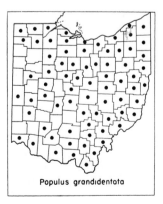

Populus grandidentota

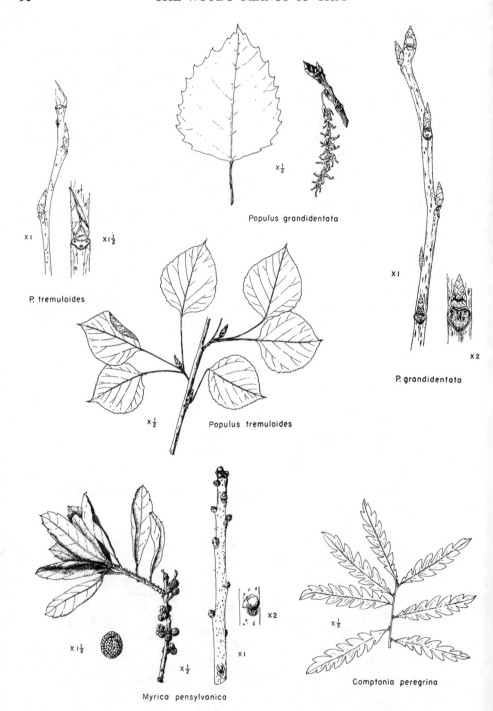

Populus grandidentata

P. tremuloides

P. grandidentata

Populus tremuloides

Myrica pensylvanica

Comptonia peregrina

10. **Populus tremuloides** Michx. Quaking Aspen
One of our widest ranging species, extending from Newfoundland across Canada to Alaska, and southward in the mountains of the West and through the northern tier of states and New England in the East. Common through northern Ohio, local southward to the Ohio River.

MYRICACEAE

A family of some 50 species of wide geographic distribution, variously classified into one, two, or three genera. All are shrubs or small trees, with resinous-dotted, usually fragrant leaves. Flowers unisexual, borne in short scaly aments. Two species in Ohio.

Leaves entire or with a few low teeth toward apex; the small globular verrucose fruits heavily covered with wax...*Myrica pensylvanica*
Leaves pinnatifid, the sinuses between lobes narrow; the small nutlets subtended by elongate bractlets, the whole fruiting ament bur-like..........................*Comptonia peregrina*

MYRICA L.

1. **Myrica pensylvanica** Loisel. Bayberry
M. carolinensis of Gray, ed. 7, and B. & B., ed. 1.
A handsome shrub with dark green foliage; conspicuous in winter because of the grayish-white waxy "berries" which persist until spring. The granular wax is melted off for bayberry candles. Local in northeastern Ohio, on beaches and in bogs.

COMPTONIA L'Her

1. **Comptonia peregrina** (L.) Coult. Sweet-fern
C. asplenifolia Ait., *Myrica asplenifolia* L. in part.
A shrub of open oak woodlands, pastures, and roadsides, often in sandy and infertile soils, where it sometimes forms large patches. The typical variety widely distributed north of our range—from eastern Canada to Manitoba—and extending southward in the Appalachian upland. In Ohio, confined to a few northern counties. The glistening yellow resinous dots and pinnatifid leaves distinguish this from all of our other shrubs. Two (or three) varieties may be distinguished, of which only the typical occurs in Ohio.

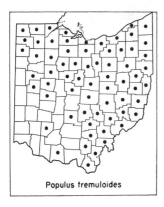

Populus tremuloides

Myrica pensylvanica

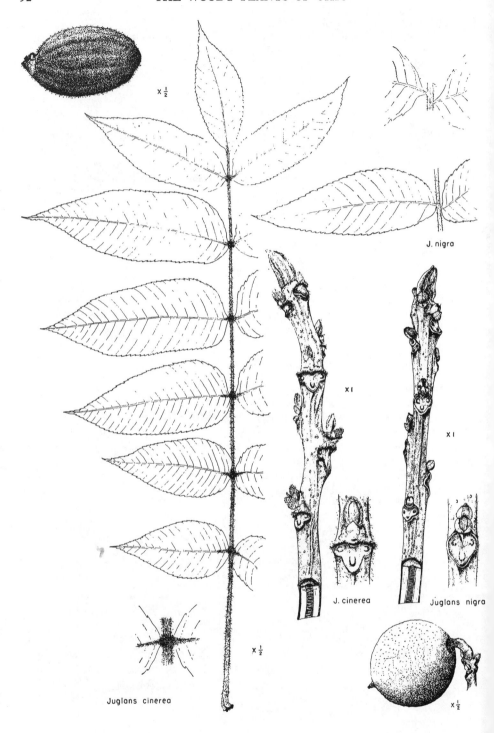

x ½

J. nigra

x 1

x 1

J. cinerea

Juglans nigra

x ½

Juglans cinerea

x ½

JUGLANDACEAE

A relatively small family (about 50 species) of trees or shrubs, mostly of the north temperate zone. The leaves of all species are alternate, pinnate, and without stipules. Monoecious, the staminate flowers in elongate catkins, the pistillate inconspicuous, solitary, or in small clusters. Fruit large, a fibrous, fleshy, or woody outer layer (exocarp) enclosing a nut. Two of the six genera are well represented in North America.

JUGLANS L. Walnut

Two of the 15 species of walnut are indigenous to eastern United States and to Ohio; four more occur in the West. A Japanese walnut has been planted in limited quantity on Waterloo State Forest. The English or Persian walnut (*J. regia* L.), which is native from southeastern Europe to Himalaya and China, is occasionally planted in our area; it has fewer leaflets than our native species, usually 7–9. The chambered pith, which can be seen clearly if a twig is cut longitudinally, is a distinctive character of the genus. The fragrance of the foliage is due to the glandular hairs of the lower leaf-surface. The generic name is derived from the Latin *Jovis glans*, the nut of Jupiter.

Pith dark- or chocolate-brown; nuts short-cylindric and pointed, the green hull viscid-pubescent; terminal bud elongate, 12–20 mm. long; leaf-scar with a small circular or elongate downy pad on its upper margin...1. *J. cinerea*
Pith cream-colored to very light brown; nuts subglobose, the green hull not viscid-pubescent; terminal bud shorter, broadly ovate to spherical, usually less than 10 mm. long; leaf-scar without downy pad; leaves with or without a terminal leaflet....................2. *J. nigra*

1. **Juglans cinerea** L. Butternut
   A tree of more northern range than the black walnut, and reaching its largest size in the Northeast. The bark of branches and small to medium size trunks is smooth and light gray, later divided by dark furrows. A medium size tree of mesic ravine slopes, creek bottoms, and river banks, widely distributed in Ohio, but less often seen in the west-central and northwestern part of the state where rich mesic sites are infrequent. Foliage more clammy and fragrant than that of black walnut; leaflets usually 9–19, a terminal leaflet normally present.

Comptonia peregrina

Juglans cinerea

2. Juglans nigra L. BLACK WALNUT

Juglans nigra

A large forest tree occasionally reaching a height of 150 ft. and trunk diameter of 4–6 ft., more often about 100 ft. in height, with straight clear trunk about one-half the height of the tree. Leaflets usually 11–23 (or 10–22), the terminal leaflet often very small or absent. A tree of rich soil, almost always present in mixed mesophytic forest communities, and frequent on high-level bottomlands. Ranging almost throughout the Deciduous Forest, except at the north; general in Ohio except in the prairie peninsula, where it is infrequent.

CARYA Nutt. (*Hicoria* Raf.) HICKORY

A genus of about 20 species, two of which are in eastern Asia, the others in eastern North America. Many of the species are valuable timber trees, and some, especially pecan, yield edible nuts of considerable commercial value. Leaves of all species pinnately compound. The best characters for distinguishing the hickories are the mature leaves, mature fruits, terminal winter-bud, and bark of the trunk, hence specimens collected late in the season are most satisfactory. It should be noted, however, that the outer bud-scales of some species are deciduous in late fall or early winter, that the winter-bud may therefore be different from the apparently mature bud of stems still retaining foliage. Natural hybrids occur and complicate identification; trees with intermediate characters, some of which are first-generation hybrids, others the progeny of back-crosses with one of the parent species, cannot always be placed in specific taxonomic "pigeon-holes." Most of the species are variable, especially in leaf-shape; named varieties of several species are distinguished by some authors. The staminate flowers of all species except pecan are in peduncled fascicles of three (rarely 4 or 5) elongate catkins. The generic name is from the Greek *Karua*, the ancient name of walnut.

a. Terminal buds with scales valvate in pairs.
  b. Leaflets 7–19, long-pointed, coarsely and often doubly serrate, the lower lateral falcate, the terminal stalked; fascicles of staminate aments nearly sessile; nut thin-shelled, sweet, nearly cylindric; husk splitting nearly to base and sometimes persistent on branch during winter; terminal bud with yellow jointed hairs and pale tomentum.......1. *C. illinoensis*
  bb. Leaflets usually 7 or 9, rarely 5 or 11, long-pointed (except on sprouts and saplings), serrate except toward base, slightly falcate, the terminal nearly sessile; fascicles of staminate aments peduncled; nut often broader than long, very bitter; husk splitting about one-half way to base; terminal bud covered with golden-yellow scale-like particles..........
        2. *C. cordiformis*
aa. Terminal buds with scales imbricate.
  b. Margin of leaflets strongly ciliate, and with a dense tuft of hairs on one or both sides of each tooth; leaflets usually 5, glabrous or pubescent; terminal bud 1–2 cm. long, tawny to blackish-brown or dark gray, apices of outer scales prolonged; inner bud-scales greatly enlarging and pink or purplish in spring, the outer tawny- to golden-pubescent on outer surface; twigs dark; bark of trunk light gray, shaggy, separating into long loose plates; husk 3–12 mm. thick, nut 4-angled......................................3. *C. ovata*
  bb. Margin of leaflets without tufts near apices of teeth, sometimes ciliate when young. Lower surface of leaves densely pubescent; leaflets 7–9, when young the margins ciliate; terminal bud 1–3 cm. long; inner bud-scales greatly enlarging when growth starts; twigs stout; fruit large, husk thick, nut not strongly ridged above.

Bark of trunk shaggy, separating into long loose plates; twigs orange or orange-tan (at least on one side); leaf-rachis nearly glabrous or pubescent with scattered hairs and indistinct fascicles; terminal bud 2–3 cm. long, brown, the apices of outer scales prolonged into widely spreading points; enlarged inner bud-scales golden-brown, satiny pubescent outside, glabrous within; husk 8–12 mm. thick.....4. *C. laciniosa*

Bark not shaggy, deeply furrowed; twigs brown, pubescent; leaf-rachis pubescent (often densely) with evidently fascicled hairs; terminal bud 1–2 cm. long, the outer scales deciduous in autumn, enlarged inner bud-scales pubescent on both faces; husk 4–5 mm. thick...5. *C. tomentosa*

Lower surface of leaves glabrous or nearly so (except in a pubescent variety) or veins pubescent; leaf-rachis usually glabrous, leaflets 5–7, the margins not ciliate; terminal bud 8–12 mm. long, inner bud-scales scarcely enlarging; twigs slender; husk thin, 1–3 mm.

Winter-buds and upper internode of twig more or less densely dotted with small yellow scales; leaflets usually 7, dotted beneath; bark close or separating into small scales or plates; husk minutely warty, splitting to base........6. *C. ovalis*

Petiole, rachis, and lower leaf-surface glabrous or nearly so...................
vars. *ovalis, obcordata, obovalis, odorata*

Rachis and (usually) petiole and lower leaf-surface densely pubescent; fruit sometimes with stipe...var. *mollis*

Winter-buds and upper internode not yellow-dotted, glabrous; twigs bright brown; leaflets usually 5, inconspicuously dotted beneath; bark close, not scaly; husk smooth, splitting only toward apex, or later to base along one suture, often constricted to stipe-like base (i.e. pear-shaped).......................7. *C. glabra*

Petiole, rachis, and lower leaf-surface glabrous or nearly so...................
vars. *glabra* and *megacarpa*

Rachis and (usually) petiole and lower leaf-surface densely pubescent...var. *hirsuta*

(For additional characters, see Manning, 1950)

WINTER KEY

a. Buds yellow, densely glandular dotted; bud-scales valvate in pairs, the outer often leaf-like...
2. *C. cordiformis*

aa. Buds not yellow.

b. Bud-scales valvate in pairs, outer pair brown, tomentose and glandular, early deciduous, inner densely tawny-pubescent; lateral buds broad-ovate...............1. *C. illinoensis*

bb. Bud-scales not valvate in pairs, but successively inwrapping.

c. Terminal bud large, over 1 cm. long.

Outer scales of terminal bud deciduous in autumn or winter.

Outer scales deciduous in autumn; inner scales yellowish, silky pubescent, and slightly glandular; twigs pubescent.........................5. *C. tomentosa*

Outer scales deciduous in winter; inner scales ciliate, sometimes finely puberulent; twigs glabrous...6. *C. glabra*

Outer scales of terminal bud persisting until spring, keeled; bud ovoid to narrowly ovoid.

Twigs orange-brown; buds brown, very large, 2–3 cm. long, apices of outer scales prolonged into long stiff widely spreading points..............4. *C. laciniosa*

Twigs greenish-, reddish-, or grayish-brown; buds blackish-brown, smaller, 1–2 cm. long; apices of outer scales prolonged into points.............3. *C. ovata*

cc. Terminal bud smaller, mostly less than 1 cm. long; outer bud-scales deciduous in winter.

Bark of trunk dark gray, smooth or fissured on old trunks, not exfoliating; inner bud-scales ciliate, sometimes finely puberulent....................6. *C. glabra*

Bark of trunk close or exfoliating in narrow plates, less shaggy than in *C. ovata*; terminal bud and upper internode of twig more or less dotted with small yellow scales..7. *C. ovalis*

1. **Carya illinoensis** (Wang.) K. Koch Pecan

C. Pecan (Marsh.) Engl. & Graebn., *Hicoria Pecan* Britt.

Pecan is a tree of alluvial bottomlands of the lower Ohio and Mississippi river basins and of the west Gulf Coastal Plain. In Indiana, it is frequent in the Wabash bottoms and occurs in scattered locations as far east as Clark and Jackson counties. In Ohio,

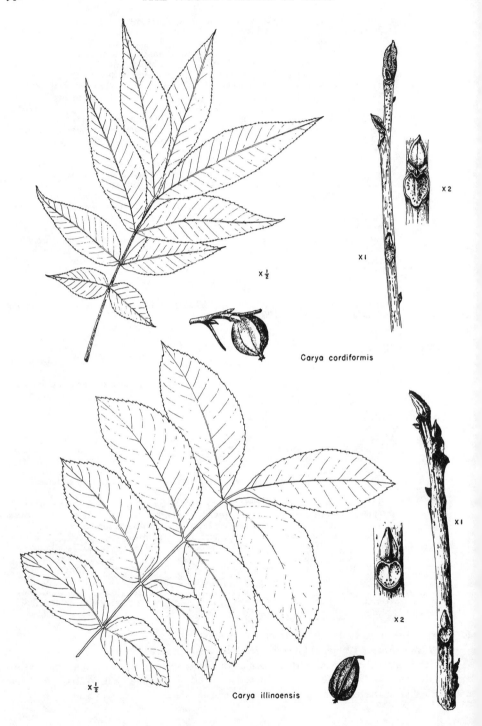

x½

x1

x2

Carya cordiformis

x½

x1

x2

Carya illinoensis

a few trees grew in the Miami Valley in Butler County; because of the irregular spacing and different ages of the trees, the species is thought to be indigenous there. Other reported occurrences (Ross and Greene counties) cannot be checked and no data on habitat are available. As the species has been planted, these doubtful records are not mapped. A large tree in eastern Kentucky (Knott Co.), in a situation which makes it appear indigenous, is known to have been grown from a nut brought from Arkansas. Wild pecans have smaller and thicker-shelled nuts than the horticultural forms now grown. Pecan has been planted on Waterloo State Forest.

2. **Carya cordiformis** (Wang.) K. Koch   BITTERNUT HICKORY
Hicoria cordiformis (Wang.) Britt., *Hicoria minima* (Marsh.) Britt.

Bitternut has a wider range than any other hickory, extending farther north and slightly farther west. It is a tree of mesic situations, and general in Ohio. It is easily recognized from mid-summer until growth starts in the spring by the valvate yellow-scurfy terminal bud.

3. **Carya ovata** (Mill.) K. Koch   SHAGBARK HICKORY
Range almost as extensive as that of bitternut hickory. Widely distributed in Ohio, and commonly associated with oaks in both dry and wet situations. The shaggy bark of this species (and *C. laciniosa*) is a conspicuous feature, especially striking on tall straight trunks of forest-grown trees. The nuts of both species are equally good. Several varieties may be distinguished, but these seem to have little taxonomic value; they are based on differences in size and shape of nuts, and kind and density of pubescence of leaves. The dense tufts of hairs on leaf-margin near teeth are found in all varieties.

4. **Carya laciniosa** (Michx. f.) Loud.   BIG SHELLBARK.   KINGNUT
A large forest tree, more circumscribed in range than our other hickories; largely confined to the Ohio and Mississippi river basins in central United States. The largest Ohio tree of this species, 11 ft. 6 inches in circumference and 130 ft. in height, is in Gallia County (Ohio Forestry Ass.). Specimens from widely separated parts of Ohio, but less frequent than *C. ovata*. The glabrous orange-tan twigs and large richly colored buds are distinctive features of this species. In leaf, distinguished from *C. ovata* by the absence of the dense tufts of hairs in ciliate margin near teeth, and (less definitely) by the 7, instead of 5, leaflets.

Carya cordiformis   Carya ovata   Carya laciniosa

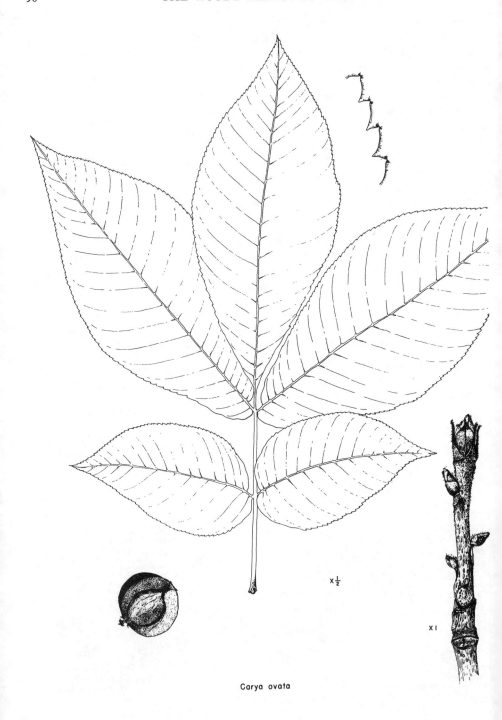

x ½

x I

Carya ovata

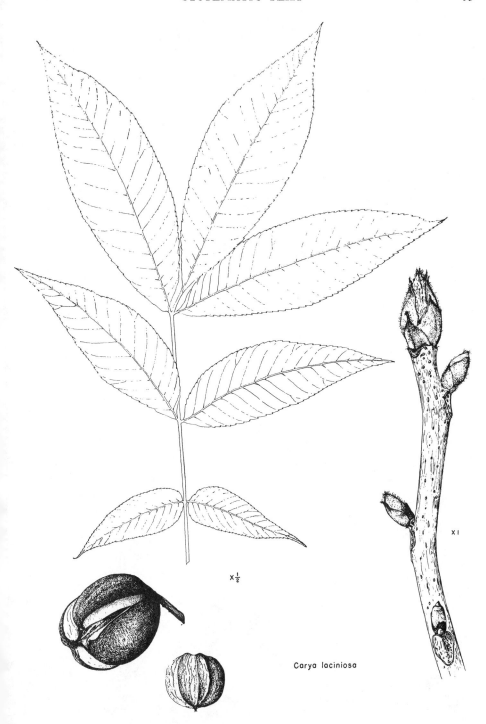

$x\frac{1}{2}$

$x\,l$

Carya laciniosa

x½

x2

x1 x1

C. ovalis

Carya glabra

x2

x½

Carya tomentosa

x½

Carya ovalis

x1

5. **Carya tomentosa** Nutt. Mockernut Hickory
Carya alba (Mill.) K. Koch, *Hicoria alba* (L). Britt.
Similar in range to *C. ovata*, but not extending as far northwestward. Foliage generally more fragrant when handled than that of other hickories. Nut thick-shelled, sweet and edible. The evidently fascicled hairs on leaf-rachis are a distinctive feature of mockernut hickory. The outer bud-scales of the terminal bud are deciduous in fall or early winter; thus the buds differ greatly in appearance from those of the other hickories with large buds, *C. laciniosa* and *C. ovata*. If, however, the outer bud-scales are still attached (late summer and fall), the bud resembles that of *C. ovata*, but the diverging tips of the outer scales are densely brown hairy, almost brush-like, and the scales elsewhere bear fascicled hairs.

6. **Carya glabra** (Mill.) Sweet Pignut
Herbarium specimens of this species and *C. ovalis* cannot always be distinguished; mature fruits with husks are necessary. Furthermore, many trees are intermediate in character. In a recent publication of the Forest Service (Little, 1953) *C. ovalis* is given as a synonym of *C. glabra*, "since it is doubtful whether the two species can be maintained, and also whether many persons will wish to do so." It is said that the two cannot be separated "except with completely mature fruit collected in November" (Manning, 1950). Characters given in our key to separate the two species are not constant, but are helpful in distinguishing the majority of specimens. The var. *megacarpa* Sarg. differs from typical *C. glabra* in its larger, obovoid fruit with thicker husk; var. *hirsuta* Ashe (considered a variety of *ovalis* by Fernald, with "rachis, and often petiole and lower surface of leaflets densely pubescent" and obovoid fruit (Manning, 1950) is recorded for Coshocton, Cuyahoga, and Highland counties.

7. **Carya ovalis** (Wang.) Sarg. Sweet Pignut
C. microcarpa Nutt. (*C. ovalis* var. *ovalis*)
A polymorphic species in which four varieties are distinguished which are doubtless present in Ohio. These differ in size and shape of nuts and fruit, and by Gleason (1952) are considered as "apparently only phases of its fluctuating variability." In such a variable species, and one which intergrades with the preceding, the recognition of varieties with essentially similar range is unwise, except for the specialist; var. *mollis* (Ashe) Sudw. (similar in pubescence to *C. glabra* var. *hirsuta*) is recorded from Adams and Coshocton

Carya tomentosa

Carya glabra

Carya ovalis

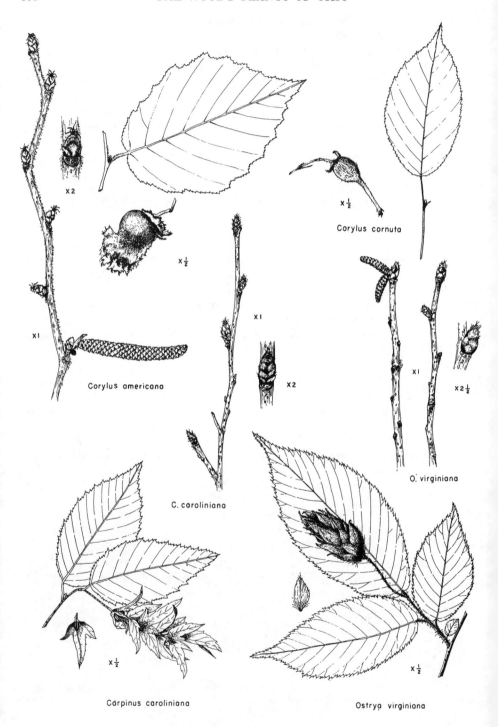

x 2

Corylus cornuta

x ½

x ½

x 1

x 1

Corylus americana

x 2

C. caroliniana

x 1

O. virginiana

x 2½

Cárpinus caroliniana

x ½

Ostrya virginiana

x ½

counties; var. *obcordata* (Muhl. & Willd.) Sarg. from Butler, Franklin, Lucas, and Trumbull counties. The maps for this and the preceding species are based upon specimens determined by W. E. Manning. Intermediates between the two species, and specimens which (because incomplete, lacking mature fruit) cannot with certainty be identified, add seven counties (Ashland, Geauga, Meigs, Morrow, Muskingum, Vinton, Warren) to the composite picture of distribution.

CORYLACEAE (BETULACEAE)

A family of over 100 species classified into six genera, one of which (with only two species) is limited to China, the others widespread in the northern hemisphere (species of *Alnus* south to Peru) of the New and Old World, some extending into arctic and alpine regions. Monoecious (rarely dioecious) trees or shrubs, with staminate flowers in catkins, the pistillate in clusters, short spikes, or catkins. Fruit a 1-celled, 1-seed nut or nutlet, or, in *Betula*, a very small samara.

CORYLUS L. Hazel. Hazelnut

Shrubs of North America, Europe, and Asia; about 15 species, two eastern American, both in Ohio. Several European and Asiatic species are in cultivation, including a purple-leaved form of the European hazel or filbert (*C. avellana* L.) The sweet and edible ovoid or subglobose nuts of the native species enclosed in a leafy cup or involucre of two enlarged bractlets partially grown together and apically cut into acute irregular lobes. Generic name from the Greek *corys*, a helmet.

> Twigs and petioles usually stipitate-glandular and more or less pubescent; mature involucre of two broad pubescent and often glandular-bristly leafy bracts longer than nut, cut irregularly along upper margin, and united toward base; leaves irregularly and doubly serrate, broadly ovate or oval...1. *C. americana*
> Twigs and petiole without glands except near nodes, young twigs slightly hairy; mature involucre densely bristly, composed of united bracts prolonged beyond nut into a long slender beak; leaves coarsely serrate, usually slightly smaller and narrower.............2. *C. cornuta*

1. **Corylus americana** Walt. Hazel

A common and widespread shrub, often forming large patches. Its range extends from Maine to Saskatchewan and south to Georgia and Oklahoma. Frequent and general in Ohio, in a variety of habitats: roadsides and thickets, mesic woods, dry rocky slopes, borders of prairie patches. Staminate catkins mature in February and early March. Length of glandular hairs and density of pubescence vary greatly.

Corylus americana

Corylus cornuta

2. **Corylus cornuta** Marsh. Beaked Hazel
   C. rostrata Ait.

Although the range of this species is more extensive than that of the last—given as Newfoundland to British Columbia, south to Georgia, Tennessee, Missouri, eastern Kansas and Colorado—it is very rare and local in much of this area, and absent from some of the included states (Indiana, Kentucky). In Ohio, known only from the slopes of Grand River in Ashtabula County, and north-facing bluffs of Scioto Brush Creek in Adams County where it grew in a hemlock-beech woods (site destroyed by relocation of N. & W. Railway).

OSTRYA Scop. Hop-hornbeam

Small to medium-size trees of the northern hemisphere. In addition to our species, two others occur in the Southwest. Staminate catkins slender and pendulous, evident in winter, pistillate slender, at first upright, in fruit inclined downward. The common name suggests the resemblance of the fruiting catkin to the hop (*Humulus*); at this time each small smooth nut is enclosed in a bladdery ellipsoid involucre, these overlapping one another in a cone-like cluster. Name from the Greek *ostrua*, a tree with very hard wood.

1. **Ostrya virginiana** (Mill.) K. Koch Hop-hornbeam

Foliage resembling that of elm or birch; bark of trunk scaly, splitting into thin narrow shaggy plates. Ranging throughout the Deciduous Forest; general and abundant in Ohio, where it grows in a variety of sites: woods (dry, wet, and mesic), dry rocky slopes and ledges, rocky shores (Erie Co.), and ravines. The forma *glandulosa* is distinguished by the stipitate-glandular new branchlets and petioles; however, all gradations occur between those densely glandular to those sparsely glandular or eglandular. Eglandular specimens are more frequent in southwestern Ohio, glandular in the Allegheny Plateau. Another variation—not described in manuals—appears to be geographically more significant. Typical *O. virginiana* has the "involucral sacs bristly-hairy at the base;" a form with the entire surface (especially at tip) pubescent with shorter stiff hairs, and on the principal veins, with bristly hairs like those at the base, is concentrated in the southwestern quarter or less of the state, with intermediate forms more or less frequent in the southern half of the Allegheny Plateau. As the hairs are easily detachable, the mature or almost mature strobile cannot be handled without itching or irritation caused by slight piercing of the skin by the bristly hairs. All combinations of glandular and bristly-hairy characters are found. Eglandular specimens with involucral sacs bristly-hairy throughout

Ostrya virginiana

Carpinus caroliniana

are almost confined to southwestern Ohio. Specimens with glandular petioles and "typical" sacs are most frequent in northeastern Ohio. The forms are not separated on our map.

CARPINUS L. Ironwood. Hornbeam

Small trees or tall shrubs of the northern hemisphere, extending into Central America. About 25 species, only one in eastern North America. The European hornbeam, *C. betulus* L., is commonly planted. Staminate and pistillate catkins developing from scaly buds, the staminate pendulous, the fruiting catkins pendulous, made up of toothed bracts subtending the small several-veined nuts.

1. **Carpinus caroliniana** Walt. Ironwood. Hornbeam. Blue-beech.

Small trees with close gray bark with a few ridges and furrows like taut muscles. Range similar to that of *Ostrya*. Common and widely distributed in Ohio, in moist and mesophytic woods. Two varieties are recognized by Fernald (1950) and distinguished as follows:

Bracts of fruiting aments blunt or pointed, entire or with 1–2 blunt teeth; leaves oblong to narrowly oblong-ovate, acute, serrate with low teeth (1 mm. or less high); southern. .var. *caroliniana*
Bracts of fruiting aments acute, usually with 1–5 sharp teeth; leaves oval or narrowly ovate and caudate-tipped, serrate with sharp slender teeth often 1–3 mm. high; published range includes Ohio. . . . . .var. *virginiana* (Marsh.) Fern.

Both leaf-shape and bract characters vary, but often not together. The species as a whole is highly variable, and varieties (as named) cannot be distinguished satisfactorily in Ohio; a few specimens might be assigned to var. *caroliniana*, more to var. *virginiana*. Gleason (1952) does not recognize varieties and Little (1953) reduces var. *virginiana* to synonomy.

BETULA L. Birch

A genus of about 40 species, most of which are northern in distribution; river birch (*B. nigra*) is an exception. Some are large trees, some are erect or depressed shrubs. Staminate flowers in elongate pendant catkins which develop from stiff short catkin-buds produced the previous season. Pistillate catkins short and oval or cylindric, ascending or later pendulous, at maturity cone-like, composed of closely imbricated 3-lobed bracts subtending the fruits. Fruit a winged nutlet or samara, the very small elliptic body bilaterally winged and tipped with two spreading stigmas. The bark of each species of birch is distinctive and in most cases furnishes a ready means of field identification. In addition to the following, the European birch (*B. pendula* Roth, *B. alba* L. in part) and the white birch of Europe (*B. alba* L., or *B. pubescens* Ehrh.) are planted; the former is occasionally adventive in northeastern Ohio. A number of natural hybrids are known.

a. Bog shrubs with dark close bark and soft-downy or tomentose twigs; leaves pubescent beneath, orbicular to broadly oval or obovate, coarsely dentate, those of fertile branches 1–3 (–4) cm. long (larger on vegetative sprouts), with 4–6 pairs of lateral veins; scales of the fruiting catkins about as wide as long, with widely divergent lateral lobes. . . . . .1. *B. pumila*
aa. Trees (sometimes bushy), with young branchlets glabrous or pubescent; leaves mostly more than 3 cm. long.
   b. Leaves with 7–12 pairs of lateral veins; bark not white.
      Leaves broadly cuneate at base; bark and twigs not aromatic; bark curly, salmon-color or pinkish-tan; young branchlets pale-tomentose; leaves pale-pubescent on both sides when young, soon glabrous and more or less lustrous above, remaining pubescent on petioles and veins beneath, rhombic-ovate, finely but sharply doubly-serrate except. below where converging toward petiole, 3–8 cm. long, with 7–9 pairs of lateral veins; scales of fruiting catkins tomentose, with 3 nearly equal narrow-oblong lobes above middle. . . . . .2. *B. nigra*

Leaves cordate, subcordate, or convexly tapering to base, with 9–11 pairs of lateral veins; bark and twigs aromatic.

Bark silvery- or yellowish-gray, curly exfoliating; twigs faintly aromatic; leaves convexly tapering or slightly cordate at base, ovate-lanceolate to elliptic, or rarely oval, thin, coarsely and irregularly sharply toothed, 6–11 cm. long, glabrous and dull above, pubescent on veins beneath; scales of fruiting catkins more or less pubescent, cut to or below the middle, lobes ciliate....................................3. *B. lutea*

Bark dark, close, resembling that of cherry; twigs strongly aromatic, wintergreen-flavored; leaves cordate or subcordate at base, ovate-oblong, sharply and doubly serrate, 6–12 cm. long, glabrous and more or less lustrous above, appressed pubescent on veins beneath; scales of fruiting catkins glabrous, lobes short and divergent.....
4. *B. lenta*

bb. Leaves with 3–7 pairs of lateral veins; bark white or whitish.

Bark chalky-white, rather close, not readily exfoliating; leaves deltoid or deltoid-ovate, the base truncate, the apex long-acuminate, coarsely serrate except toward base, glabrous, glutinous when young; fruiting catkins 1–3 cm. long, their scales puberulent, the lateral lobes divergent, much larger than the short terminal lobe....5. *B. populifolia*

Bark creamy- or pinkish-white (warm-brown on young trunks), readily exfoliating; leaves ovate to elliptic-oblong, serrate, cuneate to rounded at base, acute to acuminate at apex, pubescent in the vein-axils beneath; fruiting catkins 3–6 cm. long, their scales pubescent to glabrous, with divergent rhombic lateral lobes...........6. *B. papyrifera*

WINTER KEY

a. Shrubs; twigs soft-downy; buds pubescent with ciliate scales, oblong to oval-oblong, blunt, leaf-buds 2–3 mm. long, slender, with 2–3 exposed scales, catkin-buds 5 mm. long, the several true bud-scales much larger than the numerous exposed catkin-scales...........1. *B. pumila*

aa. Trees; twigs glabrous or slightly pubescent (or densely pubescent in the planted *B. pubescens*).

b. Bark of trunk and larger branches white (except in old trees).

Buds small, less than 5 mm. long; twigs very resinous-warty; bark close, dull white, not separating into thin papery layers; lower trunk horizontally marked with black; conspicuous black triangular patches at insertion of branches............5. *B. populifolia*

Buds 5–10 mm. long; bark separable into thin papery layers.

Bark often not spontaneously peeling; lower trunk deeply furrowed and vertically marked with black; black triangular patches beneath insertion of branches.

Twigs resinous-glandular..*B. pendula*

Twigs densely pubescent, glandless...............................*B. pubescens*

Bark peeling into thin papery layers, often pale pinkish or yellowish; trunk marked by inconspicuous horizontally elongated lenticels..................6. *B. papyrifera*

bb. Bark of trunk and larger branches not white.

Buds divergent, 5–10 mm. long; twigs with strong wintergreen flavor; bark not peeling, resembling that of wild black cherry....................................4. *B. lenta*

Buds more or less appressed; bark peeling or curling.

Buds 5–8 (–10) mm. long; bark of young trunks and branches close, and silvery- or yellowish-gray; on older trunks freely peeling and curling; twigs with slight wintergreen flavor...3. *B. lutea*

Buds less than 6 mm. long, pubescent at least toward tips of scales; bark of young trunks and branches light reddish-brown, shining, freely peeling and curling, then salmon-color or pinkish-tan.......................................2. *B. nigra*

1. **Betula pumila** L. Swamp Birch

A northern bog shrub occurring locally as a glacial relict in west-central and northern Ohio. Our only shrubby species, recognized by the coarsely dentate broad-oval to orbicular leaves, short internodes, and small blunt winter-buds. A number of varieties have been segregated; ours are referable to the typical.

2. **Betula nigra** L. River Birch

River birch is a beautiful streamside tree of the South, extending northward on the Atlantic slope to Long Island, and, locally, to northeastern Massachusetts and adjacent New Hampshire, and in the interior, in the drainage basin of the Mississippi River to southern Minnesota and adjacent Wisconsin. It enters northwestern Indiana in the

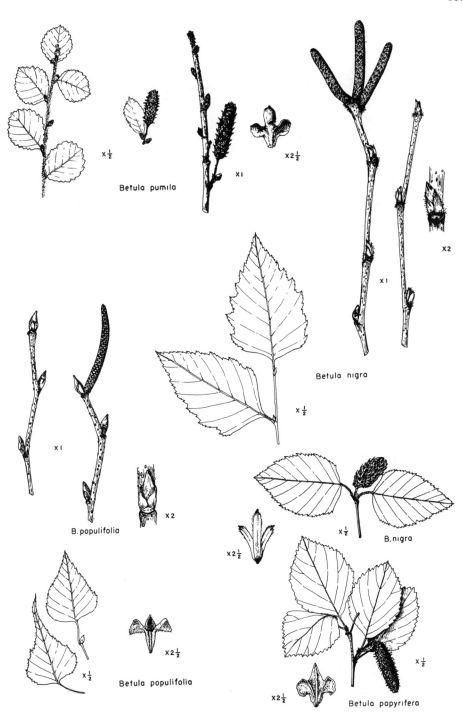

Betula pumila × ½

× 1

× 2½

Betula nigra

× 1

× 2

B. populifolia

× 1

× 2

Betula nigra × ½

B. nigra × ½

× 2½

Betula populifolia × ½

× 2½

Betula papyrifera × ½

× 2½

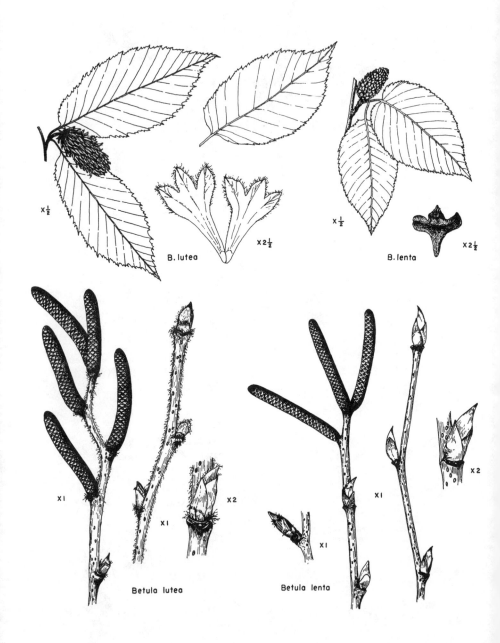

$\times \frac{1}{2}$

B. lutea

$\times 2\frac{1}{2}$

$\times \frac{1}{2}$

B. lenta

$\times 2\frac{1}{2}$

X1

X2

X1

Betula lutea

X1

X1

X2

Betula lenta

Kankakee (Illinois River) drainage, and southern Indiana in Ohio River tributaries eastward to the "flats" of Jefferson and Jennings counties. In Ohio, it is limited to a more or less wedge-shaped area extending northward from its area of more continuous distribution in eastern Kentucky. Its absence from the 100-mile stretch of the Ohio Valley between Jefferson County, Indiana, and eastern Adams County, Ohio (where it is in Scioto River drainage) is interesting, and perhaps correlated with an earlier drainage pattern, or, with acidity of habitat.

This is the only species of birch whose seeds ripen in spring or early summer, and the only semi-aquatic species (it sometimes grows where the land is inundated for several weeks at a time). Its habitat, and the curly salmon-color or pinkish-tan bark of branches and young trunks make it an easily recognized species.

3. **Betula lutea** Michx. f. YELLOW BIRCH

Yellow birch is a dominant tree of the Hemlock-White Pine-Northern Hardwoods Forest; outside of that area it is local and infrequent except in deep ravines and at higher elevations of the Appalachian Mountains. In Ohio it occurs most frequently in the northeastern part of the state, but extends southward along the rugged western border of the Allegheny Plateau. Two varieties of similar range, both of which occur in Ohio, are sometimes distinguished:

Mature bracts of the fruiting catkins 5–8 mm. long, the cuneate basal portion 1–2.5 mm. long. .
                                                                                     var. *lutea*
Mature bracts 8–13 mm. long, the elongate basal portion 2.5–6 mm. long. .var. *macrolepis* Fern.

The irregularity of teeth (large teeth at ends of veins alternating with several small ones), giving a jagged appearance to leaf-margin, the tendency toward more and longer hairs on petiole sometimes extending onto groove of midrib, and the slightly shorter and usually less acute winter-buds are characters which help to distinguish yellow birch from the next species. The silvery-yellowish bark which exfoliates in thin curly plates is characteristic, except of aberrant trees known as forma *fallax* Fassett which have dark bark; these may be mistaken for *B. lenta*.

Hybrids between *B. lutea* and *B. pumila* occur, two of which are represented in Ohio: × *B. purpusii* Schneid. from Lucas County, with numerous spur-branchlets, short-peduncled ovoid-cylindric fruiting aments with large bracts resembling those of yellow birch; and another from Stark County, with cylindric fruiting catkins on longer peduncles.

Betula pumila

Betula nigra

Betula lutea

4. **Betula lenta** L.   Cherry Birch.   SWEET BIRCH

A tree of the Appalachian Highland, ranging from southern Maine and the St. Lawrence Valley southward to Georgia and Alabama.   In Ohio, it is local in distribution, principally in a band along the hilly western margin of the Allegheny Plateau, where it grows on ravine slopes.   Specimens without fruiting catkins cannot be identified with certainty (cf. *B. lutea* forma *fallax*); combinations of characters shown by some specimens from northeastern Ohio suggest hybridization with the preceding species.   The bark resembles that of wild black cherry, hence the common name, cherry birch.   This character and the pronounced wintergreen flavor of the twigs serve to identify this species in the field.   Birch oil, distilled from twigs and small branches, is a source of wintergreen flavoring.   Black birch is not a desirable common name to use for this species as it sometimes results in confusing it with river birch, *B. nigra*, whose specific name means black.

5. **Betula populifolia** Marsh.   GRAY BIRCH.   WIRE BIRCH

A tree of the Northeast, whose range extends westward into Ontario and New York, and southward to eastern Pennsylvania and Delaware; very local beyond this range, occurring in the Northern Blue Ridge in Virginia, at the south end of Lake Michigan in northwestern Indiana, and in northern Ohio, where it is very local and perhaps adventive. Recognized by its chalky-white bark which is close and firm, with dark triangular markings at the base of branches; easily confused with *B. pendula*, which is adventive in northeastern Ohio.

6. **Betula papyrifera** Marsh.   PAPER BIRCH
B. alba L. var. *papyrifera* (Marsh.) SPACH.

A tree of the Northern Coniferous Forest and Hemlock-White Pine-Northern Hardwoods Forest; local south of these regions.   Occurs very sparingly in the "oak openings" of Lucas County.   Easily recognized by its creamy-white exfoliating bark.   The birchbark canoe of the Indians were made from the bark of this tree, which is sometimes called canoe birch.   A number of varieties are recognized; only the typical enters our area.

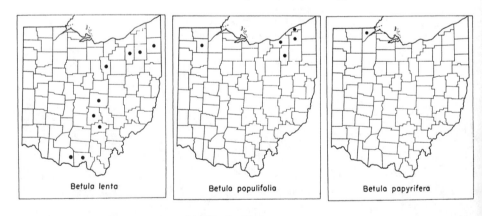

Betula lenta   Betula populifolia   Betula papyrifera

ALNUS B. Ehrh.   ALDER

A genus of about 30 species, most of which are in the north temperate zone; our species are tall shrubs of swamps, wet woods, and stream margins.   The black alder of Europe, *A. glutinosa* (L.) Gaertn. (*A. vulgaris* Hill) is sometimes planted and occasionally occurs

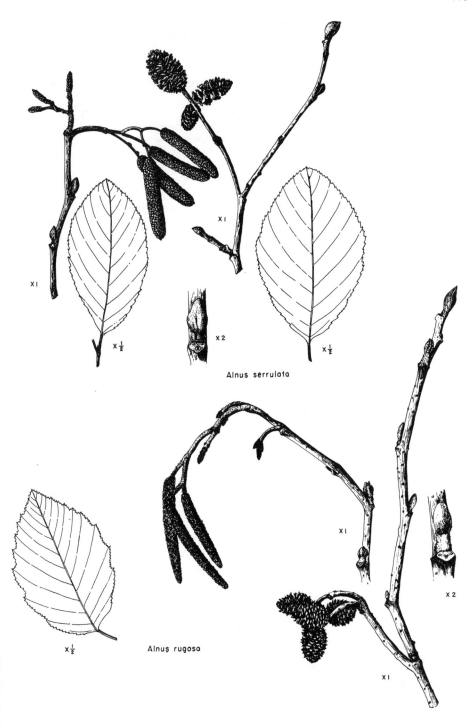

Alnus serrulata

Alnuş rugosa

as an escape; it is a tree with round crown, conspicuous in very early spring because of the abundant pendant staminate catkins. Two native species in Ohio, with flowers developing from naked buds long before the expanding leaves; leaf-buds few-scaled and stalked. Nomenclature confused; interpretation below is that of Fernald (1950).

Leaves usually doubly serrate or serrate-dentate, usually undulate, ovate to oval or sub-elliptic and broadest near middle or below, slightly or not at all glutinous; cross-veins beneath prominent, stipules lanceolate, soon deciduous; axis of young inflorescence without right-angle bends; lenticels of trunks whitish, linear-transverse................................1. *A. rugosa*
Leaves simply serrulate, usually not undulate, obovate to obovate-elliptic, broadest at or above middle, glutinous when young, cross-veins weak; stipules oval, soon deciduous; axis of young inflorescence with right-angle bends; lenticels of trunks fewer, shorter and darker, or obscure 2. *A. serrulata*

1. **Alnus rugosa** (DuRoi) Spreng. Speckled Alder
     A. incana of Gray, ed. 7, B. & B., Schaffner (1932).
     This alder of northern range is considered to be distinct from the European *A. incana* (L.) Moench, with which it has long been confused. Confined to glaciated northern Ohio. A variable species; two varieties and several forms may be recognized, but are not separated on the map.

     Leaves beneath green or tawny, not glaucous...................................var. *rugosa*
     Leaves whitened or glaucous beneath........................var. *americana* (Regel) Fern.

Each has a form with the leaves soft or velvety beneath.

2. **Alnus serrulata** (Ait.) Willd. Common Alder
     A. rugosa of Gray, ed. 7, B. & B., Schaffner (1932).
     More southern in range than the preceding; the common species of central and southern United States. In Ohio, more or less frequent in the Allegheny Plateau area from the Ohio River to Lake Erie; the Hamilton County record is based on a single small clump on a wooded stream-bank. It is found in wet soil in swampy woods and on stream-banks, apparently in slightly acid soil; it is generally absent from limestone areas. Two varieties and four forms are recognized; ours are referable to var. *serrulata*, with leaves definitely obovate, more or less cuneate at base, and glabrous or soon glabrate beneath, or to var. *serrulata* forma *noveboracensis* (Britt.) Fern., in which the leaves are densely and permanently pubescent beneath. The difference in relative position of staminate and pistillate inflorescences (as shown in the figures) will aid in distinguishing the two species of *Alnus*.

Alnus rugosa

Alnus serrulata

FAGACEAE

A large family of about 600 species of trees or shrubs, classified into five or six genera, three of which are represented in our flora. Monoecious, the staminate flowers in ascending or pendant catkins or small pendant heads; pistillate inconspicuous (the 3 styles relatively prominent), developing into a one-seeded nut wholly or partly surrounded by the enlarged and hardened bracts.

FAGUS L. Beech

A genus of some ten or twelve smooth-barked deciduous trees of the north temperate zone. Flowers axillary, developing with the leaves; staminate in heads hanging on slender peduncles, pistillate in pairs at the ends of short stout peduncles and surrounded by bractlets, the inner of which enlarge and cohere at the base to form the familiar 4-valved soft-prickly involucre surrounding the two sharply 3-angled edible nuts. The European beech, *F. sylvatica* L., is often planted; it includes a number of horticultural forms.

1. **Fagus grandifolia** Ehrh. Beech

Large tree most easily recognized by the smooth gray bark, the long slender lustrous winter-buds, and triangular nuts in a 4-valved involucre. Widely distributed, ranging through all of the more mesophytic forest regions of the East, but limited westward by dryness. Throughout Ohio, and probably in every county, but local in the area of the Prairie Peninsula. Variable in leaf shape and serration, in amount and character of pubescence of lower leaf-surface, and in prickles of the involucre, the variations correlated with geographic races or subspecies (Braun, 1950, p. 566; Camp, 1950).

CASTANEA Mill. Chestnut

Trees or shrubs of temperate regions of the northern hemisphere. Staminate flowers creamy-white, in elongate interrupted spreading or ascending catkins; pistillate usually three together in a prickly involucre, borne at the base of the staminate catkins or in leaf-axils, developing into large one-seeded edible nuts surrounded by the "bur"—an involucre densely covered on the outside by long sharp branched bristles or spines and lined with short silky hairs. Two of the ten species are represented in our flora. The European chestnut (*C. sativa* L.) and the Chinese chestnut (*C. mollissima* Bl.) are sometimes planted; spines of these species, especially the latter, are heavier than those of the native chestnut.

Fàgus grandifolia

Castanea dentata

Castanea pumila

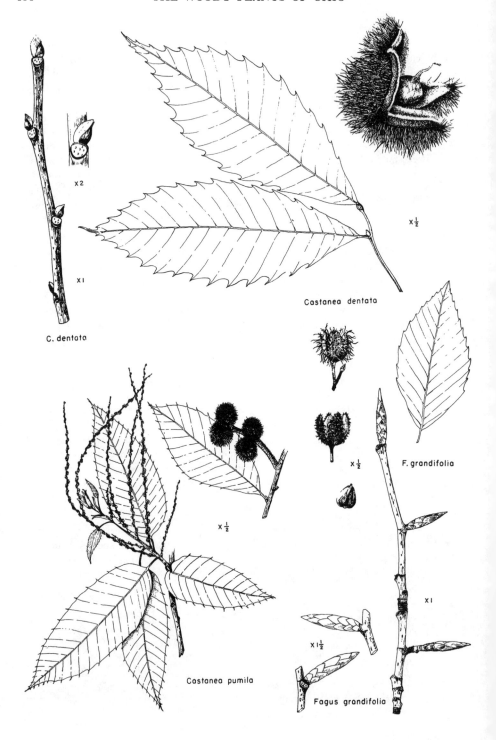

X2

X1

C. dentata

Castanea dentata

X½

F. grandifolia

X½

Castanea pumila

X1½

X1

Fagus grandifolia

Trees; leaves glabrous, long-acuminate, sharply and coarsely serrate; winter-buds glabrous;
  fruiting involucre 5–6 cm. across...1. *C. dentata*
Shrubs; leaves pubescent beneath, short-acuminate to rounded at apex, serrate or with teeth
  reduced to bristles; winter-buds pubescent; fruiting involucre 2.5–3.5 cm. across..2. *C. pumila*

1. **Castanea dentata** (Marsh.) Borkh. Chestnut

Large rough-barked tree, sometimes 100 ft. or more in height and 3–4 ft. in diameter,
occasionally 7 ft. Because of the ravages of the chestnut blight, introduced on Japanese
chestnut (first reported in New York in 1904), no large trees remain, and only occasional
standing trunks suggest the size once attained. Chestnut has a remarkable power of
sprouting from stumps, and formerly sometimes formed dense coppices with as many as
300 sprouts from a single stump; even after trees have been killed by blight, coppice sprouts
shoot up and grow for a few years until they, too, are killed. Few live sprouts are now seen.
Chestnut formerly was a dominant tree of the Oak-Chestnut Forest region, and locally
abundant through all of the Appalachian Highland and much of the Interior Low Plateau.
In Ohio, almost confined to the Allegheny Plateau, but ranging locally westward on the
Lake plains; the Clermont County record is based on a group of several trees on a steep
ravine slope cut in till of Illinoian age. Usually a tree of non-calcareous soil, reaching its
best development in rich mesic sites, but often on dry ridges and rocky slopes.

2. **Castanea pumila** (L.) Mill. Chinquapin. Chinkapin

A shrub (or sometimes a small tree) of southern range—Florida and Texas north to
southern New Jersey, Kentucky, and Arkansas. The local Ohio occurrences may be due
to introduction by flood-waters of the Ohio.

QUERCUS L. Oak

A very large genus—estimates of number of species range from 200 to 400—including
three subgenera, two of which, *Lepidobalanus*, the white oaks, and *Erythrobalanus*, the red
and black oaks, are represented in our area. Oaks are widely distributed in the north
temperate zone (few extending far northward) and higher elevations of the tropics. Staminate flowers abundant, in slender aments; pistillate arranged singly or in small groups,
each enclosed in a scaly bud-like involucre from which the 3-lobed stigma projects, and
which later develops into the familiar cup at the base of or partially surrounding the acorn.
The acorns, although seldom used as human food because of the tannin, are an important
source of food for many mammals and some birds. Several species yield tan-bark; commercial cork is a product of the thick corky bark of the cork oak, *Q. suber* L., of the
Mediterranean region. A few exotic species of oaks are planted, of which the English oak,
Q. robur L., is most often seen; it differs from white oak in its smaller and short-petioled or
subsessile leaves, and long-peduncled fruits. Likewise, a few American species from south
of our range are sometimes planted; of these, the willow oak, *Q. phellos* L., with linear
or narrowly oblong leaves, is most often seen; water oak, *Q. nigra* L., with small cuneate-
obovate leaves entire to shallowly three-lobed at summit remaining green in sheltered
situations into the winter, is planted in parks and arboretums where it may reproduce
spontaneously; turkey oak, *Q. laevis* Walt. (*Q. catesbaei* Michx.), is occasionally used in
state forest plantings; all are bristle-tip oaks. The last should not be confused with the
European Turkey oak, *Q. cerris* L., which belongs to the subg. *Lepidobalanus*.

In addition to the species known to be native to Ohio (some represented by two or
more varieties or forms), hybrids occur more or less frequently; such individuals are often
difficult to identify. There is evidence, especially in some of the bristle-tipped oaks, that

genes derived from ancient (perhaps early post-Wisconsin) crosses contaminate some individuals; this is most noticeable in northern Ohio. Specimens of oaks are best collected in September or October, when mature foliage, well-developed winter-buds, and mature fruit are available, all of which are often essential for correct determination. Shade and sun leaves, and leaves from lower and upper branches differ markedly in some species, sometimes enough that they might be referred (incorrectly) to different varieties or forms (see illustrations of *Q. muehlenbergii*, *Q. alba*, and *Q. montana*).

A. Leaves or their lobes without bristle tips; fruit maturing the first year; inner surface of shell of nut glabrous. . . . .*Lepidobalanus*, the White Oaks
AA. Leaves or their lobes bristle-tipped, i.e., midrib and, in lobed leaves, the principal veins excurrent as bristles; fruit maturing the second year; inner surface of shell of nut tomentose. .
                                                                    Erythrobalanus, Red and Black Oaks

A. The White Oaks

a. Leaves more or less deeply lobed.
  b. Leaves glabrous or essentially glabrous beneath when mature.
        Petioles more than 1 cm. long; leaves gradually narrowed to the base, bright green above, paler beneath. . . . . .1. *Q. alba*
        Petioles less than 1 cm. long; leaves usually auricled at base, dark green above, bluish beneath; planted species. . . . . .2. *Q. robur*
  bb. Leaves persistently pubescent beneath, pubescence perceptible to the touch.
        Leaves roughened above with scattered fascicled hairs, densely grayish-pubescent beneath, some of the hairs stellate; usually deeply 5-lobed, the middle lobes largest; acorns about 1–2.5 cm. long, with hoary-tomentose but not fringed cup ½–⅓ height of acorn. . . . . .3. *Q. stellata*
        Leaves dark green and glabrous above, pale green to silvery-white (or rarely rufous) pubescent beneath; leaves broadest above the middle, the upper half usually shallowly lobed, the lower deeply cut, the blade narrow between upper and lower halves; acorns 2–5 cm. long, with deep cup heavily fringed along upper edge. . . . .
                                                                    4. *Q. macrocarpa*
aa. Leaves coarsely toothed (sometimes shallowly and irregularly lobed in no. 5).
  b. Margin wavy and irregularly toothed or slightly lobed; leaves densely pubescent beneath with matted horizontally-branched hairs and some erect hairs; fruiting peduncles 3–7 cm. long, each usually bearing 2 acorns. . . . . .5. *Q. bicolor*
  bb. Margin more regularly toothed, straight parallel lateral veins ending in the teeth; fruit sessile or short-stalked.
        Leaves obovate or obovate-oblong to oblong, the apex (on fruiting branches) rounded, or at least not acute, teeth spreading; acorn large, 2–4 cm. long, cup 2–2.5 cm. in diameter.
              Leaves yellowish-green beneath, thinly pubescent or nearly glabrous, glabrous above except on midrib, the teeth rounded; fruit short-peduncled, acorn ellipsoidal, enclosed for about half its length by a thin hoary-pubescent cup roughened toward base by thickened or knob-like scales with triangular free tips. . . . . .6. *Q. montana*
              Leaves green beneath, woolly-pubescent and velvety to the touch, glabrous above or with some fascicled hairs, the margin regularly and sometimes deeply toothed or undulate; fruit sessile or subsessile, acorn ovoid to ellipsoidal, enclosed about one-third its length in a thick cup broad and flat at base, hoary-tomentose, its scales free and thickened on back, the acute scales forming a short rigid fringe around cup. . . . . .7. *Q. michauxii*
        Leaves oblong-lanceolate to obovate (shade leaves), apex (on fruiting branches) acute to acuminate, teeth slightly incurved and often callous-tipped; acorn small, 1–1.5 cm. long.
              Leaves with 8 or more veins and teeth on a side; tall trees, or sometimes dwarf in dry rocky situations. . . . . .8. *Q. muehlenbergii*
              Leaves with fewer than 8 veins and teeth on a side; shrubs, 1–3 m. tall. . . . .
                                                                    9. *Q. prinoides*

AA. The Red and Black Oaks

a. Leaves deeply or shallowly lobed, the lobes and teeth bristle-tipped.
  b. Leaves glabrous beneath, except that some species have axillary tufts of hairs.

Leaf-sinuses extending about half-way to midrib (deeper on upper branches), longest lobes shorter than to slightly exceeding width of uncut middle part of blade (sometimes more deeply cut in no. 11).

Leaves thin but firm, generally dull above, glabrous beneath (except for small axillary tufts), the lobes ascending and slightly tapering; cup saucer-shaped to shallowly cup-shaped (especially in northern forms), with tightly appressed red-brown scales; winter-buds shiny dark brown..........10. *Q. borealis* (*Q. rubra*)

Leaves thick or subcoriaceous, lustrous above, glabrous beneath except for rusty hairs in the axils, or often loosely stellate-pubescent; petioles often yellow; cups turbinate to deeply cup-shaped, with loosely imbricated pale scales, the upper forming a fringe; winter-buds strongly angled, tawny- or grayish-pubescent....
11. *Q. velutina*

Leaf-sinuses extending more than half-way to the midrib, longest lobes 2–6 times width of uncut middle part of leaf.

Cups deep or turbinate, the upper scales loosely imbricate, ,forming a fringe; winter-buds strongly angled, tawny- or grayish-pubescent; leaves thick, lustrous above...11. *Q. velutina*

Cups with upper scales appressed, not forming a fringe; winter-buds not strongly angled.

Leaf-blades 10–15 (–20) cm. long, with conspicuous axillary tufts beneath, those of upper branches more deeply lobed than lower; winter-buds gray, glabrous; acorn oblong-ovoid, with thick saucer-shaped cup (deeper in one variety), grayish-puberulent scales, and incurved rim; tree of rich or calcareous woodlands...................................12. *Q. shumardii*

Leaf-blades 8–15 cm. long, with small axillary tufts of rusty hairs; winter-buds dark reddish-brown, pale-pubescent above middle; acorn ovoid, concentrically furrowed around apex; cup deep, covering ⅓–½ acorn, scales reddish-brown, thin, usually glabrous; tree of dry usually acid soils...........13. *Q. coccinea*

Leaf-blades 8–15 cm. long, often broadly cuneate at base, axillary tufts large; winter-buds light brown, scales puberulous toward tip and sometimes ciliate; acorn small, nearly hemispheric, with flat saucer-shaped cup; tree of wet soil
14. *Q. palustris*

Leaves similar to those of preceding; axillary tufts small, pale; winter-buds bright-brown, lustrous, with ciliate scales; acorn slender ellipsoidal; tree northern...15. *Q. ellipsoidalis*

bb. Leaves pubescent beneath.

Leaves broadly obovate, shallowly 3-lobed at broad apex, thick, rusty-tomentose with stellate hairs beneath; winter-buds angled, rusty-pubescent; acorn-cup deep, scales reddish-brown, loosely imbricated; tree of barrens and dry ridges..16. *Q. marilandica*

Leaves elliptic to oblong in outline.

Leaves brownish- or rusty-pubescent beneath with stellate hairs; winter-buds strongly angled, tawny- or grayish-pubescent; acorn-cup turbinate to deeply cup-shaped, with loosely imbricated pale scales, the upper forming a fringe....
11. *Q. velutina*

Leaves grayish or ashy beneath with minute dense stellate hairs.

Leaves somewhat drooping; leaf-lobes narrow, often falcate; winter-buds chestnut-brown, puberulous or pilose, scales pale-ciliate; cup deep saucer-shaped, scales appressed, obtuse or truncate, pale-pubescent; tree southern..
17. *Q. falcata*

Leaf-lobes triangular, usually 5; winter-buds ovoid, obtuse, dark; acorns numerous, small; shrub of dry soil.............................18. *Q. ilicifolia*

aa. Leaves entire, the midrib excurrent as a bristle-tip.

b. Leaves oblong-lanceolate or narrowly elliptic, firm, lustrous above, pubescent beneath, persistent into winter.......................................19. *Q. imbricaria*

bb. Leaves linear to narrowly oblong, thin, finely reticulate-veined, dull green and usually glabrous beneath; deciduous in autumn................................20. *Q. phellos*

WINTER KEY

a. Some of the stipules persistent in the terminal cluster of buds.

b. Lateral buds appressed; stipules shorter than buds or long; twigs stout, often corky-ridged
4. *Q. macrocarpa*

bb. Lateral buds divergent; twigs not corky-ridged.

    c. Stipules twice as long as buds, narrowly linear; twigs, buds, and stipules densely tomentose...*Q. cerris*

    cc. Stipules about as long as buds; twigs glabrous or nearly so; buds and stipules tomentose
                                                           5. *Q. bicolor*

aa. Stipules not persistent.
    b. Buds at tip of twig 5 mm. or more in length.
      c. Buds conspicuously hoary- or rusty-pubescent.
          Buds obtuse to broad-acute, finely hoary pubescent, strongly angled...........
                                             11. *Q. velutina*
          Buds acute, tomentose with long hairs.
             Buds angled, conical-fusiform, rusty tomentose...........16. *Q. marilandica*
             Buds scarcely angled, brown and hoary tomentose in upper half..7. *Q. michauxii*
      cc. Buds glabrous or slightly pubescent, not strongly angled.
          Terminal bud obtuse, broadly ovoid, glossy red-brown; lateral buds almost as large as terminal.
             Acorn on long slender stalk, 4–12 cm. long......................2. *Q. robur*
             Acorn sessile or nearly so.......................................2. *Q. petraea*
          Terminal bud acute, ovoid to conical; lateral buds smaller than terminal.
          Buds nearly or quite glabrous.
              Buds bright dark brown.......................10. *Q. borealis* (*Q. rubra*)
             Buds gray to light brown; trunk often buttressed at base...12. *Q. shumardii*
          Buds pubescent above middle, or bud-scales ciliate.
              Buds brown, brownish-tomentose at tip; lower scales often whitish-margined; acorn plainly stalked, scales of cup closely appressed........6. *Q. montana*
             Buds grayish, tomentose; lower scales whitish in upper half; acorn sessile, apex with concentric rings or arcs; scales of cup free at tip...............
                                           13. *Q. coccinea*
    bb. Buds at tip of twig 5 mm. or less long; ament buds often larger.
      c. Twigs pubescent or downy at least toward tip.
        d. Bud-scales glabrous, or pubescent only on the margins; shrub, fruiting freely....
                                           18. *Q. ilicifolia*
        dd. Buds rusty pubescent or downy.
          Buds blunt pointed..3. *Q. stellata*
          Buds sharp pointed..17. *Q. falcata*
      cc. Twigs glabrous.
        d. Twigs very slender; buds long-conical, sharp pointed..............20. *Q. phellos*
        dd. Twigs stouter.
          Buds obtuse.
           Twigs red, or red and green, often with a white bloom; buds red, round-ovoid...1. *Q. alba*
          Twigs and buds brown or gray.
             Shrub or low tree not over 15 ft. in ht.; buds light brown...9. *Q. prinoides*
             Tree or sometimes a shrub; buds grayish brown.......8. *Q. muehlenbergii*
             Tree, northern and local; buds brown, pubescent at tip, the scales ciliate...
                                        15. *Q. ellipsoidalis*
          Buds acute.
           Buds grayish brown; bud-scales scarious-margined and sometimes slightly hairy...8. *Q. muehlenbergii*
          Buds dark brown, sharply pointed or bluntly acute; bud-scales smooth or with a few scattered hairs on margins...................14. *Q. palustris*
          Buds bright brown; bud-scales plainly erose or serrate........19. *Q. imbricaria*

Identifications of oaks based only on twig characters are often doubtful; additional characters should be used whenever possible.

The White Oaks, Lepidobalanus

1. **Quercus alba** L. WHITE OAK

A large and valuable forest tree of the eastern half of the United States and southern Canada, occurring almost throughout the Deciduous Forest. Probably present in all counties of Ohio. Found in a variety of habitats: mesic woods, wet flats, dry hillsides. Formerly an abundant tree of the Illinoian Till Plain ("flats") of southwestern Ohio,

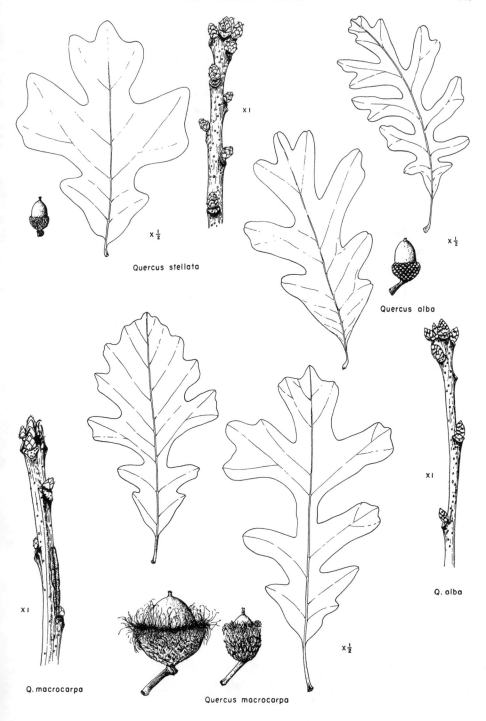

Quercus stellata

Quercus alba

Q. alba

Q. macrocarpa

Quercus macrocarpa

where trees 3–4.5 (or 5) ft. D.B.H. were frequent. The largest Ohio white oak (in Hocking Co.) measures 20 ft. 6 inches in circumference (Ohio Forestry Assn.). The light gray fissured and scaly bark (that of chinquapin oak similar) and rather evenly lobed smooth leaves distinguish this species from other oaks. The depth of the sinuses of the leaves varies greatly; in the typical form they are deep, extending from one-half to two-thirds the distance to the midrib; in forma *latiloba*, leaves are cleft less than half-way to mid-rib; in forma *repanda*, sinuses are very shallow. Leaves of a single tree vary greatly in depth of lobing, shade-leaves being less deeply cut than those exposed to sun in the higher parts of the crown.

2. QUERCUS ROBUR L. ENGLISH OAK

A variable species of Europe, North Africa, and western Asia; frequently planted, and occasionally spontaneous. Leaves commonly somewhat similar to those of white oak, but smaller, with truncate, cordate, or auricled base, bluish-green in color, long-persisting in the fall. The durmast oak, *Q. petraea* (Matt.) Lieblein (*Q. sessiliflora* Salisb.), is similar, but has leaves green beneath, and fruit sessile or short-stalked.

3. **Quercus stellata** Wang. POST OAK

A tree of more southern and western range than white oak, extending north of the Ohio River through southern Ohio on and adjacent to the Allegheny Plateau, in the "knobs" of Indiana, and in the less wet and poor soils of the Illinoian drift area of Illinois. Commonly associated with black-jack oak within the range of the latter; on dry ridge-tops of the Allegheny Plateau; on compact soils of the wedge-shaped rolling oak belt which extends north from the "Bluegrass region" just to the west of the Plateau.

The gray bark, somewhat darker and more deeply fissured than that of white oak, the stiff branching, and the leaves stellate-pubescent beneath and with large obovate middle lobe distinguish this species.

4. **Quercus macrocarpa** Michx. BUR OAK; MOSSY-CUP OAK

A large tree ranging from the Appalachian Mountains westward and northwestward to central Texas, eastern Montana, and Saskatchewan; rare and local eastward and southward of this area. In Ohio, most frequent west of the Allegheny Plateau; originally an important constituent of the Great Black Swamp area of northwestern Ohio, as well as of the oak-hickory area of west-central Ohio. Bur oak occupies a variety of habitats ranging from wet to dry; often found on high-level alluvial bottoms of small streams.

Quercus alba

Quercus stellata

Quercus macrocarpa

The massive form, heavy twigs often ridged with strips of cork, the persistent long stipules in the cluster of terminal buds, the deeply lobed leaves broadest above the central deep sinus, and the beautifully fringed deep cups of the large acorns (cups sometimes 5–6 cm. in diam.) make this an easily recognized species.

The forma or variety *olivaeformis* (Michx.) Gray is distinguished by the more slender acorns about half as thick as long and a third to half covered by the cup.

5. **Quercus bicolor** Willd.   SWAMP WHITE OAK

More northern in range than any other of our oaks except *Q. ellipsoidalis*, occurring mainly north of the Ohio and Potomac rivers. Found throughout Ohio in swamps and bottom-lands, but infrequent in the southeastern part; associated with pin oak on the "flats" of southwestern Ohio. The name, meaning two-colored, refers to the contrast between the green upper and white or whitish pubescent under surface of leaves. This character and the irregularly wavy or slightly lobed margin of the broadly obovate leaves, and the long-peduncled acorns (2–6 cm.) distinguish the swamp white oak.

6. **Quercus montana** Willd.   ROCK CHESTNUT OAK.   MOUNTAIN OAK.
Q. prinus L.

A tree of the Appalachian Highlands and non-calcareous sections of the Interior Low Plateau; a dominant in the Oak-Chestnut Forest region, and in certain communities of the Mixed Mesophytic and Western Mesophytic Forest regions. In Ohio, it is common in the southern part of the Allegheny Plateau, less frequent northward, and westward along Lake Erie (a comparable westward extension north of Lake Erie through southern Ontario).

The name *Quercus prinus* has been shunted back and forth between two species of oaks. This has led and will continue to lead to endless confusion, for the name may in one publication refer to the swamp chestnut oak (*Q. michauxii*), in another to rock chestnut oak (*Q. montana*). Fernald (1950), Gleason (1952), and Little (1953) use *Q. prinus*. In the Department of Agriculture Yearbook, TREES (1949), in Dean and Chadwick's OHIO TREES, in Deam's FLORA OF INDIANA (1940), in Rehder (1940), Sargent (1922), and many other publications between 1915 and 1950, *Q. montana* was used for rock chestnut oak and *Q. prinus* for swamp chestnut oak. Unaccompanied by the common name, *Q. prinus* is an ambiguous name. For this reason, we are retaining the specific name *montana* for this oak.

Chestnut oak is the dominant tree of the oak or shale barrens of southerly and westerly

Quercus bicolor

Quercus montana

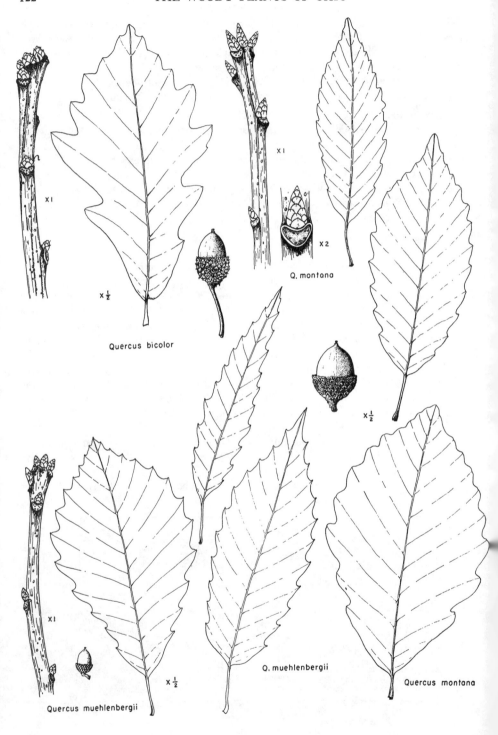

X I

X ½

Quercus bicolor

X I

X 2

Q. montana

X ½

X I

X ½

Quercus muehlenbergii

Q. muehlenbergii

Quercus montana

slopes on the Ohio Black Shale; in such situations, it is a medium-size tree of irregular form. In more mesic sites, it develops into a large tree with straight trunk, and mingles with other tree species. The heavy and very slowly decomposing leaf-litter of pure or almost pure stands of this oak is responsible for the sparse ground vegetation. The acorns are large (2.5–3.5 or 4 cm. long), usually more than twice the size of those of *Q. muehlenbergii* with which it is sometimes confused, and the teeth of the leaves are rounded instead of pointed; the bark is dark and, in old tree, deeply furrowed.

7. QUERCUS MICHAUXII Nutt. SWAMP CHESTNUT OAK

Q. prinus L. in many publications.

A tree of wet and poorly drained soils, common in the "flats" of southeastern Indiana, but not known in comparable situations of southwestern Ohio. Its general range includes the Atlantic and Gulf coastal plains and Mississippi embayment, beyond which it extends to the flats of southeastern Indiana and locally to swamps of the Mississippian Plateau in south-central Kentucky. Planted on a few of the Ohio state forests.

8. **Quercus muehlenbergii** Engelm. CHINQUAPIN OAK; YELLOW OAK

Q. acuminata (Michx.) Sarg., *Q. prinoides* Willd. var. *acuminata* (Michx.) Gl.

A wide-ranging species, extending from the Piedmont and Hudson Valley westward to Kansas, Oklahoma, and Texas, and very locally to New Mexico. Most frequent on calcareous soils, and often in dry rocky situations. Widely distributed in Ohio, but most common on the limestone soils of southwestern Ohio, a fact which the distribution map does not show.

The light gray bark resembles that of white oak; the small, dark, and sometimes striped acorn is sweet and edible if roasted. Leaves from branches fully exposed to the sun, or from higher parts of the crown are narrow, thick, and shiny, bearing little resemblance to the broad, dull, and thin shade leaves. The latter might be confused with leaves of *Q. montana*, as are also leaves of *f. alexanderi* (Britt.) Treal.

9. QUERCUS PRINOIDES Willd. DWARF CHINQUAPIN OAK

A shrub only 1–3 m. tall, whose range is similar to that of *Q. muehlenbergii*. Westward, it is scarcely distinguishable from that species. Both are considered varieties of *Q. prinoides* by Gleason (1952), in which case the dwarf chinquapin oak is called var. *prinoides*. Leaves with only 4–7 (5–8) pairs of veins. Dwarfish individuals of *Q. muehlenbergii* are sometimes incorrectly determined as *Q. prinoides*. No Ohio specimens; those reported as such are dwarfed forms of *Q. muehlenbergii*.

Quercus muehlenbergii

Quercus borealis

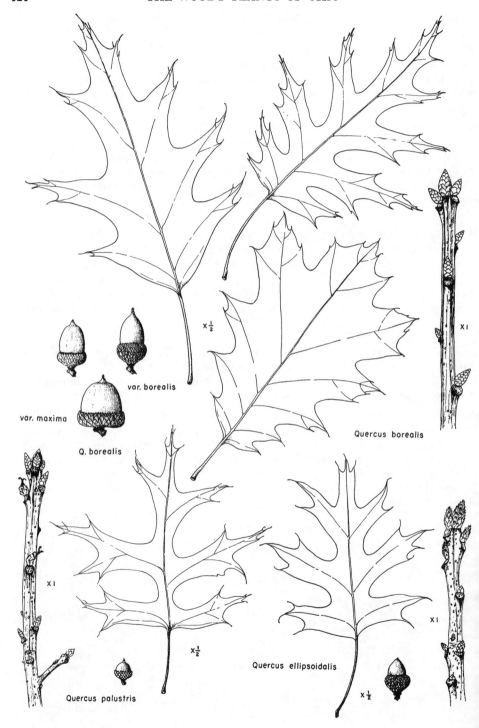

var. borealis

var. maxima

Q. borealis

Quercus borealis

Quercus palustris

Quercus ellipsoidalis

The Red and Black Oaks, Erythrobalanus

10. **Quercus borealis** Michx. f. RED OAK

Q. rubra L. of most authors.

Red oak ranges almost throughout the Deciduous Forest, including most of the area of the Hemlock-White Pine-Northern Hardwoods Forest. It occurs throughout Ohio in a variety of habitats; generally absent from poorly-drained or swampy situations.

The nomenclature of this species is confused. Gleason (1952) states "the name *Q. rubra* was originally applied to a mixture of several species; to avoid confusion the name is therefore rejected." Most publications between 1915 and 1950 (see under *Q. montana*) use *Q. borealis* for red oak; some authors used the name *Q. rubra* for southern red oak or Spanish oak (*Q. falcata*). Following Gleason, we are using the name *borealis* for red oak.

Regardless of nomenclature, two varieties are recognized which have many intermediate forms differing in size of acorn and depth of cup:

Q. borealis var. **borealis**—*Q. rubra* var. *borealis* (Michx f.) Farw.

Q. borealis var. **maxima** (Marsh.) Ashe—*Q. rubra* var. *rubra*

The second of these is the red oak most widespread in Ohio; the first is northeastern in range, and in its typical form scarcely enters our area. Many intermediates occur in the northern part of the state, and some specimens may be referred to var. *borealis*; others may carry genes of *Q. ellipsoidalis*. Most Ohio red oaks are *Q. borealis* var. *maxima*. No distinction is made on the map.

Best recognized by the relatively large leaves with sinuses extending about half-way to the midrib (deeper on leaves from high in the crown), the saucer-shaped or flattish acorn cups, and the shiny brown winter buds.

11. **Quercus velutina** Lam. BLACK OAK

Ranging almost throughout the Deciduous Forest, usually in dry sites. Widely distributed in Ohio, but most abundant on dry slopes of the Allegheny Plateau, and in the sandy soil of ancient beach ridges near Lake Erie. The relatively large glossy leaves lobed about half-way to the midrib (more deeply in some forms), the usually yellow to orange-brown petioles, the free scale-tips forming a fringe around the cup, and the angled, hoary-tomentose winter-buds distinguish black oak. The inner bark of twigs and young branches is orange, whence the common name yellow oak, sometimes applied to this species.

Very variable in leaf shape and pubescence of under surface of leaves; a number of

Quercus velutina

Quercus shumardii

Quercus coccinea

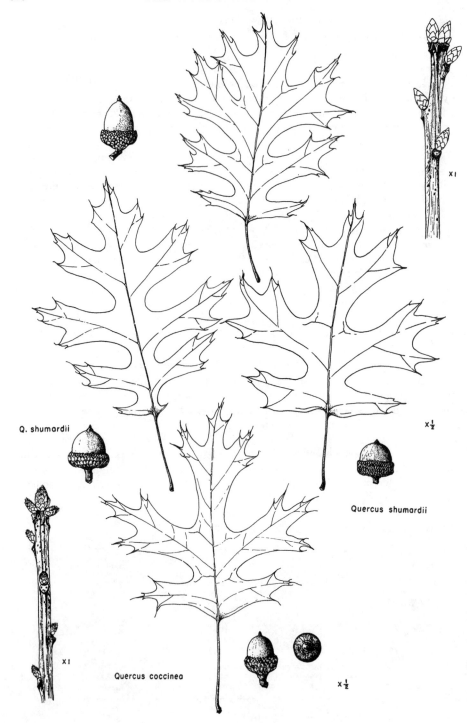

Q. shumardii

x 1

Quercus shumardii

x ½

Quercus coccinea

x ½

x 1

poorly characterized forms are based on these variable features. Least typical in northern Ohio, where some specimens have very small, or somewhat ellipsoid acorns, or scarcely free scale-tips, characters which suggest genes from *Q. ellipsoidalis.*

12. **Quercus shumardii** Buckl. SHUMARD RED OAK
    Q. texana of ed. 7, not Buckl.
    A tree of southern range, extending northward on the Atlantic slope to Virginia, and in the Mississippi Valley to Missouri, southern Illinois, Indiana, and southwestern Ohio (perhaps more widely distributed in the western part of the state than our map shows). A tree of stream terraces, well-drained alluvial terraces, and ravine slopes where it mingles with other mesophytes. Somewhat resembles *Q. coccinea*, but does not grow in the same habitats; can be distinguished by the gray or brownish-gray glabrous winter-buds, the conspicuous axillary tufts on the lower side of the leaf, and the shallower cup of the acorn. The var. *schneckii* (Britt.) Sarg., differing in its deeper cup, is recognized by some authors.
    A handsome tree, turning bright red in late fall; the young leaves in spring susceptible to late frost damage.

13. **Quercus coccinea** Muenchh. SCARLET OAK
    Less wide-ranging than red oak, generally confined to the Central Deciduous Forest. In Ohio, it is common on the non-calcareous soils of the Allegheny Plateau, infrequent in other parts of the state. A tree of ridge-crests and bluffs, dry south and west slopes, and (in the northern part of Ohio) low sandy ridges. Frequently confused with other bristle-tipped oaks—*Q. velutina*, *Q. ellipsoidalis*, and *Q. shumardii*. Distinguished from *Q. velutina* by the deeper and more rounded sinuses of the leaves, the absence of the fringe around the cup formed by free scale-tips, the brown non-angular winter-buds pubescent only above the middle, and the brilliant fall coloration; from *Q. shumardii* by the winter-buds (gray and glabrous in that species), and the fruit; from *Q. ellipsoidalis* by the relatively thicker acorn. Concentric rings or discontinuous arcs around the apex of the acorn help to identify *Q. coccinea*. Range and habitat differ from that of *Q. shumardii*, which is a more mesic species.
    A variety, *tuberculata* Sarg., distinguished by the thickened tuberculate scales of the cup, is sometimes recognized; it occurs with the typical in Pike and Scioto counties.

14. **Quercus palustris** Muenchh. PIN OAK
    More limited in range than most of our oaks; in general, north-central in distribution. Widely distributed in Ohio in wet bottomlands, borders of swamps, and most abundant

Quercus palustris

Quercus ellipsoidalis

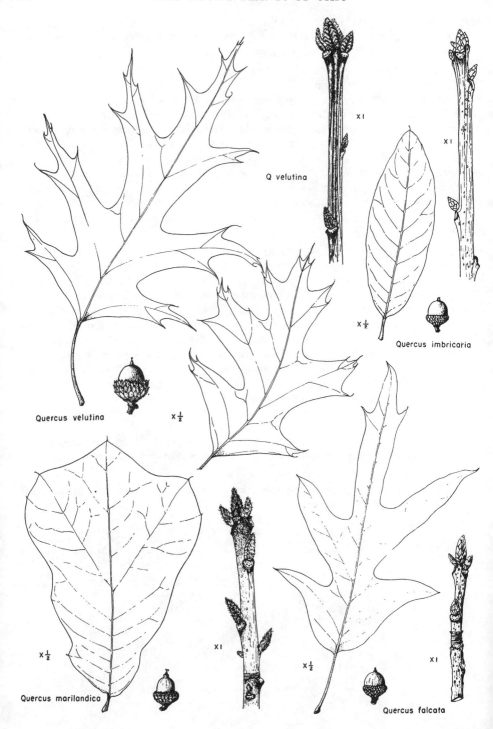

Q velutina

x 1

x 1

x ½

Quercus imbricaria

Quercus velutina   x ½

x ½

Quercus marilandica

x 1

x ½

x 1

Quercus falcata

on the "flats" or Illinoian Till Plain of southwestern Ohio where it occurs in pure stands of second- or third-growth forest. Easily recognized from a distance by the branch habit— lower branches inclined downward, middle branches horizontal, upper ascending. The nearly spherical acorns are smaller than those of other Ohio oaks. The common name refers to the many short spur-like twigs on slender branches.

15. **Quercus ellipsoidalis** E. J. Hill. JACK OAK. NORTHERN PIN OAK

A tree of limited range, largely confined to the middle and western parts of the Great Lakes area. In Ohio, found in a few localities in the northwest. Often confused with *Q. coccinea* and *Q. velutina*; sometimes called "black oak" which adds to danger of confusion.

16. **Quercus marilandica** Muenchh. BLACKJACK OAK

Mainly south of the Ohio and Potomac rivers, having much the same range as *Q. stellata*, with which it is commonly associated. More limited in distribution in Ohio than that species; confined to a few of the southern counties. The open irregular crown, very coarse twigs, broadly obovate thick leaves only obscurely lobed at the summit, and rusty-pubescent winter-buds make recognition of typical forms easy. Some forms have leaves approaching those of *Q. velutina* in outline.

17. **Quercus falcata** Michx. SPANISH OAK. SOUTHERN RED OAK

Q. triloba Michx., *Q rubra* of some authors (see note under *Q. borealis*), *Q. digitata* Sudw. A southern species, reaching Ohio only in its southernmost part. Variable; several varieties and forms recognized. Ohio specimens are referable to the typical variety.

18. QUERCUS ILICIFOLIA Wang. BEAR OAK

A shrub or small tree 1–5 m. tall, intricately branched and often forming large patches. Common in the Allegheny Mountains, and said to extend westward to Ohio (Gleason, 1952); no specimens.

19. **Quercus imbricaria** Michx. SHINGLE OAK

Ranging almost throughout the Central Deciduous Forest, and most abundant in the Middle West. Widely distributed in Ohio, but seldom abundant; occupies a variety of sites. The only entire-leaved oak native to Ohio. Leaves thick and glossy, often hanging on through the winter. The specific name means overlapping, and refers to the use of the wood for shingles by early settlers in Illinois.

Quercus marilandica

Quercus falcata

Quercus imbricaria

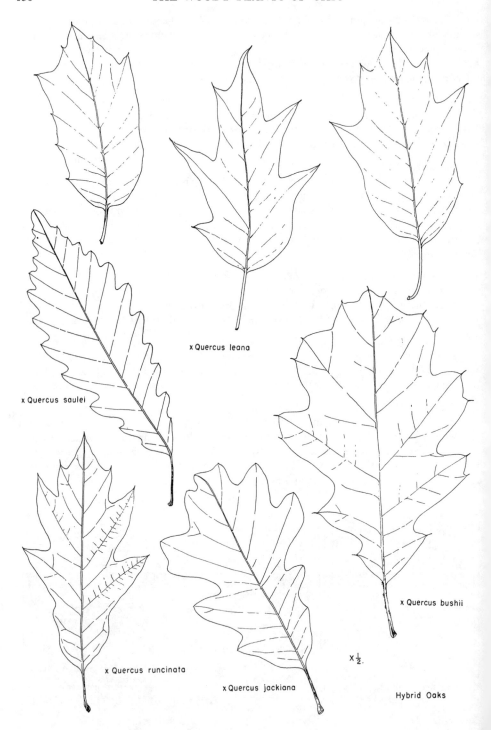

x Quercus leana

x Quercus saulei

x Quercus bushii

x Quercus runcinata

x Quercus jackiana

$\times \frac{1}{2}$.

Hybrid Oaks

20. Quercus phellos L. Willow Oak

A southern species whose range and habitat are similar to that of swamp chestnut oak. It does not extend as far north as Indiana or Ohio; planted in limited quantities on a few of the state forests, and reported from Jackson and Scioto counties, but there are no specimens to support these records. Leaves thin, and narrower than those of *Q. imbricaria*; twigs very slender.

Hybrid Oaks

Interspecific hybrids within each subgenus of oaks occur frequently; hybrids between species in different subgenera do not occur. It is believed that most of the hybrid oaks which are found are first generation (F₁) hybrids; this is explained as due to the very small percentage of germination and growth of the millions of acorns produced by a given tree in its lifetime and hence the even smaller percentage that might produce a hybrid individual. Again, cross-pollination of the hybrid and one of the parent species (back-cross) and growth of an acorn to maturity could occur but seldom. Hybrids appear most often where one of the parent species is uncommon. When a strange or apparently abnormal oak is found it is well to note what species occur (or did occur in the recent past) in the vicinity. First-generation hybrids display characters of both parents and knowing the possibilities may help to determine the probably identity of the tree in question. Although first-generation hybrids are most frequent, there is little experimental evidence to prove the supposed rarity of individuals of later generations. During postglacial migrations when ranges of species were changing, hybridization doubtless took place, as now, and progeny of such crosses, in regions from which one or both parents have departed may resemble one parent in most characters; such forms give the taxonomist much difficulty (Palmer, 1948; Gleason, 1952).

Hybrids known to occur in Ohio, and others known to occur in adjacent states (and both of whose parents occur in Ohio) are listed below; characters are mostly from Trelease (1924) and Palmer (1948). Others known hybrid oaks are listed by Palmer and by Gleason (1952).

Hybrid Oaks Known to Occur in Ohio

× **Q. bebbiana** Schneider (*Q. alba* × *macrocarpa*)
   Hardin.
   A hybrid "showing various degrees of transition between the parent species." Winter-buds tomentulose, stipules persistent; leaves elongated, usually cuneate; acorn oblong, large, fringe short or lacking.

× **Q. bushii** Sargent (*Q. marilandica* × *velutina*)
   Adams, Gallia, Highland, Ross, Scioto.
   Winter-buds fusiform, brown-hairy; leaves obovate, with 3 or 5 subtruncate lobes (resembling *velutina*); scales of acorn-cup thin, appressed, creamy-tomentulose.

× **Q. exacta** Trelease (*Q. imbricaria* × *palustris*)
   Ashland, Delaware.
   Difficult to distinguish from × *Q. runcinata*; leaves somewhat smaller and thinner, and with fewer primary veins.

× **Q. fernowi** Trelease (*Q. alba* × *stellata*)
   Scioto.
   Winter-buds subellipsoid, 3–5 mm. long; twigs canescent; leaves oblanceolate, lobes about 5 on a side.

× **Q. hawkinsiae** Sudw. (*Q. borealis* × *velutina*)
   Brown, Hamilton, Ottawa.
   Distinguished from *borealis* by the somewhat pubescent scales of the winter-buds and the more persistent tomentum on under surface of leaves.

× **Q. hillii** Trelease (*Q. bicolor* × *macrocarpa*)
   Ashtabula.

"A relatively abundant and widely distributed hybrid." Buds tomentulose, stipules persistent; leaves elongate, usually cuneate, white beneath; acorn stalked, ovoid, cup with short fringe.

× **Q. jackiana** Schneid. (*Q. alba* × *bicolor*)
Brown.
Leaves lobed, more or less intermediate in outline between the parent species, but resembling those of white oak; under surface more or less densely tomentose (as in *bicolor*); fruit long-stalked or nearly sessile, cup deep; winter-buds blunt, round-ovoid, 2–3 mm. long.

× **Q. leana** Nutt. (*Q. imbricaria* × *velutina*)
One of the most frequent natural oak hybrids; known in Ohio from more than a score of widely distributed counties (see map); type specimen from Cincinnati about 1836.

X Quercus leana

A translation of Nuttall's original description, published in 1859, follows: Leaves membranaceous, very long petioled, oblong-oval, with rounded or subcordate base, sinuate-pinnatifid, glabrate, the wide lobes acuminate and bristle-tipped; fruit short pedicelled solitary or in twos; cup hemispheric, the scales ovate obtuse, the acorn subglobose, striped, rather small, with umbo short conic. (From Michaux, North American Sylva, Vol. IV, Nuttall, Vol. I.)

× **Q. mutabilis** Palmer & Steyerm. (*Q. palustris* × *shumardii*)
Hamilton, Lawrence.
Tree with aspect of *Q. palustris* and leaves resembling *Q. shumardii*; acorn-cups shallow.

× **Q. runcinata** (A. DC.) Engelm. (*Q. borealis* × *imbricaria*)
Adams, Hamilton, Highland, Jackson, Lawrence, Pickaway, Scioto, Summit, Tuscarawas.
Winter-buds red-brown; leaves lance-oblong or oblanceolate to broadly elliptic, usually rather deeply lobed, lobes often 5, in texture similar to *Q. imbricaria*.

× **Q. saulei** Schneid. (*Q. alba* × *montana*)
Belmont, Lawrence, Warren (?), Washington.
Said to be "one of the commonest hybrids among the white oaks." Winter-buds conic-ovoid, 3–5 mm. long, acute; twigs glabrous; leaves resembling those of rock chestnut oak in outline, but incised more deeply, the ovate or oblong lobes rounded at apex.

× **Q. tridentata** (A. DC.) Engelm. (*Q. imbricaria* × *marilandica*)
Lawrence.
Leaves somewhat 3–4 lobed toward apex, persistently gray-stellate beneath, the pubescence lcose and soft; winter-buds scarcely angled.

× **Q. coccinea** × **imbricaria**
Muskingum.
Leaves relatively narrow, often with 3 lobes on each side, separated by shallow sinuses.

Hybrid Oaks Which May Occur in Ohio

× Q. DEAMII Trelease (*Q. alba* × *muehlenbergii* or *alba* × *macrocarpa*)
Bartlett (1951) discusses the probable parentage of this hybrid, which is honored by the "Deam Oak Monument" northwest of Bluffton, Indiana, established to preserve the type tree. A specimen of the original collection, and leaves from a number of offspring, are shown by Bartlett in five plates.

× Q. PALEOLITHICOLA Trelease (*Q. ellipsoidalis* × *velutina*)
Young leaves, branchlets, and winter-buds more pubescent than in *Q. ellipsoidalis*; acorn-cups usually turbinate, with rather loose, pubescent scales.

ULMACEAE

Trees or shrubs, widely distributed in temperate and tropical regions; about 15 genera and over 150 species. Two genera are represented in our flora; another, *Planera*, with one species, is found in swamps of the Southeast. The Asiatic *Zelkova serrata* (Thunb.) Mak. is occasionally planted; it resembles an elm but has smooth gray bark and bears small oblique short-stalked drupes which mature in autumn. Leaves 2-ranked, alternate, simple and pinnately veined, usually inequilateral at base; flowers inconspicuous, perfect or unisexual, borne in cymes or fascicles, or solitary; fruit a nutlet, samara, or drupe.

ULMUS L. ELM

Trees of the north-temperate zone; three or four species in the Ohio flora, several European and Asiatic species—especially wych elm, *U. glabra* Huds. (*U. montana* With.), English elm, *U. procera* Salisb. (*U. campestris* Mill.), Siberian elm, *U. pumila* L.—commonly planted as street or shade trees. Flowers in lateral clusters, in our species produced in late winter or early spring before the leaf-buds begin to grow; two southern species and the Chinese elm, *U. parvifolia* Jacq., bear flowers in late summer or fall. Fruit a round or elliptic samara with centrally placed seed. Winter-buds directed to one side of the twig, i.e. axis of bud not perpendicular to leaf-scar.

Twigs rough, scarbrous pubescent; winter-buds rusty pubescent with long hairs; flowers very short-pedicelled; leaves very rough above, densely pubescent beneath; samaras nearly round, 1–2 cm. long, eciliate..1. *U. rubra*
Twigs glabrous or finely pubescent; winter-buds glabrous or nearly so; flowers long-pedicelled; leaves glabrous but usually roughish above, pubescent or nearly glabrous beneath; samaras elliptic, deeply notched at apex, about 1 cm. long, glabrous with ciliate margin..2. *U. americana*
Year-old twigs pubescent or sometimes glabrous, older twigs often with irregular corky ridges; winter-buds pubescent and scales ciliate; flowers long-pedicelled, racemose, the racemes short; leaves smooth and more or less lustrous above, pubescent beneath; samaras broadly elliptic, shallowly notched, about 1.5 cm. long, pubescent on surface and short ciliate...3. *U. thomasi*
Twigs very slender, glabrous or on vigorous shoots pubescent, usually with 2 broad, thin, opposite corky ridges interrupted at nodes; winter-buds small, 2 mm. or less long, glabrous; flowers long-pedicelled; leaves smooth above, pale pubescent beneath, with axillary tufts; samaras lance-ovate, apex with 2 incurved awns, slender-stalked, pubescent and long-ciliate.........
4. *U. alata*

WINTER KEY

a. Twigs very rough; buds rusty tomentose...1. *U. rubra*
aa. Twigs not very rough, glabrous or finely pubescent.
  b. Buds dark brown or blackish, conspicuously darker than the twigs.
    Leaf buds broadest below the middle, tapering from there to apex; bud-scales sparsely ciliate with long blackish hairs; leaf-buds on leading twigs 6–10 mm. long; large or small tree..*U. glabra*
    Leaf buds broadest at or a little above middle, therefore more abruptly pointed; bud-scales sparsely short pubescent with tawny hairs; leaf-buds on leading twigs 4–6 mm. long..*U. procera*
  bb. Buds brown, little darker than the twigs.
    Leaf buds very small, uppermost scarcely 2 mm. long and nearly as broad.
      Twigs prominently corky winged...............................4. *U. alata*
      Twigs not corky winged.......................................*U. pumila*
    Leaf buds, at least the uppermost 3–6 mm. long, about twice as long as broad.
      Bud scales glabrous or short pubescent; bundle-scars usually 3; tree often vase-shape; trunk dividing into several large branches.................2. *U. americana*
      Bud-scales downy ciliate, sometimes short pubescent; bundle-scars usually more than 3 (4–6); twigs sometimes corky; tree not vase-shaped, trunk continuing into crown...3. *U. thomasi*

1. **Ulmus rubra** Muhl. RED ELM. SLIPPERY ELM
   U. fulva Michx.
   Ranging through most of the Deciduous Forest, in well-drained mesic sites and rocky (calcareous) hillsides. Well distributed in Ohio, usually occurring as scattered individuals. The wood is dark brown or red, the inner bark fragrant and mucilaginous, characters which have given rise to the common names. Easily recognized by the rusty winter buds and very rough twigs and leaves.

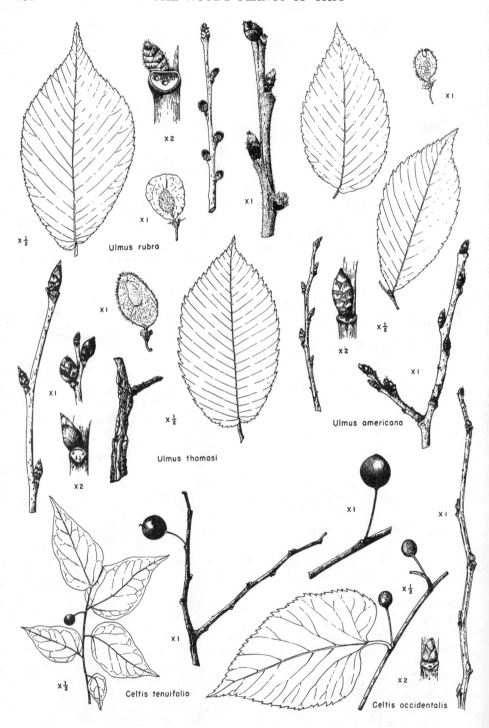

Ulmus rubra

Ulmus thomasi

Ulmus americana

Celtis tenuifolia

Celtis occidentalis

2. **Ulmus americana** L. White Elm. American Elm

   The most widespread American species of elm, extending far westward and northward; general and abundant in Ohio.   In size, exceeding all other American species; the largest known white elm, at Marietta, Washington Co., Ohio, is 26 ft. 6½ inches in circumference (Ohio Forestry Assn.).   Open-grown trees are commonly vase-form and branch low, forest-grown trees have straight columnar trunks.   Leaves smooth or scabrous above and branchlets the first year glabrous or pubescent, variations used to distinguish four named forms, all of which occur in Ohio.   The form with smooth leaves and glabrous branchlets (forma *laevior* Fern.) corresponds with the Linnean type and therefore should be regarded as the typical *U. americana* (Seymour, 1952).   Found in a variety of sites: frequent on alluvial bottomlands and along ravines, and sometimes a constituent of mesic woodlands northward.   Often an early invader of cleared land and slumped banks, but not usually persisting in dry sites.   Attacked by the Dutch elm disease and elm phloem necrosis which have killed large numbers of trees.

3. **Ulmus thomasi** Sarg. Rock Elm.   Cork Elm
   U. racemosa Thomas, not Borkh.

   Largely confined to the upper Mississippi Valley and lower Great Lakes region southward to the Ohio River.   Infrequent in Ohio.   Similar in appearance to white elm with which it is often associated in mesic sites, and not with certainty distinguished from it except by inflorescence and samaras unless the branches are corky-ridged.

4. Ulmus alata Michx.   Winged Elm

   A southern species extending north to central Kentucky, southern Indiana, southern Illinois, and central Missouri.   Said to occur "scattered generally over the southern Ohio counties" (Dean and Chadwick), but there are no Ohio specimens to validate this report. Easily recognized by the broad thin cork wings on slender twigs, and small leaves.

5. Ulmus serotina Sarg., September Elm, is local south of Ohio.

   A specimen reported from Adena woods, near Chillicothe, Ross Co., an area in which a number of southern and exotic species are planted.

Ulmus rubra

Ulmus americana

Ulmus thomasi

CELTIS L. Hackberry

   A large genus (about 75 species) of the north-temperate and tropical zones; only two species in Ohio.   Leaves somewhat 3-nerved at base; flowers axillary, the pistillate solitary or in pairs from upper axils of the new growth, the staminate clustered and toward base of

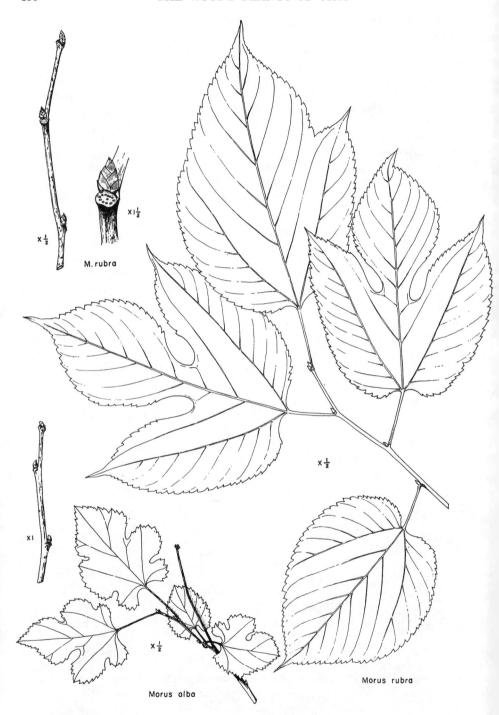

x $\frac{1}{2}$

x 1$\frac{1}{2}$

M. rubra

x $\frac{1}{2}$

x 1

x $\frac{1}{2}$

x $\frac{1}{2}$

Morus rubra

Morus alba

new growth. Fruit a subglobose drupe with sweet but thin pulp. Winter-buds with tips closely appressed to the twig.

> Leaves sharply serrate except toward base, 5–12 cm. long, ovate, inequilateral, acuminate, green on both sides or somewhat paler beneath, smooth or rough; mature drupes dark purplish- or brownish-black; trees, the bark dark with irregular corky ridges............1. *C. occidentalis*
> Leaves entire or sparingly toothed, (more teeth on vigorous shoots), 2–8 cm. long, acute, grayish-green, pubescent beneath; mature drupes orange-brown; shrubs, the bark pale gray
> 2. *C. tenuifolia*

1. Celtis occidentalis L. HACKBERRY

Ranging over much of eastern United States except the area of the Hemlock-White Pine-Northern Hardwoods, and the southern forests of the Gulf slope, where it is replaced by the sugarberry (*C. laevigata* Willd.). Extremely variable; several varieties and even species have been distinguished in the complex which seems best considered as one polymorphic species (Gleason, 1952; Little, 1953). Variations from tree to tree, or even on a single tree, may be noted. Scabrous-leaved forms may be found in dry sunny situations, and more frequently, westward. Dwarf forms in shallow soil over limestone may be mistaken for the following species. Witch's brooms are common on trees in open situations, less frequent on forest trees.

2. Celtis tenuifolia Nutt. DWARF HACKBERRY

C. pumila of most Amer. auth., not Pursh (and of Schaffner, 1932), *C. occidentalis* var. *pumila* of Gray, ed. 7, but not of ed. 8.

A grayish-green, straggly, irregularly branched shrub of southern range, on dry rocky hills. Local in Ohio on dolomite outcrops in openings and prairie patches. Two varieties, *tenuifolia* and *georgiana* (Small) Fern. & Schub. sometimes distinguished; the latter differing from the typical in its coriaceous leaves scabrous above, and pubescent twigs.

MORACEAE

A large family of some 60 genera and 2000 species, mostly tropical. The family includes a number of well known economically important and ornamental species of warm climates, as the edible fig (*Ficus carica* L.) and many other species of that genus, the breadfruit and the jackfruit, somewhat comparable in structure to the familiar osage-orange. Mostly trees or shrubs, with milky juice and alternate, simple or palmately-lobed leaves. Flowers unisexual, apetalous, usually with 4 sepals. Fruit, in ours, an assemblage of small achenes or drupes, i.e. a syncarp, resembling a blackberry.

Celtis occidentalis

Celtis tenuifolia

Morus rubra

MORUS L. Mulberry

Small trees or shrubs of temperate and subtropical regions of the northern hemisphere; cultivated for their edible fruits or for the leaves which are the chief food of silkworms. Leaves lobed or undivided, 3–5 nerved at base. The juicy edible syncarp composed of achenes covered by the succulent berry-like calyx. One native species, and one commonly spreading from cultivation.

Leaves usually cordate-ovate except on vigorous shoots, there palmately lobed (the lobes acute), serrate, soft-pubescent beneath, or pubescent only on veins and veinlets, smooth or scabrous above; fruit dark purple: winter-buds 5–7 mm. long, somewhat divergent, the bud-scales with dark margins...1. *M. rubra*
Leaves roundish in general outline, serrate and irregularly 3 or more lobed (lobes obtuse), glabrous beneath except on principal veins and in vein-axils; fruit pale to dark; winter-buds 2–4 mm. long, appressed; introduced..2. *M. alba*

1. **Morus rubra** L. Red Mulberry. Mulberry

Small or medium-size tree of eastern United States, except the area of the Hemlock-White Pine-Northern Hardwoods. Widely distributed in Ohio, but apparently infrequent in the north-central area. Leaves, although commonly cordate-ovate, vary to mitten-shape, 3-lobed, or sometimes 5-lobed. Undivided leaves somewhat resemble those of *Tilia*, from which they can be distinguished by texture and pubescence; twig and bud characters are distinctive, buds of *Morus* with more than 3 exposed scales, and bundle-scars arranged in an ellipse in the leaf-scar. Variable in the amount of pubescence, on which character named forms and varieties have been based.

2. Morus alba L. White Mulberry

From Asia; long cultivated and frequently seen as an escape. A number of varieties and horticultural forms, among them *M. alba* var. *pendula* Dipp., with slender pendulous branches and usually lobed leaves, and *M. alba* var. *tartarica* (L.) Ser., with bushy crown and usually dark fruit. Reported from more than three-fourths of the Ohio counties, throughout the state.

BROUSSONETIA L'Her.

An Asiatic genus of two species, one of which is locally naturalized in eastern United States.

1. Broussonetia papyrifera (L.) Vent. Paper-mulberry
   Papyrus papyrifera (L.) Kuntze

A smooth-barked tree with pubescent twigs, and leaves similar to those of mulberry, but grayish and densely soft-pubescent beneath. The name, which means paper-bearing, refers to the use of the inner bark in making paper. The syncarp is highly colored and showy: globose, about 2 cm. in diam., orange, with protruding red fruits. Young twigs and branches killed back in severe winters, hence found as an escape only southward. In Ohio, reported from Brown, Hamilton, and Washington counties.

MACLURA Nutt. Osage-orange

A monotypic American genus.

1. Maclura pomifera (Raf.) Schneid. Osage-orange
   Toxylon pomiferum Raf. ex Sarg.

Native in a narrow belt extending from coastal Texas northward into Oklahoma and Arkansas, barely including northwest Louisiana. Frequently planted for hedges throughout the eastern half of the United States (except extreme north) and spreading by root-shoots.

Persistent, but doubtfully naturalized, although sometimes spontaneous from seed. Widely distributed in Ohio, and probably in every county.

Spiny tree (often clipped to appear shrub-like); leaves glossy, ovate to oblong-lanceolate, long-pointed; dioecious, with axillary inflorescences, the staminate short loose racemes, the pistillate dense and globular, developing into a spherical multiple fruit about 1 dm. in diam., green or yellowish-green, roughened with rounded protuberances; composed of enlarged receptacle and calyces enclosing achenes. Bark of trunk orange in the furrows, bark of roots brilliant orange. Wood hard and durable.

LORANTHACEAE

Woody, shrub-like plants, parasitic on trees. A family of about 1300 species in 36 genera, largely tropical. Leaves usually opposite, in some genera scale-like; flowers inconspicuous in our species; fruit berry-like. Two genera represented in the eastern American flora: *Phoradendron* and *Arceuthobium*. The latter parasitic on conifers, with leaves reduced to small scales, is northern and does not reach our area. The mistletoe of folklore and literature is *Viscum album* L., a native of Europe and northern Asia.

PHORADENDRON Nutt.

A genus of warm-temperate and tropical America; about 300 species, only one of which extends to cool-temperate latitudes. The name is from the Greek *phor*, thief, and *dendron*, tree.

1. **Phoradendron flavescens** (Pursh) Nutt.
   MISTLETOE

Parasitic on the branches of many species of trees; in Ohio, reported on *Ulmus americana, Prunus serotina, Acer saccharum,* and *Nyssa sylvatica.* A thickly branched evergreen shrub forming somewhat spherical clumps sometimes almost 1 m. in diameter; conspicuous in winter when the host trees are leafless. Branches and thick obovate leaves opposite; flowering and fruiting spikes axillary; fruit globose, berry-like, whitish, the flesh gelatinous and sticky, poisonous. Migration is entirely by the agency of birds. Growth of seedlings is slow; haustoria ("roots") penetrate through the bark into the wood of the host tree, from which mistletoe derives part of its nourishment. Much more abundant southward, ranging northward into southern Ohio in the unglaciated area and margin of the Illinoian glacial area.

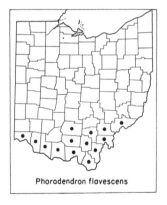

Phoradendron flavescens

BERBERIDACEAE

A family of herbs and shrubs, of which none of the latter is native to Ohio. *Berberis* is represented here by two introduced species.

BERBERIS L. BARBERRY

A large genus of spiny shrubs, most of which are native to Asia. Flowers yellow, perfect, with 6 petaloid sepals, 6 smaller petals, each with 2 glandular spots at its base, in axillary, elongated or umbel-like racemes, or solitary. Fruit a one- to several-seeded

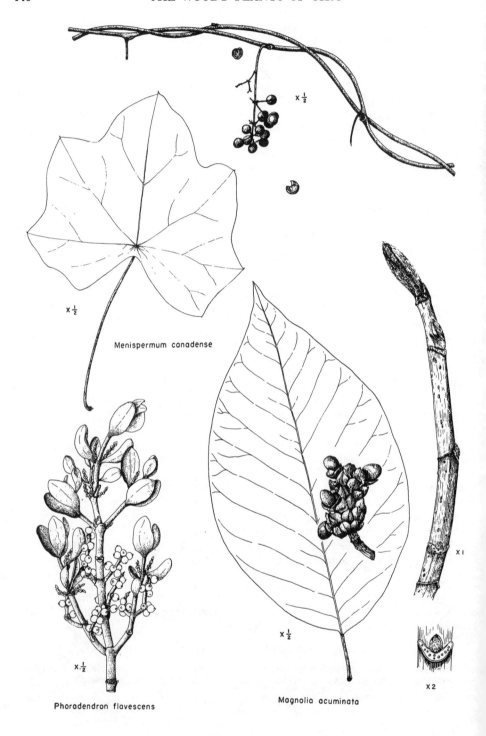

$x\frac{1}{2}$

Menispermum canadense

$x\frac{1}{2}$

Phoradendron flavescens

Magnolia acuminata

$x\frac{1}{2}$

$x\frac{1}{2}$

$x 1$

$x 2$

berry. Most species of the genus are alternate hosts of wheat rust or black stem rust. The generic name is derived from the Arabic name of the fruit, *berberys*.

Spines mostly simple; branches brown; leaves entire, obovate to spatulate, small, 1–3 cm. long; flowers solitary or 2–4 in fascicles...1. *B. thunbergii*
Spines mostly branched; branches gray; leaves bristle-toothed, obovate to oblong, 2–4 cm. long; flowers in racemes...2. *B. vulgaris*

1. BERBERIS THUNBERGII DC. JAPANESE BARBERRY
Commonly planted, and often appearing as an escape. Reported from almost half our counties. Immune or highly resistant to wheat rust.

2. BERBERIS VULGARIS L. COMMON BARBERRY
Native of Europe, formerly much planted and naturalized or appearing as an escape. Now uncommon; an alternate host to wheat rust, hence largely eradicated in the Barberry Eradication Campaign. Reported from about 20 Ohio counties, but no recent collections.

LARDIZABALACEAE, a small family of eastern Asia and Chile, is represented by the twining shrub, *Akebia quinata* (Houtt.) Dcne., which occasionally spreads from cultivation. Easily distinguished by its digitate leaves with 5 stalked entire leaflets.

MENISPERMACEAE

Woody climbers with palmately veined leaves, and small flowers in racemes or panicles; chiefly tropical in distribution. About 65 genera including 400 species, many of which are poisonous.

MENISPERMUM L. MOONSEED

A genus of only two species, one eastern American, the other eastern Asian. Leaves peltate near the margin; flowers in panicles whose peduncles arise above the leaf-axil; sepals and petals 4–8; fruit a subglobose drupe with scar of stigma to one side and toward base. The stone is laterally flattened, reniform or crescent-shaped, and thickened around margin, to which character the generic name, from the Greek *men*, moon, and *sperma*, seed, refers.

1. **Menispermum canadense** L. MOONSEED
Half-woody climber whose stems commonly freeze to the ground northward, but in southern Ohio persist except in very severe winters. Drupes bluish-black, resembling frost grapes, but bitter and poisonous. Usually only one drupe develops from a single flower (which has 2–4 pistils); occasionally 2 or more drupes may be seen crowded on a single receptacle. Widely distributed in Ohio, often in mesic woods and thickets.

Menispermum canadense

MAGNOLIACEAE

Trees or shrubs of warm-temperate and tropical America and Asia, only a few extending into cold-temperate latitudes. Flowers usually large, terminal, regular, perfect, with 3 sepals, 6 or more petals, numerous stamens and pistils. Stamens and pistils spirally arranged, the carpels developing into follicles, samaras, or berries, the whole forming a

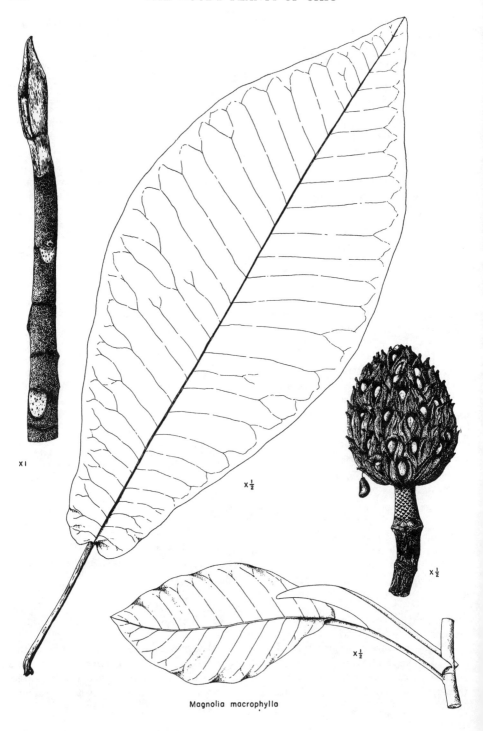

x1

x½

x½

x½

Magnolia macrophylla

cone-like aggregate fruit. Of the 10 genera, 2 are represented in the temperate American flora. The family is one of ancient lineage, well represented in floras of Upper Cretaceous times.

MAGNOLIA L. Magnolia

Of the nine species of Magnolia of temperate eastern America, three are native to Ohio. Two others are often planted, sweetbay (*M. virginiana* L.) and, in the southern part of the state, the evergreen southern magnolia (*M. grandiflora* L.). Leaves large, entire, stipulate, the stipule-scars entirely encircling the twig; the stipules, which function as bud-scales, enlarge with growth of shoots and leaves, and for a short time, are conspicuous, either while partly adnate to petiole or as they turn yellowish and droop before dropping off. Flowers perfect, large and terminal, with 9–13 similarly colored sepals and petals; fruit cone-like, composed of carpels splitting on the back at maturity allowing escape of red berry-like seeds which remain attached for a time by a slender thread, which is a vascular bundle of the seed stalk.

In the American species, the flowers are produced with or after the leaves; the early-blooming starry magnolia, *M. stellata* (Sieb. & Zucc.) Maxim., and the hybrid × *M. soulangeana* Soul., both of which are frequently planted, are Asiatic. Magnolia has been well represented in fossil floras of temperate latitudes from Upper Cretaceous to Recent time, in North America, Europe, and Asia. None persisted in Europe through the Pleistocene; present-day magnolias are restricted to eastern North America and eastern Asia. Of the American magnolias, the range of the cucumber-tree, *M. acuminata*, extends farthest north (into southern Ontario and western New York).

Leaves very large, up to 1 m. in length, cordate or broadly auricled at base, green above, white and finely pubescent beneath; buds and young twigs downy; flowers white..1. *M. macrophylla*
Leaves large, 3–6 dm. long, tapering from above the middle to an acute base, green on both sides but paler beneath; buds and twigs glabrous or nearly so; flowers cream-colored.......
2. *M. tripetala*
Leaves 1–2.5 dm. long, broadest near the middle, rounded to acute at base, green on both sides, finely soft-pubescent beneath; buds silky-pubescent; twigs glabrous or nearly so; flowers greenish-yellow...3. *M. acuminata*

WINTER KEY

Twigs and buds bright green; buds long-pointed.............................*M. virginiana*
Twigs and buds not bright green.
Buds large, 25–50 mm. long; twigs stout.
    Buds densely downy; twigs greenish-purple with fine whitish down.....1. *M. macrophylla*
    Buds smooth and glaucous; twigs reddish or greenish-brown, glabrous except near end-bud
                                                  2. *M. tripetala*
Buds smaller, 15–25 mm. long.
    Buds long hairy, the hairs not appressed...................................*M. stellata*
    Buds silky-hairy, the hairs closely appressed.
        Leaf-scars narrowly crescent to U-shaped; tall tree..................3. *M. acuminata*
        Leaf-scars triangular to broadly crescent-shaped; tall exotic shrub.......*M. soulangeana*

1. **Magnolia macrophylla** Michx. Bigleaf Magnolia

A tree attaining a height of 15–18 m. and trunk diameter of 3–4 dm. Bigleaf magnolia not only has the largest leaves of any species of Magnolia, but also the largest entire leaves of any of our trees. The broadly auricled large leaves, white beneath, distinguish this species. Flowers opening in June, fragrant, very large, petals 12–20 cm. long; fruit ovoid or nearly globose, pubescent, 6–8 cm. long, rose-colored when ripe. Local through much of its range, which is chiefly south of Ohio; often in deep ravines or gorges in mesic communities, sometimes in drier sites. In Ohio, known only from the Rock Run area, Jackson County.

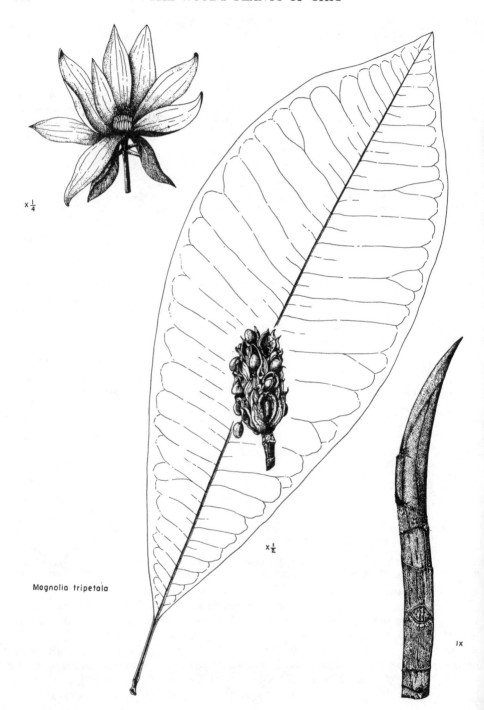

x ¼

x ½

Magnolia tripetala

1x

2. **Magnolia tripetala** L. UMBRELLA MAGNOLIA

A tree attaining about the same size as bigleaf magnolia, often with a number of stems around the base of the main trunk. The large leaves, except on rapidly growing shoots, are approximate near tip, suggesting the common name. Flowers opening in late May, ill-scented, large, petals 10–14 cm. long; fruit oblong, 7–10 cm. long. Widely distributed in the Appalachian Highland as far north as southern Pennsylvania and southern Ohio. In Ohio occurring in four counties along the more rugged western border of the Unglaciated Allegheny Plateau.

3. **Magnola acuminata** L. CUCUMBER-TREE

A large forest tree, commonly reaching a height of 75–80 feet and trunk diameter of 2–4 feet. A Morrow Co. tree measures 17 ft. 7 inches in circumference and 84 ft. in height (Ohio For. Assn.). Leaves much smaller than those of our other two species, more or less regularly scattered along the twigs; flowers opening in May, greenish-yellow, with narrow petals 5–8 cm. long; fruit oblong, 6–8 cm. long. Chiefly Appalachian in range, and extending locally westward to the Knobs of Indiana, the southern Illinois hills, and the Ozark region of Missouri and Arkansas. In Ohio, more frequent in the northern part of the Allegheny Plateau. Generally occurring singly or in small groups; logging operations have greatly reduced its numbers.

LIRIODENDRON L. TULIPTREE

A genus of two closely related species, one eastern American, the other in China. Leaves long-petioled, truncate or broadly retuse at apex, and with one or sometimes two broad angular lobes on a side—the whole leaf thus 4-angled. Flowers cup-shaped, with 3 green sepals, soon reflexed, and 6 yellow petals, each with a deep orange patch toward base. Fruit cone-shaped, composed of samaras arranged on the elongate axis.

Like *Magnolia*, *Liriodendron* is a genus of ancient lineage, abundantly represented in Upper Cretaceous and Tertiary time in various parts of the northern hemisphere. The two existing closely related species are the last survivors of a once larger genus.

1. **Liriodendron tulipifera** L. TULIPTREE. YELLOW-POPLAR

The tallest eastern American hardwood species, sometimes attaining a height of 200 ft. and a trunk diameter of 8–10 ft.; the largest Ohio tree (in Vinton County) is 4.5 ft. D.B.H. (Ohio Forestry Assn.). In eastern Erie Co., "many of the largest trees in the primeval forest were of this species" (E. L. Moseley, 1894, herb.slip). Widely distributed

Magnolia macrophylla

Magnolia tripetala

Magnolia acuminata

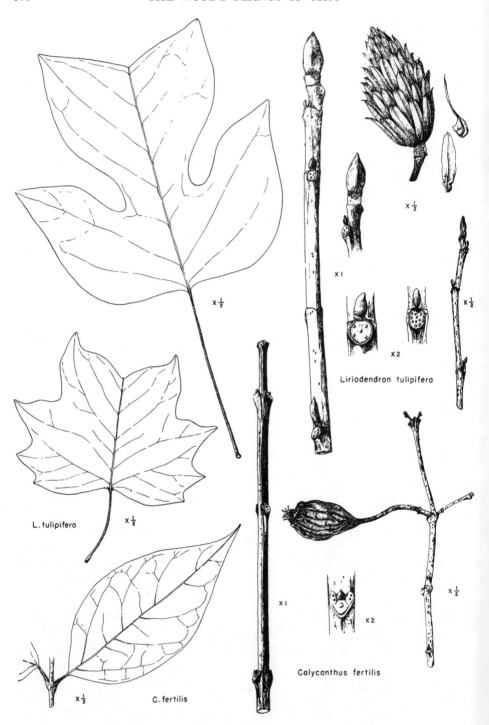

$\times \frac{1}{2}$

$\times 1$

$\times \frac{1}{2}$

$\times \frac{1}{2}$

$\times 2$

Liriodendron tulipifera

L. tulipifera $\times \frac{1}{2}$

$\times 1$

$\times 2$

Calycanthus fertilis

$\times \frac{1}{2}$

$\times \frac{1}{2}$ C. fertilis

in the Deciduous Forest east of a line extending from Lake Michigan southward to the Ohio River and thence along the Mississippi River; limited westward by drier climate. A dominant in some communities of the Mixed Mesophytic Forest region and of the Oak-Chestnut region. Widespread in Ohio, but less frequent in the west-central and northwestern parts of the state—the climatically drier parts. The unusual squarish shape of leaves, the handsome tulip-like flowers, and the dry cone-like aggregate fruit are distinctive. The flowers are emphasized by the scientific name: the generic derived from the Greek *lirion*, lily, and *dendron*, tree (for the lily-like or tulip-like flowers); the specific from the Latin, meaning tulip-bearing. The very misleading common name, yellow-poplar, is the lumberman's term. The tuliptree is the state tree of Indiana and of Kentucky.

Liriodendron tulipifera

Calycanthus fertilis

CALYCANTHACEAE

A small family of only six species, four in America (*Calycanthus*) and two in China (*Chimonanthus*).

CALYCANTHUS L.   CAROLINA ALLSPICE

Represented in southeastern United States by two or three species; another (*C. occidentalis* Hook. & Arn.) in California. Leaves opposite, entire; sepals and petals similar, indefinite in number, inserted, with the numerous stamens, around the rim of a hollow receptacle to which the lower parts of the sepals are adnate, and within which the pistils are borne. The aggregate fruit suggesting a rose-hip, but large, 5–7 cm. long, the ripened achenes (often mistaken for seeds) ovoid or ellipsoid, about 8–10 mm. long. The generic name, from the Greek *kalux*, calyx and *anthos*, flower, refers to the hollow receptacle.

1. **Calycanthus fertilis** Walt.   SWEET SHRUB

A shrub of the southern part of the Appalachian Highland, known in Ohio only from Meigs County. Flowers dark red to reddish-brown, not as fragrant as those of the commonly planted strawberry-shrub, *C. floridus* L., from which it may be distinguished by the glabrous lower leaf surface or, in forma *nanus* (Loisel.) Schelle, glaucous lower leaf surface.

ANNONACEAE

A large family of about 70 genera and perhaps 1000 species in the tropics of both hemispheres. Only one species, *Asimina triloba*, inhabits the cool-temperate zone.

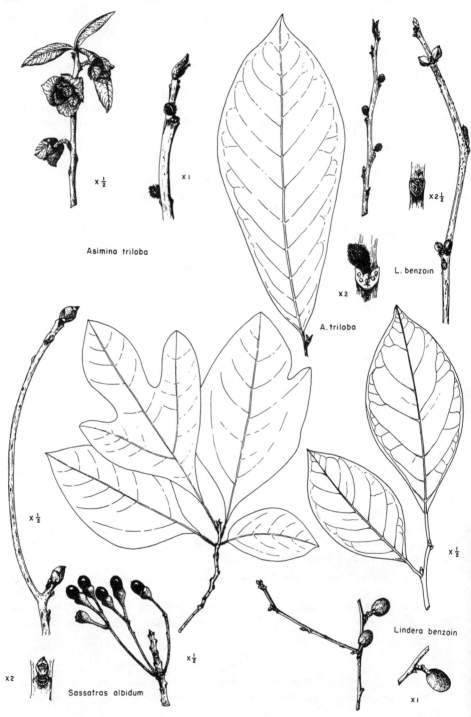

$\times \frac{1}{2}$ $\times 1$

Asimina triloba

$\times 2\frac{1}{2}$

L. benzoin

$\times 2$

A. triloba

$\times \frac{1}{2}$

$\times \frac{1}{2}$

$\times 2$

Sassafras albidum

$\times \frac{1}{2}$

Lindera benzoin

$\times 1$

ASIMINA Adans. Pawpaw

The generic name is derived from the American Indian name for pawpaw. The genus is exclusively American, the other 7–9 species in lower latitudes.

1. **Asimina triloba** (L.) Dunal. Pawpaw

A small tree easily recognized by its large obovate entire leaves (suggestive of those of *Magnolia tripetala*), silky-pubescent dark brown buds, dark purple-red flowers (with 3 sepals and 6 petals in two whorls) which appear before the leaves, and large ellipsoidal yellowish-green fleshly edible fruit, in which the large seeds are embedded. A pawpaw develops from one of the carpels of the flower; the several pawpaws in a cluster are derived from a single flower. In mesic woods of ravine slopes and sometimes in cleared areas, especially pastures (the ill-scented, when bruised, foliage is not palatable to cattle) throughout Ohio, but local northward.

Asimina triloba

LAURACEAE

Only two of the approximately 50 genera (including about 1000 species) occur in the Ohio flora. Most of the species of this family are tropical or subtropical, among them avocado, laurel (the true laurel of the Mediterranean region), camphor, etc. The California-laurel or Oregon-myrtlewood, *Umbellularia californica* (Hook. & Arn.) Nutt., of our West Coast is one of the few temperate zone species. All are aromatic trees or shrubs with alternate simple (entire or lobed) leaves, clustered, usually yellow, flowers, and 1-seeded berry or (in ours) drupe.

SASSAFRAS Nees

Our sassafras of eastern North America and two Asiatic species are the last survivors of a long line of species of Sassafras dating back, in the fossil record, to the close of Lower Cretaceous time, or almost to the beginning of the record of dicotyledonous trees. Leaves, almost throughout these millions of years, have been similar to those of existing species: 3-lobed or mitten-like, or sometimes entire. Dioecious, the flowers with 6-parted perianth (calyx); staminate with 9 stamens in 3 rows, those of the inner whorl each bearing a pair of stalked glands at base; pistillate with rudimentary stamens and ovoid ovary which develops into a blue drupe about 1 cm. long, supported on a club-shaped red pedicel. Flowers in early spring, before the leaves.

1. **Sassafras albidum** (Nutt.) Nees Sassafras

  S. officinale var. *albidum* Blake

  Sassafrass albidum var. **molle** (Raf.) Fern.

  S. officinale Nees & Eberm., *S. variifolium* (Salisb.) Ktze.

Sassafras within our area is represented by two varieties, the typical, with leaves and young twigs glabrous, is widespread and var. *molle*, with puberulent twigs, leaves densely pubescent when young and permanently pubescent beneath, is local. The extremes are distinct, but intermediate forms occur.

Sassafras is usually a small tree, with abundant root-shoots, hence may form extensive patches; occasionally 80–90 ft. in height and with trunk diameter of 6 ft. The largest

Ohio sassafras, in Lake County, is about 60 ft. tall, with trunk diameter of over 4.5 ft. (Ohio Forestry Assn.). Often a pioneer in fields and on open eroded slopes, in poor, usually non-calcareous soils. Found almost throughout Ohio, but infrequent northward and in the calcareous soils of western Ohio. Oil of sassafras, used to perfume soap and as a base for many perfumes, is distilled from the roots. The tree is aromatic throughout. A tea, more popular in the South than elsewhere, is brewed from the roots. An over-dose of the oil may act as a narcotic.

LINDERA Thunb.

About 60 species in temperate and subtropical eastern and southern Asia; only two in North America, one widespread, the other local and southern.

1. **Lindera benzoin** (L.) Blume SPICE-BUSH
   Benzoin aestivale Nees
   Aromatic shrub with obovate to elliptic entire leaves, yellow flowers in almost sessile lateral clusters opening in very early spring before the leaves, and bright red ellipsoidal drupes (6–10 mm. long) which often persist after the leaves have fallen. An attractive shrub of rich moist soil in woods; found throughout Ohio. Killed back severely—sometimes almost to the ground—by extremely low winter temperatures, but sprouting and recovering rapidly.

   Two varieties are recognized, based on differences of pubescence:

   Leaves and young branches glabrous...var. *benzoin*
   Leaves pubescent beneath, at least on veins, and young branchlets pubescent.............
   var. *pubescens* (Palmer & Steyerm.) Rehd.

   The latter is more southern in range.

Sassafras albidum

Lindera benzoin

SAXIFRAGACEAE

A large family of about 75 genera, often divided into five families or subfamilies (six if *Penthorum* is placed here). Our shrubs belong to the subfamilies HYDRANGEOIDEAE and RIBESIOIDEAE—HYDRANGEACEAE and GROSSULARIACEAE of Schaffner's catalog (1932). The subfamily Saxifragoideae, to which the family Saxifragaceae is often limited, includes only herbaceous plants. Best represented in the temperate and arctic zones.

PHILADELPHUS L. Mock-orange. Syringa

A genus of some 50 or 60 species of the north-temperate zone. Several are native south of Ohio; the commonly planted European *P. coronarius* L. has been reported as an escape in several counties, and the American *P. inodorus* L. from 2 counties, in Pike Co. in woods where doubtfully native. Flowers in terminal cymes or short racemes, showy, with 4 sepals, 4 white petals, numerous stamens, and 4-celled ovary developing into a capsule splitting into 4 valves. Leaves opposite, entire or toothed.

HYDRANGEA L. Hydrangea

A genus of North and South America and eastern Asia; one species native in our area, two others in the southeastern states, several exotic species planted. Shrubs with opposite, petioled leaves without stipules. Flowers perfect, small, in terminal panicles or corymbs with marginal flowers often sterile and showy because of enlarged white calyx. Capsules small, dehiscent at apex, with many minute seeds.

1. **Hydrangea arborescens L.** Wild Hydrangea

Hydrangea arborescens

A shrub of rich moist soil, common in ravines and on shaded bluffs; sometimes forms large patches on wet slumping banks in woods. A species of the Central Deciduous Forest; in Ohio, almost limited to the southern half or two-thirds of the state. The commonly planted snowball-like Hydrangea of early summer is *H. arborescens* forma *grandiflora* (E. G. Hill) Rehd., which was "found wild in Ohio before 1900" (Rehder, p. 286).

Several varieties may be distinguished, each with a sterile-flowered form:

Leaves glabrous beneath, or pubescent on veins.
Leaves ovate to suborbicular, rounded to cordate at base.................var. *arborescens*
Leaves ovate-oblong to lance-oblong or -elliptic, gradually narrowed toward base..........
　　　　　　　　　　　　　　　　　　　　　　　　　var. *oblonga* T. & G.
Leaves densely grayish-pubescent over whole surface beneath....................................
　　　　　　　　　　　　　　　　　var. *deamii* St. John (*H. cinerea* Small)

These varieties intergrade and it is not always possible to assign a specimen to one of them; many are intermediate between var. *arborescens* and var. *oblonga*. More or less typical var. *arborescens* is more frequent in Ohio than other forms; var. *deamii*, more western in range, is not known to occur.

RIBES L.

This large genus of the colder and temperate parts of the northern hemisphere and Andean region of South America is divided into subgenera or by some authors into distinct genera—*Ribes*, the currants, and *Grossularia*, the gooseberries. *Ribes nigrum* L., the black currant from Europe, and most of the American currants are carriers of white pine blister rust, and hence their eradication from areas where white pine grows has been attempted.

Low, sometimes prickly shrubs with palmately veined and lobed leaves, fascicled on short lateral shoots and evidently alternate on leading twigs. Flowers 5-parted, regular

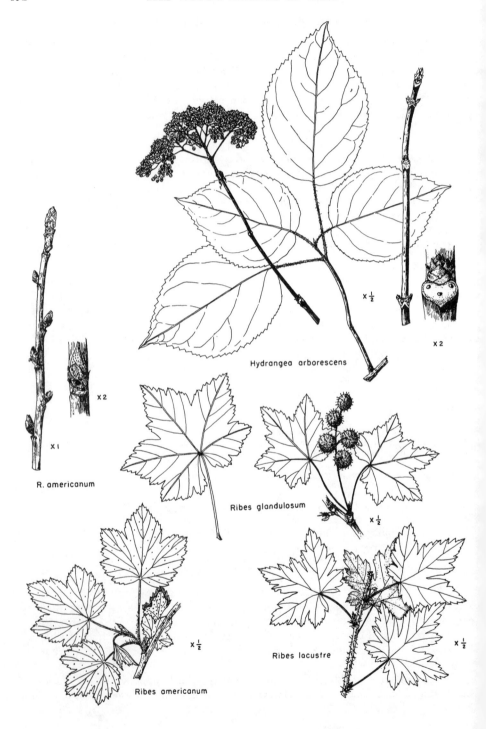

Hydrangea arborescens

x ½

x 2

R. americanum

x 2

x 1

Ribes glandulosum

x ½

Ribes lacustre

x ½

Ribes americanum

x ½

or nearly so, usually perfect; ovary inferior, developing into a berry crowned by the shriveled remains of the calyx. In addition to the native species, a number are cultivated as ornamentals or for their edible fruit; some of these occasionally occur as escapes, but none seems to be naturalized in Ohio.

a. Flowers in racemes, usually 5 or more; pedicels jointed at summit; stems without spines (except no. 3)...THE CURRANTS
 b. Leaves resinous-dotted beneath, flowers whitish to greenish-yellow.
  Racemes many-flowered; bracts about 5 mm. long, longer then pedicels; calyx glabrous, 8–10 mm. long, tubular-companulate; native......................1. *R. americanum*
  Racemes few-flowered; bracts shorter than pedicels; calyx pubescent, 5–6 mm. long, broadly campanulate; cultivated, from Europe........................2. *R. nigrum*
 bb. Leaves not resinous-dotted beneath.
  Ovary and fruit glandular-bristly; flowers not yellow; plants ill-scented.
  Stems prickly; leaves deeply 3–5 lobed, cut much more than half-way to base; racemes spreading or drooping; flowers greenish or purplish; berries purplish-black
                              3. *R. lacustre*
  Stems not prickly; leaves 5 (–7) lobed about half-way to base; racemes ascending; flowers white or pinkish; berries red...........................4. *R. glandulosum*
  Ovary and fruit not glandular-bristly; flowers yellow or yellowish; plants not ill-scented.
  Flowers golden yellow, fragrant; berries black or yellow; petioles pubescent with short whitish hairs (dense or sparse), the same kind of pubescence continuing along leaf-margin and veins of the upper leaf-surface; western, often planted..5. *R. odoratum*
  Flowers yellow-green; berries bright red, juicy; cultivated, from Europe..6. *R. sativum*
aa. Flowers solitary or in clusters, 1–4 (rarely 5); pedicels not jointed; stems usually with nodal spines..THE GOOSEBERRIES
 b. Ovary and fruit with glandless prickles; nodal spines 1–3 (rarely lacking), internodes with or without prickles; calyx lobes shorter than tube, greenish, soon reflexed; stamens not exserted, about equaling petals.................................7. *R. cynosbati*
 bb. Ovary and fruit not prickly.
  Stamens long-exserted, fialments capillary, 1–1.5 cm. long, nearly twice length of greenish-white spreading sepals; berries glabrous; nodal spines larger than preceding, up to 2 cm...8. *R. missouriense*
  Stamens not exserted, equaling or shorter than calyx lobes.
  Calyx glabrous, ovary and berry glabrous; petioles with compound elongate trichomes scattered among unbranched hairs; native.........................9. *R. hirtellum*
  Calyx pubescent; ovary pubescent or glandular, berry pubescent, glandular-bristly, or smooth; from Europe.....................................10. *R. grossularia*

WINTER KEY

a. Twigs without nodal spines or internodal prickles; leaf-scars rather broad.
 Resin-glands present (sometimes inconspciuous) on twigs and bud-scales.
  Glands minute; wood ill-scented.................................1. *R. americanum*
  Glands large..2. *R. nigrum*
 Resin-glands absent.
  Twigs densely or sparsely minutely pubescent........................5. *R. odoratum*
  Twigs glabrous or nearly so.
   Buds more or less purple-red..................................4. *R. glandulosum*
   Buds gray-puberulent...6. *R. sativum*
aa. Twigs usually with nodal spines, with or without internodal prickles; leaf-scars very narrow.
 Buds straw-colored, glossy.
  Nodal spines usually large, up to 2 cm., red......................8. *R. missouriense*
  Nodal spines little larger than the abundant internodal prickles; twigs straw-colored, glossy...3. *R. lacustre*
 Buds brownish.
  Twigs pale gray, epidermis exfoliating.............................9. *R. hirtellum*
  Twigs yellowish, becoming gray.
   Bud-scales with keel, more or less silky......................7. *R. cynosbati*
   Bud-scales not keeled, glabrous or nearly so.................10. *R. grossularia*

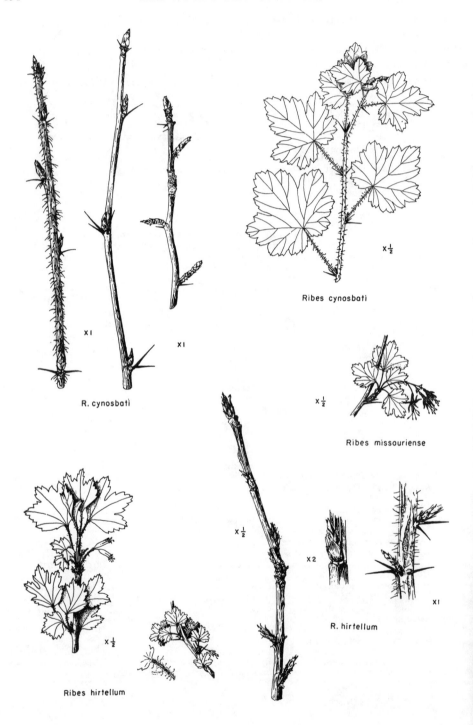

Ribes cynosbati

x½

x1

x1

R. cynosbati

x½

Ribes missouriense

x½

x½

x2

x1

R. hirtellum

Ribes hirtellum

1. **Ribes americanum** Mill. WILD BLACK CURRANT
   R. floridum L'Her.
   The only wild currant widely distributed in Ohio. Erect shrub, about 1 m. tall, without spines or bristles. The yellow resinous dots of the lower leaf-surface and large greenish-yellow flowers distinguish this from other currants.

2. RIBES NIGRUM L. BLACK CURRANT
   Occasionally planted and rarely escaped; reported only from Hamilton County.

3. **Ribes lacustre** (Pers.) Poir. SWAMP CURRANT. BRISTLY BLACK CURRANT
   The purplish-black bristly fruit, densely bristly internodes, glandular bristly inflorescence, deeply lobed leaves and skunk odor of bruised leaves and fruit help to distinguish this northern species which barely enters our range.

4. **Ribes glandulosum** Grauer SKUNK CURRANT
   A northern species known in Ohio only from Ashtabula County. The common name refers to the odor of bruised foliage and berries, a feature of *R. lacustre*, also, from which this is readily distinguished by the smooth stems, and red berries.

5. RIBES ODORATUM Wendland f. BUFFALO CURRANT. GOLDEN CURRANT
   R. aureum of ed. 7, not Pursh
   An American shrub whose natural range is west of the Mississippi River. Often planted for its golden yellow fragrant flowers, and occasionally spreading from cultivation. Reported from 12 Ohio counties, but perhaps not everywhere an escape.

6. RIBES SATIVUM Syme. GARDEN or RED CURRANT
   R. vulgare of ed. 7, not Lam., *R. rubrum* of Schaffner, 1932.
   Cultivated, and occasionally escaped. Many horticultural varieties.

7. **Ribes cynosbati** L. WILD GOOSEBERRY. DOGBERRY
   Grossularia cynosbati (L.) Mill.
   More generally distributed in Ohio than any other species of *Ribes*; frequent in woods, both dry and wet. Two varieties may be distinguished in our range:

   Leaves soft-pubescent...var. *cynosbati*
   Leaves soon glabrate, or sparingly pilose on veins beneath...............var. *glabratum* Fern.

Ribes americanum

Ribes lacustre

Ribes glandulosum

8. **Ribes missouriense** Nutt.   MISSOURI GOOSEBERRY

R. gracile Pursh, not Michx., *Grossularia missouriensis* (Nutt.) Cov. & Britt.

The general range of this species is to the west of Ohio: Indiana to Minnesota and South Dakota, south to Kansas, Missouri, and Tennessee. Known in Ohio from two Hamilton County collections.

9. **Ribes hirtellum** Michx.   SMOOTH GOOSEBERRY

Ribes (Grossularia) oxyacanthoides in part, *Grossularia hirtella* (Michx.) Spach.

A northern species; the best of the edible native gooseberries; fruit glabrous. Three varieties are recognized, only the typical entering our area.   *R. hirtellum* is similar to the far-northern *R. oxyacanthoides*, but the fruiting canes are usually less bristly and the leaf-blades not glandular.   The long branched or plumose hairs scattered among unbranched hairs on petioles are a specific character.

10.   RIBES GROSSULARIA L.   GARDEN GOOSEBERRY

Grossularia reclinata (L.) Mill.

A large number of garden forms in cultivation which vary in size, color, and pubescence of fruit.   Occasional as an escape; reported from 5 Ohio counties.

Ribes cynosbati

Ribes missouriense

Ribes hirtellum

HAMAMELIDACEAE

A family of shrubs and trees, best developed in warm-temperate and tropical regions, especially eastern and tropical Asia; about 20 genera and 100 species; two genera, each with one species, in our area.

HAMAMELIS L.   WITCH-HAZEL

Tall shrubs, rarely tree-like; leaves short-petioled, pinnately veined, sinuate-dentate, oblique at base; buds naked, stalked; flowers perfect, complete, in small short-peduncled axillary clusters; fruit a 2-valved, 2-seeded capsule; seeds shining-black, slender-ellipsoidal, 7–10 mm. long.   Five or six species, two or three in eastern America, the others in eastern Asia.   The vernal witch-hazel, *H. vernalis* Sarg., a shrub of the Ozark region, and *H. mollis* Oliver, from China, are sometimes planted.

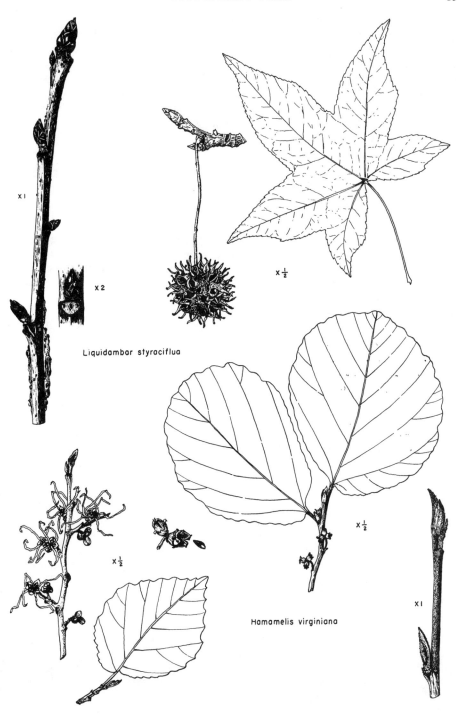

Liquidambar styraciflua

Hamamelis virginiana

1. **Hamamelis virginiana** L. WITCH-HAZEL

Most abundant in mesic woods of ravine slopes and bottoms, often in drier sites; occurring singly or in groups, each plant usually with several stems, i.e., shrub-like, although in height often tree-like. The smooth gray bark and asymmetric wavy-margined leaves are distinctive. Flowers with 4 linear petals, pale to bright yellow, in late fall when the leaves are turning or after leaf-fall, sometimes continuing through December. The capsules mature one year after flowering; upon splitting, seeds are shot explosively to several feet. The linament "witch-hazel" is distilled from branches of this shrub.

Two varieties are recognized:

Mature leaves glabrous beneath or thinly pubescent on veins.................var. *virginiana*
Leaves densely stellate-pubescent and reddish-brown or whitened beneath..var. *parvifolia* Nutt.

The typical variety is widely distributed in Ohio, but infrequent in the "prairie peninsula"; var. *parvifolia* is recorded in Ohio from Vinton and Lorain counties.

LIQUIDAMBAR L. SWEETGUM

Trees with leaves palmately veined and lobed; buds with overlapping bud-scales with shining resinous coating; flowers usually monoecious and without perianth, in globular heads, the staminate heads in terminal soon deciduous racemes, the pistillate heads solitary on long peduncle, developing into a hard spherical fruit 3–5 cm. in diameter, composed of many small crowded 2-beaked capsules each bearing 1–2 perfect seeds and many abortive sawdust-like seeds.

The resin, storax, is derived from *L. orientalis* Mill., a native of Asia Minor; the American species exudes a resin from cracks in the bark which is said to equal the imported storax. The generic name is a combination of the Latin *liquidus*, liquid, and the Arabic *ambar*, alluding to the fragrant gum.

1. **Liquidambar styraciflua** L. SWEETGUM

A large forest tree of southern range, extending northward into southern Ohio and southern Indiana and westward into Oklahoma and Texas; also in the mountains of Mexico and Central America. In Ohio, sweetgum is a tree of alluvial flats in the southern part of the Scioto and adjacent Ohio River drainage, and of the wet upland flats of the Illinoian Till Plain of southwestern Ohio, where trees over 9 ft. in circumference were frequent. In the latter situation it is associated with pin oak, or red maple, beech, and white oak

Hamamelis virginiana

Liquidambar styraciflua

in old-growth forests, and is a common pioneer in secondary succession. The largest known Ohio sweetgum, in Gallia County, is 14 ft. 9½ inches in circumference (Ohio Forestry Assn.). Sweetgum has been planted on ten of our state forests, some far north of the natural range of the species. Where planted on broad swampy stream flats (as on Zaleski State Forest in Vinton Co.) it appears as if native. A valuable timber tree (known to lumbermen as red-gum), and a fine ornamental, coloring brilliantly in fall.

PLATANACEAE

Deciduous trees of one genus.

PLATANUS L. Sycamore. Plane-tree

A genus of about 10 species: one, *P. orientalis* L., in the Old World (southeast Europe to India); one, × *P. acerifolia* (Ait.) Willd., a hybrid of early (before 1700) and doubtful origin (probably *P. orientalis* × *P. occidentalis*), commonly planted under the name London Plane as a street tree; one, *P. occidentalis* L. of eastern United States; the others in southwestern United States and Mexico. Large trees, the bark (except on the lower part of old trunks) exfoliating in autumn in thin brittle sheets leaving younger branches smooth and white, older ones blotched with dark persistent fragments of old bark. Leaves palmately lobed, with sheathing stipules which encircle twig, thus leaving a ring scar; buds enveloped in enlarged petiole-base until leaf-fall, hence almost surrounded by leaf-scar. Flowers monoecious, in separate spherical long-pedunculate heads, the staminate early deciduous. Fruit a syncarp made up of elongated achenes each surrounded by a ring of bristly hairs. Fruit-balls usually solitary in our species, 2 or occasionally more in the London Plane, and 3 or more, racemose, in other species.

1. **Platanus occidentalis** L. Sycamore

The largest of all eastern trees in girth, but not in height. The largest for which measurements are given— 48 ft. in circumference—grew in Daviess Co., Indiana (Deam, 1953); the largest Ohio tree—42 ft. 7 in. circumference at 4½ ft. above ground (D. & C.)—is recorded for the Muskingum River valley in northern Washington County. A more recent record (Ohio Forestry Assn.) mentions a Pickaway County sycamore 32 ft. 10 in. as the largest in the state.

A tree of floodplains, streamsides, and banks where there is local seepage, but not where subjected to long periods of submergence. Fruit maturing late in the season but remaining attached and intact until spring.

Platanus occidentalis

In the spring the ball loosens, and the individual achenes, each bearing a tuft of hairs, are carried away by the wind. The leaves, although glabrous at maturity except on the veins beneath, when young are heavily coated with sharp-pointed branching hairs (as are also petioles, stipules, and young branches) which soon become detached and blow about in the wind, often collecting on the ground in fuzzy wads at the margins of open spots.

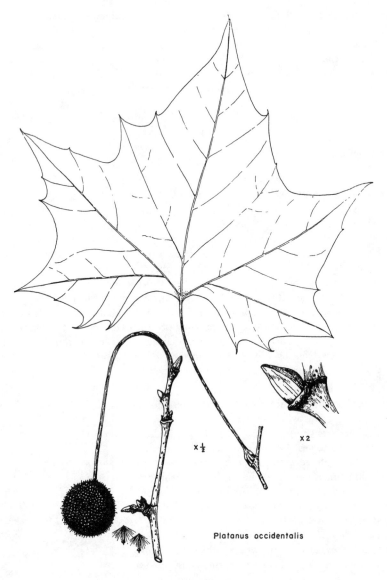

Platanus occidentalis

ROSACEAE

A large family of about 100 genera and 3000 species, herbs, shrubs, or trees, widely distributed and almost cosmopolitan, most abundant in the north temperate zone.   Sometimes divided by fruit characters into several families, better considered as subfamilies as all are linked by floral characters.   Members of the Rose Family can generally be distinguished easily by floral characters: 5-merous perigynous flowers, the sepals, petals (sometimes wanting), and numerous stamens on the margin of the flat to urn-shaped or conical receptacle.   Leaves alternate in all native species.

The family is of great economic importance because of the many ornamentals and species bearing edible fruits; it contains few weeds.

PHYSOCARPUS Maxim. NINEBARK

Deciduous shrubs whose bark peels into thin strips, hence the common name. The generic name, from the Greek *physa* (or *phusa*), bladder or bellows, and *karpos*, fruit, refers to the inflated capsules or follicles. A genus of 12 or 13 species, best represented in the West; two species in the Southeast; one in northeast Asia; and one in our range.

1. **Physocarpus opulifolius** (L.) Maxim. NINEBARK

A common and widespread shrub frequently found on stream bluffs and terraces; often planted. Easily recognized by the exfoliating bark, the somewhat lobed roundish leaves, small white flowers (with dark anthers) in umbel-like corymbs, and inflated, often reddish follicles in clusters of 3 (3–5).

Two varieties are distinguished:

Fruit glabrous or nearly so..var. *opulifolius*
Fruit stellate-pubescent.................................var. *intermedius* (Rydb.) Robins.

The typical variety is more eastern in distribution than var. *intermedius*, but the two intergrade. Most Ohio specimens are referable to the typical variety; one, from Defiance County, to the more western var. *intermedius*.

SPIRAEA L. SPIRAEA

Deciduous shrubs of the northern hemisphere; nearly 100 species, about half in cultivation. A few of the latter are occasionally reported as escapes, and the old-fashioned bridal-wreath, *S. prunifolia* Sieb. & Zucc., reported from six Ohio counties, persists and spreads after cultivation. Shrubs with elongate simple leaves (in our species) and small white or pink flowers with widely spreading petals (flowers flat).

Leaves glabrous or nearly so beneath; sepals spreading, not reflexed.
Branchlets of inflorescence and receptacle puberulent; stems yellowish- to reddish-brown;
leaves oblong to oblanceolate; sepals usually obtuse; flowers white..............1. *S. alba*
Branchlets of inflorescence and receptacle glabrous; stems red-brown or purplish-brown;
leaves generally elliptic; sepals usually acute; flowers white or pale pink......2. *S. latifolia*
Leaves woolly beneath with white or tawny tomentum, rugulose above; sepals reflexed; flowers
deep rose-pink...3. *S. tomentosa*

1. **Spiraea alba** DuRoi MEADOW-SWEET. MEADOW SPIRAEA

Shrub, about 1 m. in height. Widely distributed in Ohio, more frequent northward,

Physocarpus opulifolius

Spiraea alba

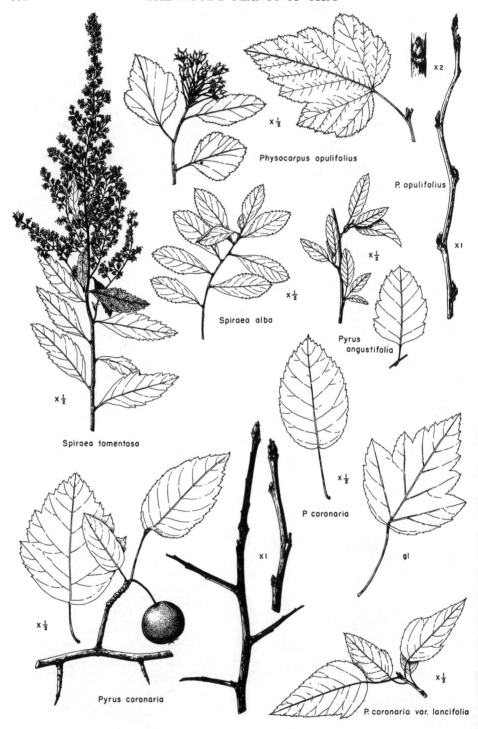

Physocarpus opulifolius

P. opulifolius

x 2

x ½

x I

Spiraea alba

x ½

Pyrus
angustifolia

x ½

Spiraea tomentosa

x ½

P coronaria

x ½

gl

Pyrus coronaria

x I

x ½

P. coronaria var. lancifolia

x ½

local southward; in wet, usually acid soil, ditches, pond and bog borders, and on the Illinoian Till Plain of southwestern Ohio, in wet meadows and forest openings. *S. salicifolia* L., a similar species of cool-temperate latitudes of the Old World, has pink flowers; it is sometimes planted, and in the East, has been reported as an escape.

2. **Spiraea latifolia** (Ait.) Borkh. MEADOW-SWEET

A shrub of the Northeast, in "moist or dry, usually upland or rocky soil, old fields and meadows." Known in Ohio only from Lake County.

3. **Spiraea tomentosa** L. HARDHACK. STEEPLE-BUSH

The rose-pink flowers in dense narrow panicles, and the leaves, woolly beneath, distinguish this handsome native shrub. This species ranges farther south in interior United States (to Ga., Tenn., and Ark.) than the two preceding; in Ohio, it is most frequent in the Allegheny Plateau, extending westward on the Lake plain and locally on the Illinoian Till Plain, where it occurs in meadows in the pin oak flats.

Two varieties have been distinguished, which differ in density of inflorescence:

Pedicels hardly visible in the densely-flowered inflorescence; mature flowers and fruits 12–17 per cm. on branches. . . . . var. *tomentosa*
Pedicels easily visible in a less dense inflorescence; flowers and fruits 8–11 per cm. on branches
var. *rosea* (Raf.) Fern.

The densely-flowered var. *tomentosa* is found in the northeastern part of the range of the species, the var. *rosea* on the Coastal Plain and in the interior. The eastern limit of range of var. *rosea* is in eastern Ohio; the western limit of range of typical var. *tomentosa* is in eastern New York and Pennsylvania; intermediates prevail in the intervening area. Most of our Ohio specimens are var. *rosea*; three (one each from Gallia, Lorain, and Stark Counties) are var. *tomentosa*; a few are intermediates, one of which (from Portage County) approaches var. *tomentosa*. Colonies, rather than individuals, should be studied, as there is some variation from plant to plant in a colony. All may be referred to as *S. tomentosa*.

Characters used in the key, and statements concerning range taken from Salamum, 1951.

SORBARIA A. Br. FALSE SPIRAEA

Deciduous shrubs of eastern Asia, with stipulate, pinnately compound leaves and small spiraea-like flowers in large terminal panicles.

Spiraea latifolia

Spiraea tomentosa

1. SORBARIA SORBIFOLIA (L.) A. Br.
   Schizonotus sorbifolius (L.) Lindl.
   A suffruticose or nearly herbaceous plant, 1–2 m. in height. Reported as an escape in Coshocton, Harrison, Lake, and Licking counties.

PYRUS L.

A large genus of the northern hemisphere including perhaps 130 species; often subdivided into a number of genera among which are *Pyrus* (in a restricted sense), *Malus*, *Aronia*, and *Sorbus*. These may be better considered as subgenera, although in America they seem to be distinct because of the relatively few American species. Hybrids are known between *Sorbus* and *Aronia*, and between *Sorbus* and *Pyrus*. Characters which we, because of our limited contacts, may think of as distinctive, are not always so; for example, our American species of *Sorbus* have compound leaves, while some of those of the Old World have simple leaves. This character of American *Sorbus* is generally used in keys to distinguish it from the other "genera" here included in *Pyrus*. The fruit of all members of the genus is a pome, large and fleshy in subgenera *Pyrus* and *Malus*, smaller and berry-like in *Aronia* and in our species of *Sorbus*.

Only 6 species of the inclusive genus *Pyrus* are indigenous to Ohio. In addition to the native species, a few, as pear and apple, are often spontaneous; many species are planted as ornamentals; a few hybrids have been reported.

a. Leaves simple.
   b. Trees or tree-like or coarse shrubs with some branchlets spur-like and usually spine-tipped; flowers large, 2 or more cm. across; fruit large, 2 or more cm. in diam.
   Petals white; fruit pyriform, flesh with grit cells; leaves glabrous, finely serrate; terminal bud pointed, conical...1. *P. communis*
   Petals pink or pinkish; fruit globose or depressed-globose; terminal bud generally blunt-pointed, ovoid..Subg. MALUS
   Leaves not lobed, involute in bud; leaves, young shoots, pedicels, and outside of persistent calyx tomentose..2. *P. malus*
   Leaves often lobed, folded in bud; spine-tipped branchlets commonly present.......
   Native Crab apples
   Calyx (outside) and pedicels glabrous or nearly so; leaves usually glabrous beneath at maturity.
   Leaves broadly lanceolate to ovate or oval, mostly acute or acuminate, serrate or doubly serrate...3. *P. coronaria*
   Leaves oblong to lanceolate or narrowly elliptic or oblong, round-tipped or blunt, crenate-serrate or -dentate...4. *P. angustifolia*
   Calyx (outside) and pedicels densely tomentose; leaves permanently pubescent beneath...5. *P. ioensis*
   bb. Shrubs with slender branches, without thorn-like or spur-like branchlets; leaves with distinct glands along midrib above; flowers white or pinkish, small, about 1 cm. across; fruit small and berry-like, about 1 cm. or less in diam., in compound clusters..Subg. ARONIA
   Inflorescence, lower surface of leaves, and young branches glabrous; fruit black.......
   6. *P. melanocarpa*
   Inflorescence, lower leaf-surface, and young branches pubescent.
   Fruit red when ripe, 5–7 mm. in diam.; calyx-lobes conspicuously stipitate glandular
   7. *P. arbutifolia*
   Fruit dark purple or purple-black when ripe, 8–10 mm. in diam.; calyx-lobes glandless or nearly so...8. *P. floribunda*
aa. Leaves large, pinnately compound; flowers small, white, in large cymes; fruit berry-like, bright red or orange-red; terminal bud large, 10–15 mm. long, curved at tip; small trees or large shrubs...Subg. SORBUS
   b. Branchlets, inflorescence, and winter-buds villous, lower leaf-surface pubescent; winter-buds white-villous toward tip, not glutinous; introduced................9. *P. aucuparia*

bb. Branchlets, inflorescence, and winter-buds glabrous or nearly so, inner bud-scales villous; winter-buds glutinous, scales glabrous on back.
  Leaflets oblong, rounded or blunt-pointed at tip, rounded at base, coarsely toothed to middle or below; lenticels few, scattered.............................10. *P. decora*
  Leaflets lanceolate, sharp-pointed at tip, rounded to cuneate at base, finely toothed almost to base; lenticels numerous.................................11. *P. americana*

WINTER KEY

a. Bundle-scars 5 (sometimes 4); terminal bud large, 10–15 mm. long..............Subg. SORBUS
  b. Terminal bud white-hairy, at least in upper half, not glutinous...........9. *P. aucuparia*
  bb. Terminal bud not white-hairy, glutinous.
    Buds slightly glutinous, hairy toward tip, inner bud-scales villous; lenticels few.......
                                                                          10. *P. decora*
    Buds very glutinous, glabrous, or inner scales ciliate; lenticels numerous....*P. americana*
aa. Bundle-scars 3 or in 3 groups; terminal bud smaller, 3–6 (–8) mm. long.
  b. Shrubs; buds flattened and appressed, with 5 exposed pointed scales more or less glandular-denticulate...Subg. ARONIA
    Twigs and buds glabrous...6. *P. melanocarpa*
    Twigs and buds more or less pubescent.
      Buds soon glabrous...8. *P. floribunda*
      Buds retaining pubescence.....................................7. *P. arbutifolia*
  bb. Trees or coarse shrubs, often colonial; exposed bud-scales usually 4, more or less keeled above.
    Terminal bud conical, evenly tapering to a sharp point; twigs very dark brown or greenish brown; wild forms often with sharp spine-tipped branchlets...1. *P. communis*
    Terminal bud ovoid and blunt pointed, occasionally sharp pointed but its sides convex..
                                                                          Subg. MALUS
      Twigs more or less pubescent toward tip; buds tomentose at least toward tip; twigs bright or dark brown; stout spur-like branches numerous..............2. *P. malus*
      (Amount of pubescence varies in different varieties; twigs of crab apple more slender, usually less pubescent.)
      Twigs glabrous; buds glabrous.
        Buds narrowly ovoid, bright red-brown; twigs slender; short pointed or spine-like branchlets present; fruit sometimes persistent in winter, 2–4 cm. in diam........
                                                                          3. *P. coronaria*
        Buds broadly ovoid, bright yellow-brown, the scales dark margined; twigs stout; stout spur-like fruiting branches numerous; fruit often persistent in winter, small, about 1 cm. in diam...*P. baccata*

1. PYRUS COMMUNIS L. PEAR

The common pear, when wild, usually has abundant spine-tipped branchlets. Otherwise, it resembles its well known orchard parent. Reported from more than half the Ohio counties.

Subgenus MALUS APPLE

The large (usually pink) flowers in simple umbels or umbel-like clusters borne on short lateral branchlets, the styles united at base, and the large fleshy pome characterize this subgenus.

2. PYRUS MALUS L. APPLE

Malus malus (L.) *Britt.*, *Malus pumila* Mill.

Apple trees frequently persist about old house-sites and orchards, and sometimes come from seed spontaneously along fence-rows and roadsides and in old fields. *P. malus* is the parent of most of our cultivated apples; some may be derived from hybrids of this and related species. The Siberian crab is, by some authors, considered a hybrid of *P. malus* and *P. baccata*, by others it is referred to *P. baccata* L.

3. **Pyrus coronaria** L. WILD CRAB
    Malus coronaria (L.) Mill., *Malus glaucescens* Rehd. (in Schaffner, 1932)
    The wild crab is an exceedingly fragrant and ornamental species; it frequently forms large thickets which are almost impenetrable because of the close branching and abundant spine-tipped branchlets. Leaves of fertile and vegetative branchlets differ, and leaves vary from base to tip of vigorous branchlets. The flowers are similar to, but more highly colored than those of the common apple, and are borne on longer and more slender pedicels; fruits green or yellowish-green, about 2.5 cm. in diam., fragrant, and very acid; excellent for making jelly.
    A variable species in which several varieties may be recognized, some of which are segregated as species by some authors; hybrids also occur, adding to the confusion. Three varieties occur in our range:

    Leaves of flowering branches ovate to oblong-ovate.
    Hypanthium and outer surface of sepals glabrous.........................var. *coronaria*
    Hypanthium sparingly villous, sepals glabrate, leaves paler beneath...................
    var. *dasycalyx* (Rehd.) Fern.
    Leaves of flowering branches ovate-lanceolate to oblong-lanceolate, tending to be acuminate,
    those of vigorous shoots triangular-lanceolate.................var. *lancifolia* (Rehd.) Fern.

    Recognition of varieties is not always possible with incomplete material; as the varieties intergrade, it may be best to consider *P. coronaria* a polymorphic species and not segregate varieties (map). However, a form with long petioles (4–5 cm.) and triangular-ovate leaves glaucous beneath, those of vigorous shoots often deeply lobed, is *Pyrus glaucescens* (Rehd.) Bailey. Although usually included in *P. coronaria*, Ohio material seems to be fairly distinct (gl of figure); specimens from Cuyahoga, Erie, Jefferson, Knox, Lucas, Wayne, and Williams counties. The narrow-leaved var. *lancifolia* is usually more or less distinct (map); var. *dasycalyx*, sometimes regarded as a hybrid between *P. coronaria* and the more western *P. ioensis*, is found in a few western and central Ohio counties (Auglaize, Crawford, Franklin, Lucas).

4. **Pyrus angustifolia** Ait. WILD CRAB
    Malus coronaria (L.) Mill. (in Schaffner, 1932)
    This narrow-leaved wild crab with obtuse or subacute leaves is southern in distribution, and as now understood, occurs only in the southern part of Ohio. The illustration is from an Adams County specimen determined many years ago at the Arnold Arboretum. Most of our wild crabs with narrow leaves should be referred to *P. coronaria* var. *lancifolia*.

Pyrus coronaria

Pyrus coronaria var. lancifolia

Pyrus angustifolia

5. **Pyrus ioensis** (Wood) Bailey WILD CRAB
Malus ioensis (Wood) Britt.

A more western species not generally thought to occur east of Indiana; distinguished by the densely tomentose calyx and pedicels, and leaves persistently pubescent beneath, at least on veins. Ohio specimens here referred to *P. ioensis* (from Clinton, Erie, Logan, Lucas, and Morrow counties) seem to show some introgression from *P. coronaria*, with which it is known to hybridize. Perhaps they should best be thought of as "approaching" *ioensis*. Other specimens (from Crawford and Union counties) have some characteristics of *P. ioensis*, and perhaps should be referred to *P. coronaria* var. *dasycalyx*. The handsome Bechtel's Crab, with large double flowers, is a horticultural form frequently planted.

Subgenus ARONIA CHOKEBERRY

Recognized by its distinctly shrub habit, the glands along midrib, smaller flowers in compound clusters, and small berry-like fruit. The fruit is edible but puckery until cooked; it can be used for jelly.

6. **Pyrus melanocarpa** (Michx.) Willd. BLACK CHOKEBERRY
Aronia melanocarpa (Michx.) Ell.

Best distinguished by the absence of pubescence, and black fruit 7–10 mm. in diam. Our most abundant species, found in about one-half of the Ohio counties, principally in the Allegheny Plateau, pin oak flats of southwestern Ohio, and Lake district of northern Ohio. Occurs in a variety of habitats—bogs, wet meadows and swamp woods (acid soil), and in dry (acid) soil of ledges and steep slopes.

7. PYRUS ARBUTIFOLIA (L.) L.f. RED CHOKEBERRY
Aronia arbutifolia (L.) Ell.

Distinguished by the grayish or whitish pubescence of young branchlets and lower leaf surface, the usually conspicuously glandular calyx, and the red fruit 5–7 mm. in diam. No Ohio specimens are referred to this species. Usually in boggy situations.

8. **Pyrus floribunda** Lindl. PURPLE CHOKEBERRY
P. arbutifolia var. *atropurpurea* Robins., *Aronia atropurpurea* Britt., *Aronia prunifolia* (Marsh.) Rehd.

Very similar to *P. arbutifolia*, from which it can be distinguished only by the dark purple mature fruit, and the usually glandless calyx. Local in Ohio; more northern in

Pyrus melanocarpa

Pyrus floribunda

Pyrus decora

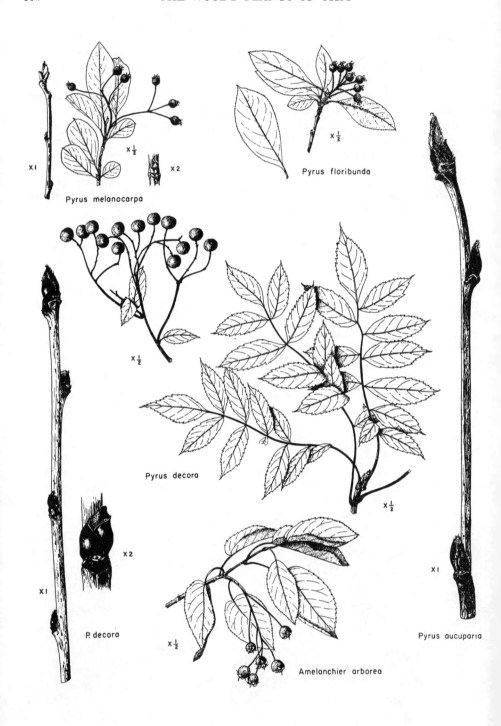

Pyrus melanocarpa

x1   x½   x2

Pyrus floribunda   x½

Pyrus decora   x½

x½

P. decora   x2   x1

Amelanchier arborea   x½

Pyrus aucuparia   x1

distribution than the other species of *Aronia*. A hybrid of this and *P. melanocarpa* was found growing with the parent species in a bog in Stark County.

Although specimens of *P. floribunda* vary in density of pubescence, there appears to be no valid reason to separate our specimens. A few (from Licking, Stark, Summit, and Wood counties) are more densely pubescent than others, and the glands on the midrib of upper leaf-surface more prominent, and might be referred to *P. arbutifolia*. However, following Gleason's interpretation, that species is not found in Ohio; it grows "on the Coastal Plain and not far inland."

Subgenus SORBUS MOUNTAIN-ASH

Small trees or coarse shrubs with odd-pinnate leaves (in our species), small white flowers in large very compound cymes, and small berry-like usually bright red pomes.

9. PYRUS AUCUPARIA (L.) Gaertn. MOUNTAIN-ASH
Sorbus aucuparia L.

A European species, often planted, and occasionally occurring as an escape in northern Ohio.

10. **Pyrus decora** (Sarg.) Hyland. MOUNTAIN-ASH
P. sitchensis of ed. 7, not Piper; *Sorbus decora* (Sarg.) Schneid., *Sorbus scopulina* Britt., not Greene.

Very similar to *P. aucuparia*, with which it is often confused. Confined to northeastern Ohio and the Lake Erie shore. *P. americana* (Marsh.) DC., which probably does not enter our area, can be distinguished by its glabrous or merely ciliate inner bud-scales, longer leaflets (3–5 times as long as wide), and smaller flowers and fruits.

AMELANCHIER Medic. SERVICEBERRY, JUNEBERRY, SHAD-BUSH

Ornamental, early-blooming small trees or shrubs of the northern hemisphere, principally in North America. Flowers (in our species) white or pale pink, in short racemes; petals narrow (obovate, oblong, oblanceolate, or linear); the small juicy pomes sweet and edible; leaves simple, serrate or dentate.

Authorities differ in the interpretation of species, which are difficult or impossible to identify without flowers, immature fruit, and mature leaves. Growth habit should be noted, as some species are colonial. Hybrids occur frequently, which add to the difficulty of determination. Plants with the aspect of one of the colonial shrub species may have the glabrous ovary of *A. arborea* or *A. laevis*; plants with the pinkish-white flowers and densely white-tomentose young leaves of *A. arborea* may have the long fruiting pedicels and entirely glabrous leaves of *A. laevis*. The interpretation of species used here is that of Gleason (1952), in which *A. spicata* includes *A. humilis* and *A. stolonifera* of Fernald's treatment (1950). Additional information, not included in the two manuals, may be found in publications by Wiegand (1912), Nielsen (1939), and Jones (1946).

The winter-buds of all species are long and slender, somewhat resembling those of beech, but with fewer exposed scales; at least some of the bud-scales are more or less three-toothed at apex, the middle tooth sometimes awn-like.

Ovary glabrous at summit; leaves acute to acuminate at apex, finely toothed; racemes spreading to pendulous; petals 1 cm. or more long; sepals reflexed in fruit; trees or erect shrubs, not colonial.
  Leaves at flowering time small and densely white-pubescent, pubescent to nearly glabrous at maturity, doubly serrate, the teeth 6–10 per cm.; pedicels more or less pubescent, 0.5–2 cm. long; winter-buds 6–13 mm. long.............................1. *A. arborea*

Leaves at flowering time about half grown, glabrous, bronzy or suffused with sienna-red, teeth callous-tipped, 6–8 per cm.; pedicels glabrous, 1–3 cm. long, the lowest becoming 2.5–5 cm. long; winter-buds 9–17 mm. long...............................2. *A. laevis*
Ovary pubescent at summit.
Stoloniferous shrub; leaves with lateral veins relatively weak, curved forward, branching toward margin, some of the smaller branches reaching teeth; racemes more or less dense, lowest pedicels becoming 7–15 mm. long................................3. *A. spicata*
Straggling shrub with one to several stems; leaves with lateral veins relatively strong, simple or few-forked, extending into teeth of leaf-margin; racemes loose and open, lowest pedicels becoming (7–) 15–30 (–40) mm. long................................4. *A. sanguinea*

1. **Amelanchier arborea** (Michx. f.) Fern. Downy Serviceberry

A. canadensis of auth., not Medic., *A. canadensis* var. *botryapium* (L.f.) T. & G.

Our only common and widespread species, probably absent only from some of the counties in the calcareous area of western Ohio; most frequent in the acid soils of the Allegheny Plateau. Flowering very early, thus conspicuous at a distance in the leafless woods; often associated with red maple. Leaves vary from almost glabrous (a few hairs on petiole) to densely felted beneath, from broad ovate-oval and short-acute to elliptic or elliptic-oblong and acuminate, usually with sides in part parallel.

2. **Amelanchier laevis** Wieg. Allegheny Serviceberry

Local in Ohio; conspicuous at flowering time because of the bronzy or sienna-red color of unfolding leaves, pedicels, and calyx. An abundant species at higher elevations in the

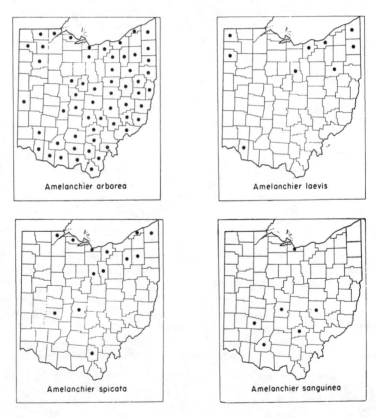

Amelanchier arborea

Amelanchier laevis

Amelanchier spicata

Amelanchier sanguinea

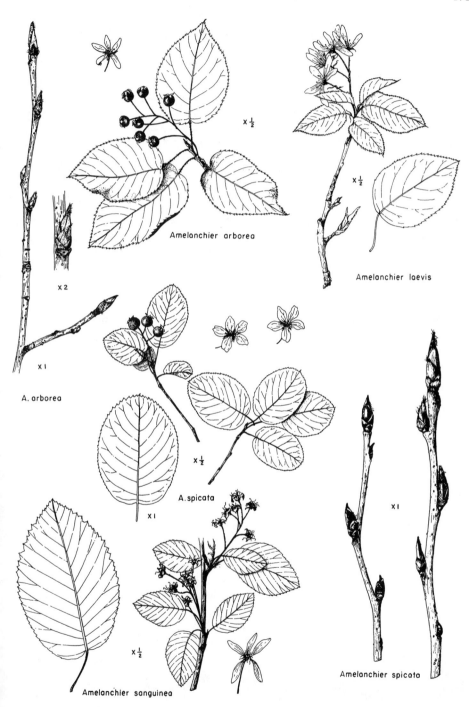

Amelanchier arborea

Amelanchier laevis

x½

x2

x1

A. arborea

A. spicata

x½

x1

x1

Amelanchier sanguinea

x½

Amelanchier spicata

Southern Appalachians. Summer specimens of glabrous-leaved forms of *A. arborea* may be mistaken for this species, which is most distinct at flowering time. Hybrids are recorded from several counties.

3. **Amelanchier spicata** (Lam.) K. Koch

A. humilis Wieg. and *A. stolonifera* Wieg. included here.

A northern species, or species-complex; widely scattered in Ohio, most abundant northward, local southward. The great range of variation in leaf-shape and serration (even on the same bush), and in size and number of teeth—characters which Fernald uses to distinguish *A. humilis* and *A. stolonifera*, but which are not always grouped the same in Ohio material—make desirable the use of the name *spicata* for our stoloniferous shrubs.

4. **Amelanchier sanguinea** (Pursh) DC.

A. amabilis Wieg. included here.

A straggling shrub or small tree with several stems, rare in Ohio. Veins relatively straight, at least some running directly to the teeth, leaves large, petals very narrow-spatulate or linear, hypanthium relatively shallow and broad (in pressed specimens of flowers sometimes compressed vertically instead of laterally). The specific name refers to the red or reddish branchlets.

CRATAEGUS L. HAWTHORN. THORN. RED HAW*

The name is from a Greek word signifying strength or invincibility; and it probably referred to the dense and thorny nature of the plants that made them immune from attack rather than to the strength of the wood, as has often been suggested. The plants are shrubs or small trees usually with crooked thorny branches. The leaves are alternate and vary greatly in shape and size in different species, and to some extent within the species. Terminal leaves of sterile shoots or of new and vigorous growth are often unlike those of the flowering branchlets. Such leaf specimens should never be collected for determination except in connection with flowering or fruiting material. The flowers are regular, normally with five petals and five to twenty stamens, and one to five styles. In most species they are several in number, borne in simple or branching corymbs or cymes, or rarely they grow singly or two or three in a close cluster. The fruit is a pome or haw with one to five bony nutlets embedded in the firm to succulent flesh. The fruits vary in shape from subglobose to oblong-cylindric, ovoid, or obovoid; they are sometimes slightly 5-angled. In our species, the color when ripe is some shade of red or more rarely yellow; but some species remain green and hard throughout the season. The fruit of some species is edible with a pleasing and distinctive flavor.

While most species of *Crataegus* are not absolutely selective as to soil, they show a decided preference for alkaline soils, and they are most abundant and varied in limestone regions; this is well illustrated by the range of *C. mollis*. The *Crataegus* flora of Ohio is a large and diverse one due to the wide range of ecological conditions found in the state and to the fact that large areas are underlaid with limestone or are covered with alkaline or neutral soils.

For convenience, the genus has been divided into a number of groups or series that differ from each other in the degree of their distinctness of characters. Fourteen series are recognized as occurring in Ohio.

---

*Text and keys contributed by Ernest J. Palmer, and figures drawn from specimens determined by him. For further general information concerning the genus, and for distribution of varieties within the species, see Ohio Jour. Sci. **56**(4): 205–216, July 1956. A few of the species, known only from one or two collections, are omitted from the key, and appear in the text following better known or more widely distributed species.

KEY TO THE SERIES*

a. Veins of the leaves running to the sinuses as well as to the points of the lobes; leaves mostly ovate or deltoid in outline, distinctly lobed; fruit subglobose, 4–10 mm. thick.

  b. Leaves often trilobate, up to 5–6 cm. long, thin, turning red or yellow in autumn, early deciduous; flowers in many-flowered compound corymbs, opening in late May or early June; fruit 4–6 mm. thick, bright red, becoming succulent, remaining on branches until late winter; nutlets 3–5 . . . . .1. CORDATAE (p. 174)

  bb. Leaves usually deltoid or broadly ovate in outline with 5–7 lobes, firm in texture, remaining green and persistent until late in season; flowers opening usually before middle of May; fruit firm or mellow at maturity, falling early; nutlets 1–2 or rarely 3 . . . .
2. OXYACANTHAE (p. 174)

aa. Veins of the leaves running only to the points of lobes or larger teeth; leaves entire or variously lobed; fruit subglobose, oval, ovoid, or pyriform, 0.5–1.5 cm. thick.

  b. Flowers single or rarely 2–3 in cluster; stamens 20–25; sepals foliaceous, pectinate; slender shrubs usually less than 1.5 m. tall . . . . .3. PARVIFOLIAE (p. 176)

  bb. Flowers more numerous, in simple or compound corymbs or cymes; stamens 5–20; sepals entire or glandular-serrate, not foliaceous; trees or stout arborescent shrubs.

    c. Leaves of various shapes, cuneate to subcordate at base, glabrous or pubescent; fruit 0.5–1.5 cm. thick; nutlets 1–5, not pitted on ventral surface.

      d. Leaves mostly spatulate or obovate, broadest above middle except rarely on terminal shoots; unlobed or sometimes more or less lobed on terminal shoots.

        e. Leaves firm to subcoriaceous, often glossy above, the veins inconspicuous or rarely slightly impressed above; flowers 1–1.5 cm. wide; fruit usually 1–1.3 cm. thick or less, with 1–2 or rarely 3–5 nutlets, remaining hard and inedible, usually dull red or green at maturity . . . . .4. CRUS-GALLI (p. 176)

        ee. Leaves firm but not subcoriaceous, dull above, the veins noticeably or conspicuously impressed above; flowers usually 1.3–2 cm. wide; fruit usually becoming mellow and somewhat edible; nutlets 3–5 . . . .5. PUNCTATAE (p. 180)

      dd. Leaves mostly rhombic, ovate, oblong-ovate, or deltoid in outline, broadest below or about middle, usually more or less lobed at ends of shoots.

        e. Foliage and inflorescence glandular, often conspicuously so; leaves usually narrowed at base, except sometimes at ends of shoots; flowers mostly 3–8 in simple or little-branched corymbs; fruit usually remaining hard, green, bronze-yellow or dull red at maturity; branching shrubs, 1.5–3 m. tall . . . . . .
6. INTRICATAE (p. 182)

        ee. Foliage and inflorescence eglandular or sometimes sparsely glandular when young; flowers usually 8–15 or more in simple or compound corymbs.

          f. Leaves mostly ovate, oblong-ovate, or elliptic in outline, abruptly narrowed or rounded at base, or sometimes deltoid or suborbicular at ends of shoots; fruit usually bright or dull red at maturity, firm to succulent; nutlets usually 3–5.

            g. Leaves of flowering branchlets mostly oblong-ovate or rhombic, sometimes broadly ovate to suborbicular and rounded at base at ends of shoots, more or less incised with shallow or rounded lobes; flowers 1.2–1.5 cm. wide; nutlets 3–5 . . . . .7. ROTUNDIFOLIAE (p. 183)

            gg. Leaves of flowering branchlets mostly ovate, rounded at base, those at ends of shoots similar but larger, slightly lobed below middle; flowers 1.5–1.8 cm. wide; nutlets usually slightly pitted on ventral surface, or sometimes plain . . . . .8. BRAINERDIANAE (p. 185)

          ff. Leaves mostly ovate, oblong-ovate, or deltoid in outline, rounded to truncate at base, or at ends of shoots broadly ovate to deltoid and often truncate to subcordate at base, more or less lobed; fruit 0.8–1.7 cm. thick; nutlets 3–5.

            g. Leaves mostly ovate, more or less incised with acute lobes terminating in acuminate, often reflexed points, thin at maturity, glabrous except for short appressed hairs on upper surface while young; flowers 1.3–1.8 cm. wide; pedicels glabrous (except in *C. lucorum*); anthers pink; fruit 0.8–1.5 cm. thick, usually becoming mellow or succulent; nutlets 3–5
9. TENUIFOLIAE (p. 186)

            gg. Leaves mostly ovate or oblong-ovate, firm to thick at maturity, more or less incised, the points of lobes not reflexed; flowers 1.3–2.5 cm. wide; stamens 8–20; fruit 0.8–1.7 cm. thick; nutlets 3–5.

*Reprinted from Ohio Journal of Science **56** (4): 205-216, July, 1956

    h. Petioles and midribs of leaves slender; flowers 1.3–2 cm. wide; filaments as long or nearly as long as petals; fruit 0.8–1.8 cm. thick, glabrous; nutlets 3–5.

      i. Leaves glabrous or short-pilose above when young, glabrous beneath (except in *C. locuples*); fruit usually firm or hard at maturity, inedible; nutlets relatively large.

        j. Leaves with short appressed hairs above while young, becoming glabrous; stamens 10 or less; fruiting calyx small and sessile.................10. SILVICOLAE (p. 187)

       jj. Leaves glabrous or essentially so from the first (except in *C. locuples*); fruit often pruinose, sometimes slightly 5-angled, with prominent elevated calyx..................
11. PRUINOSAE (p. 189)

      ii. Leaves glabrous or pubescent; stamens 10–20; fruit usually becoming mellow and edible; nutlets relatively small........
12. COCCINEAE (p. 194)

    hh. Petioles and midribs of leaves thick and prominent; filaments shorter than petals; fruit pubescent at least while young, 1.3–1.5 cm. thick, usually becoming mellow and edible; nutlets 3–5, usually 5.............................13. MOLLES (p. 195)

  cc. Leaves mostly ovate, oblong-ovate, or elliptic, abruptly narrowed or rounded at base, more or less lobed, glabrous except for short appressed hairs on upper surface while young; flowers mostly 4–14 in simple or little-branched corymbs; fruit hard or firm at maturity; nutlets 2–3 (or rarely more), deeply pitted on ventral surface............
14. MACRACANTHAE (p. 197)

WINTER CONDITION

    As the species of *Crataegus* are identified with great difficulty at any season, it is impossible to determine them satisfactorily in winter condition. Winter twigs of a few of the more common species (or series) which have more or less distinctive aspects are illustrated, but the characters shown will not serve to distinguish these from others which may be found.

    The cockspur thorn (*C. crus-galli* and others of its series) are so abundantly provided with relatively slender slightly curved thorns as to lend a distinctive winter aspect. *C. punctata*, in winter, has a silvery-gray aspect. The common *C. mollis*, some trees of which are thorny, others thornless, may often be recognized by general aspect, if a number of trees have been identified in summer and are used to illustrate winter aspect. The commonly planted Washington thorn (*C. phaenopyrum*) can be recognized, at least in early winter, by its abundant corymbs of small (4–5 mm.) fruit. The two introduced species (*C. oxyacantha* and *C. monogyna*), which are sometimes found as escapes, can be distinguished from all other species of our area by the very short straight thorns, not more than 1–2 cm. long. Members of the series *Intricatae* are typically shrubby.

    Aspect, thickness of twigs (stout or slender), length and stoutness of thorns and prominence of twig-like base of thorns, color of bark of trunk and branches, are characters which should be observed. Within a limited area, after one has become familiar with all species in summer condition, winter characters, although not diagnostic, may be found to distinguish at least some of the species.

Series I. CORDATAE

1. **Crateagus phaenopyrum** (L. f.) Medic. WASHINGTON THORN

C. cordata Ait.

    A tree up to 7–8 m. high with slender thorny branchlets and thin scaly brownish-gray bark. A native species, commonly planted and sometimes escaping from cultivation. It is one of the most desirable species for planting on account of its symmetrical shape, distinctive foliage, and abundance of small bright red haws that persist until late in the season.

Series II. OXYACANTHAE Loud. ENGLISH HAWTHORN

Leaves mostly ovate to broadly obovate, with 3–5 shallow rounded or acute serrulate lobes, narrowed or cuneate at base, or at ends of shoots abruptly narrowed or rounded at base; stamens about 20; styles 2–3; fruit usually subglobose, nutlets 2 or rarely 3. .2. *C. oxyacantha*

Leaves mostly ovate to deltoid, usually abruptly narrowed at base, or rounded or truncate at ends of shoots, more deeply incised and divided into 3–5 (–7) oblong or oblong-lanceolate lobes; stamens about 20; styles usually single or rarely 2; fruit oblong, ellipsoid or subglobose, nutlets single...3. *C. monogyna*

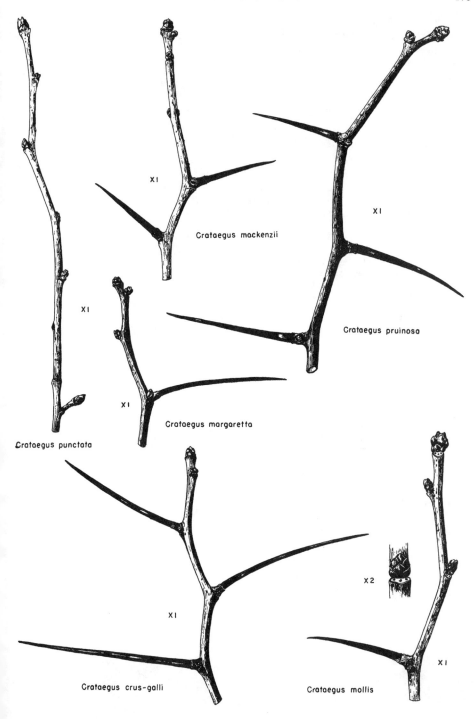

Crataegus mackenzii

Crataegus pruinosa

Crataegus margaretta

Crataegus punctata

Crataegus crus-galli

Crataegus mollis

2. CRATAEGUS OXYACANTHA L.

A small tree or arborescent shrub up to 5–6 m. high with slightly scaly bark and slender branchlets usually armed with short stout thorns. Native of the Mediterranean region, commonly cultivated and rarely escaped in our region.

3. CRATAEGUS MONOGYNA Jacq. ENGLISH HAWTHORN

A small tree similar to the last and often not distinguished from it by nurserymen; commoner in cultivation than the last, and more frequently found as an escape.

Series III. PARVIFOLIAE

4. **Crateagus uniflora** Muenchh.

C. trianthophora Sarg.

A slender shrub, usually with several spreading stems or branches and stoutish branchlets armed with slender thorns. Young branchlets pubescent at first, later glabrous. The general distribution is southern and southeastern, from the Atlantic coast to eastern Texas; in Ohio, found only rarely near the Ohio River.

Series IV. CRUS–GALLI Loud.

Foliage and inflorescence glabrous or essentially so; leaves of flowering branchlets mostly obovate, cuneate at base, not reticulately veined.
    Mature leaves deep green or dark green above, unlobed or rarely obscurely lobed at ends of shoots; young branchlets red or brownish-red becoming gray.
        Mature leaves thick or subcoriaceous (except sometimes in shade), mostly obovate or spatulate, sharply serrate above middle, pointed or rounded at apex; fruit usually 1–1.5 cm. thick, nutlets 1–2, rarely 3...5. *C. crus-galli*
        Mature leaves firm but not subcoriaceous, mostly oblong-obovate, pointed at apex, serrate with broad shallow or crenate teeth; fruit 0.6–1 cm. thick; nutlets 1–3, usually 2.......
                6. *C. pyracanthoides* var. *arborea*
    Mature leaves yellowish green, veins slightly but noticeably impressed above; terminal shoot leaves often slightly lobed; young branchlets olive or yellowish green, becoming light gray.
        Mature leaves relatively thin, mostly narrowly obovate or oblong-elliptic, from ⅜ to ⅔ as broad as long; stamens about 10 (rarely 20); anthers pink......7. *C. fontanesiana*
        Mature leaves thick or subcoriaceous, veins distinctly impressed above, mostly broadly obovate or oblong-elliptic, from ⅔ to ⅚ as broad as long; stamens about 10; anthers white or pale yellow...8. *C. hannibalensis*
Foliage, inflorescence, and young branchlets pubescent; leaves of flowering branchlets mostly 2–3.5 cm. long, reticulately veined beneath when mature; petioles 2–8 mm. long...........
                9. *C. engelmannii*

5. **Crataegus crus-galli** L. COCKSPUR THORN

C. algens Beadle, *C. arduennae* Sarg., *C. attenuata* Ashe, *C. eburnea* Ashe, *C. mollipes* Ashe, *C. trahax* Ashe.

Crataegus phaenopyrum

Crataegus uniflora

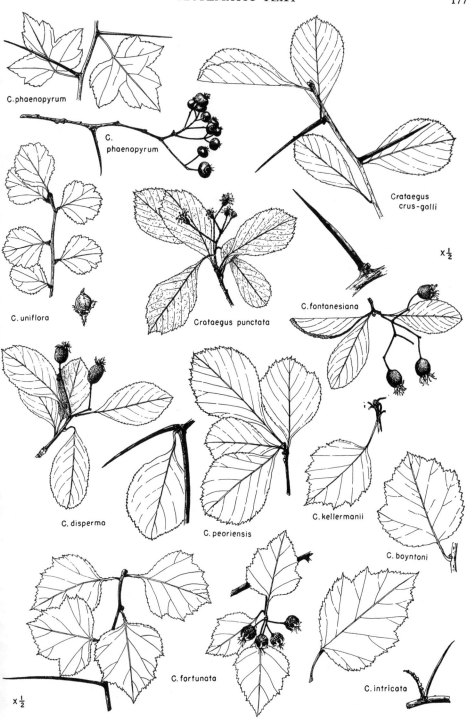

C.phaenopyrum

C. phaenopyrum

Crataegus crus-galli

C. uniflora

Crataegus punctata

C.fontanesiana

$x\frac{1}{2}$

C. disperma

C. peoriensis

C. kellermanii

C. boyntoni

C. fortunata

C. intricata

$x\frac{1}{2}$

A tree up to 10 m. high with a low broad crown of intricate wide-spreading branches, stout flexuous thorny branchlets, and dark gray-brown scaly bark; often flowering and fruiting as an arborescent shrub with an open irregular top.

Var. **crus-galli.** Leaves of flowering branchlets mostly obovate, cuneate at base, unlobed or rarely slightly lobed at ends of shoots, thick, dark green and glossy above at maturity; flowers 1–1.5 cm. wide; stamens about 10; anthers pink or pale yellow; fruit obovoid or subglobose, often slightly 5-angled, 0.8–1. cm. thick, with thin dry flesh, remaining green or turning dull red late in the season; nutlets 1–2, or rarely 3. Var. **barrettiana** (Sarg.) Palmer (*C. barrettiana* Sarg.) has thinner less glossy leaves, with veins slightly impressed above. Var. **exigua** (Sarg.) Egglest. (*C. exigua* Sarg.) has leaves at ends of shoot often slightly lobed, fruit oblong, bright red, with usually a single nutlet. Var. **leptophylla** (Sarg.) Palmer (*C. leptophylla* Sarg.) has slightly larger leaves, about 20 stamens, and 3–4 (–5) nutlets. Var. **pachyphylla** (Sarg.) Palmer (*C. pachyphylla* Sarg.) has leaves mostly broadly ovate, up to 3–3.5 cm. wide, subcoriaceous, with slightly impressed veins when mature. Var. **pyracanthoides** Ait. has smaller, narrowly oblong or linear-oblong leaves, and fruit 1 cm. or less thick.

6. **Crataegus pyracanthoides** Beadle var. **arborea** (Beadle) Palmer
   C. arborea Beadle.
   A tree up to 10 m. high with slender spreading branches, slender thorny or sometimes thornless branchlets, thin scaly bark.

7. **Crataegus fontanesiana** (Spach) Steud.
   C. genesseensis Sarg., *C. tenax* Ashe, *C. wilkinsoni* Ashe.
   A tree up to 8–9 m. high with pale slender usually thorny branchlets, slightly scaly brownish-gray bark on trunk; sometimes fruiting as a stout shrub.

8. **Crataegus hannibalensis** Palmer
   A tree 6–8 m. high with slender spreading branches, pale thorny branchlets, and slightly scaly pale gray-brown bark.
   Crataegus persimilis Sarg., with leaves somewhat similar to those of *C. hannibalensis*, but differing from that species in the flowers with 15–20 stamens, pink anthers, coarsely glandular-serrate sepals, and dark crimson fruit, is possibly a hybrid between *C. crus-galli* and *C. succulenta* or a similar species.

9. **Crataegus engelmanni** Sarg.
   A tree up to 6–7 m. high with broad flattened crown of stiff spreading branches, stoutish thorny branchlets, or often a stout irregularly branched shrub. This southern and western species has been found so far only in Richland County.
   C. ohioensis Sarg., with leaves somewhat resembling those of *C. pyracanthoides* var. *arborea*, but differing from that species in having slightly villous flowering corymbs, and in the erect persistent sepals, and 4–5 nutlets, is known only from the type locality in Franklin County. It is probably of hybrid origin.
   C. vallicola Sarg., with leaves similar to those of *C. hannibalensis*, but slightly villous along veins above while young, and slightly smaller flowers with 10–20 stamens and pink anthers.

Crataegus crus-galli

Crataegus pyracanthoides var. arborea

Crataegus fontanesiana

Crataegus hannibalensis

Crataegus persimilis

Crataegus engelmanni

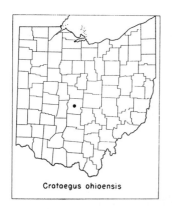

Crataegus ohioensis

Crataegus vallicola

Series V.   PUNCTATAE Loud.

a. Leaves of flowering branchlets mostly obovate or oblong-obovate, broadest above middle, gradually narrowed below to a short winged petiole.

    Leaves dull green above, serrate or irregularly dentate and more or less lobed above middle, pubescent (except in 1 var.); terminal shoot leaves often deeply incised with 2–3 pairs of acute spreading lobes; stamens about 20; fruit usually 1.2–1.5 cm. thick; nutlets 3–5, usually 5..10. *C. punctata*

    Leaves bright green and slightly glossy above, glabrous or sparsely pubescent, unlobed or obscurely lobed except sometimes at ends of shoots; stamens about 10.

        Leaves green on both sides or slightly paler beneath, those of flowering branchlets finely serrate above middle and usually unlobed, pointed or rounded at apex; terminal shoot leaves larger, pointed or acuminate at apex, more or less lobed; nutlets 1–3, usually 3
11. *C. disperma*

        Leaves decidedly paler beneath, usually obtuse or rounded at apex, often unlobed or only slightly lobed even at ends of shoots; nutlets 2–3.....................12. *C. peoriensis*

aa. Leaves of flowering branchlets mostly oval or oblong-ovate, broadest about middle; terminal shoot leaves mostly broadly ovate to suborbicular, sometimes as broad as long or broader.

    Leaves with obscure or shallow lateral lobes, not deeply cut or laciniate even at ends of shoots; flowers mostly 6–12 in glabrous or villous corymbs; stamens about 20.

        Leaves mostly oblong-ovate, abruptly narrowed or rounded at base, or at ends of shoots often broadly ovate and rounded to truncate at base; nutlets 3–5, usually less than 5.

            Foliage, inflorescence, and young branchlets glabrous; fruit red or orange, punctate, often slightly 5-angled..13. *C. kellermanii*

            Foliage, inflorescence, and young branchlets pubescent; fruit dull dark red, slightly pruinose..14. *C. indicens*

        Leaves mostly broadly ovate to suborbicular; flowers 1.6–2 cm. wide; nutlets 3–5, usually 5
15. *C. suborbiculata*

    Leaves with 4–5 pairs of sharp lateral lobes, or at ends of shoots deeply incised; flowers 1.3–1.5 cm. wide, mostly 6–7 in nearly simple villous corymbs; stamens about 20; anthers white or pale yellow..16. *C. mansfieldensis*

10. Crataegus punctata Jacq.

    A tree up to 8–10 m. high, with broad depressed crown of stout intricate wide-spreading branches and stout thorny branchlets; bark brownish gray, fissured and ridged on trunk.

    Var. **punctata.** Leaves dull yellowish green, short-villous above and along veins beneath while young; flowers 1.3–2 cm. wide; stamens about 20; anthers usually pink; fruit subglobose or short-oblong, usually 1.2–1.5 cm. thick, dull red or orange-red; nutlets 3–5, usually 5. Var. **aurea** Ait. (*C. crocata* Ashe) has fruit bright yellow at maturity, and usually pale yellow anthers. Var. **canescens** Britt. has leaves densely gray-pubescent on both surfaces. Var. **microphylla** Sarg. has smaller leaves, mostly 2–2.5 cm. long, 1.5–2 cm. wide, and smaller flowers 1–1.2 cm. wide. Var. **pausiaca** (Ashe) Palmer (*C. pausiaca* Ashe, *C. calvescens* Sarg., *C. porrecta* Ashe) has foliage and inflorescence glabrous or nearly so, and leaves slightly lustrous above; stamens 10–20, nutlets 2–3. It is perhaps a hybrid between *C. punctata* and a species of the Crus-galli series.

11. Crataegus disperma Ashe

    C. cuneiformis of Egglest. in part, *C. praestans* Sarg.

    A tree up to 7–8 m. high with slender usually thorny branchlets and often with compound thorns on trunk and older branches. Intermediate in character between *C. crus-galli* and *C. punctata*, and may have originated as a hybrid between these species, although it is now rather common and widespread.

12. Crataegus peoriensis Sarg.

    C. grandis Ashe, *C. pratensis* Sarg., *C. cuneiformis* of Egglest. in part.

    A tree similar to the last in size and appearance.

Crataegus punctata

Crataegus disperma

Crataegus peoriensis

Crataegus kellermanii

Crataegus indicens

Crataegus suborbiculata

Crataegus mansfieldensis

13. **Crataegus kellermanii** Sarg.

An arborescent shrub or small tree with slender flexuous thorny branchlets. Probably a hybrid between *C. punctata* and *C. pruinosa* or a related species.

14. **Crataegus indicens** Ashe

A tree up to 7–10 m. high with ascending branches, slender dull brown thorny branchlets and dark rough bark. Perhaps a hybrid between *C. punctata* and a species of the *Macracanthae* series. Known only from the type locality in Richland County.

15. **Crataegus suborbiculata** Sarg.

A tree up to 6–7 m. high with glabrous, often flexuous thorny branchlets and pale gray slightly scaly bark; sometimes fruiting as an arborescent shrub.

16. **Crataegus mansfieldensis** Sarg.

An arborescent shrub or small tree with slender thorny branchlets, orange-green and slightly villous while young, and pale gray bark. Of doubtful relationship and perhaps a hybrid between some species of the *Punctatae* and of the *Coccineae* series.

Series VI.　INTRICATAE Sarg.

a. Foliage and inflorescence glabrous or essentially so; fruit glabrous.
  b. Leaves of flowering branchlets relatively large, mostly 2.5–4.5 cm. wide; flowers 1.3–1.7 cm. wide; stamens about 10.
    Fruit red, russet, or green flecked with red at maturity; leaves yellowish green, firm but veins not noticeably impressed above; sepals entire or partly glandular-serrate.
      Leaves mostly ovate, broadest below middle; fruit russet or bronze-green more or less flecked with dull red.
        Leaves with 4–5 pairs of sharp spreading lateral lobes, or at ends of shoots often deeply cut and with tips of lobes reflexed; fruit obovoid or pyriform, russet or yellowish green...17. *C. intricata*
        Leaves usually with 3–4 pairs of broad shallow lateral lobes, the tips not reflexed; fruit subglobose or short-oblong, green more or less blotched with dull red........
                              18. *C. boyntoni*
      Leaves mostly oval or oblong-ovate, broadest about middle; fruit becoming bright red or orange-red...19. *C. rubella*
      Fruit bright yellow at maturity; leaves thick, veins impressed above; sepals deeply glandular-serrate...20. *C. fortunata*
  bb. Leaves of flowering branchlets relatively small, mostly 2–3.3 cm. wide; flowers 1–1.3 cm. wide; stamens 5–10, often 5...21. *C. borseyi*
aa. Foliage and inflorescence pubescent; fruit pubescent at least while young..22. *C. biltmoreana*

Crataegus intricata

Crataegus boyntoni

17. **Crataegus intricata** Lange
C. diversifolia Sarg., *C. inducta* Ashe, *C. meticulosa* Sarg.
An irregularly branched shrub 1–3.5 m. high with slender thorny branchlets.
Var. **intricata.** Leaves mostly ovate or oblong-ovate, dull bluish-green, sharply
lobed; stamens about 10; anthers pale yellow or rarely pink; fruit green or russet at maturity.
Var. **straminea** (Beadle) Palmer (*C. straminea* Beadle) differs in the yellowish green
slightly lobed oblong-ovate or elliptic leaves, pink anthers, and yellowish green or dull
orange fruit.
18. **Crataegus boyntoni** Beadle
An arborescent shrub or rarely a small tree up to 6–8 m. high with flexuous thorny or
sometimes nearly thornless branchlets.
19. **Crataegus rubella** Beadle.
Usually a shrub 1–4 m. high or rarely a small tree with slender thorny or nearly unarmed
branchlets.
20. **Crataegus fortunata** Sarg.
A shrub 2–3 m. high with crooked ascending branches and slender thorny branchlets;
most easily distinguished in autumn by the bright yellow fruit.
21. **Crataegus horseyi** Palmer
A shrub 2–4 m. high, usually with several crooked ascending stems and slender thorny
branchlets. For original description of this species, see Ohio Jour. Sci. **56**:211–12. 1956.
22. **Crataegus biltmoreana** Beadle
C. polybracteata Ashe, *C. modesta* Sarg., *C. intricata* of Egglest., not Lange.
An irregular shrub or rarely small tree 1–5 m. high with crooked ascending stems and
slender thorny branchlets pubescent while young.

Series VII.   ROTUNDIFOLIAE Egglest.

Leaves of flowering branchlets mostly short-obovate or oblong-elliptic, serrate with shallow or
crenate teeth and usually with 2–4 pairs of shallow rounded lateral lobes above middle, or at
ends of shoots broadly ovate to suborbicular, more deeply lobed or sometimes deeply incised
near base, glabrous or slightly pubescent above while young; flowers mostly 4–10 in simple
glabrous or slightly villous corymbs; stamens about 20; nutlets 2–4, usually 3. .23.   *C. margaretta*
Leaves of flowering branchlets mostly rhombic or oblong-obovate, sharply serrate and with
2–3 pairs of short acute lateral lobes above middle, or at ends of shoots often ovate, more
coarsely serrate and more deeply lobed; flowers mostly 4–7 in nearly simple compact glabrous
corymbs; stamens 10; nutlets 3–5. . . . . .24.   *C. sicca* var. *glabrifolia*

Crataegus rubella   Crataegus fortunata   Crataegus horseyi

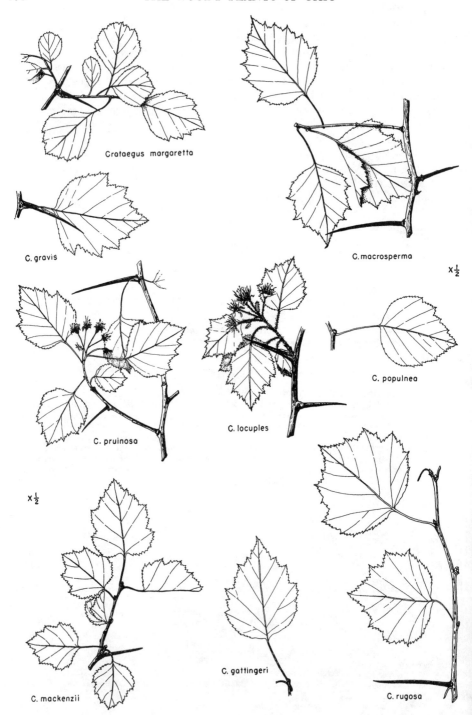

Crataegus margaretta

C. gravis

C. macrosperma

$\times \frac{1}{2}$

C. pruinosa

C. locuples

C. populnea

$\times \frac{1}{2}$

C. mackenzii

C. gattingeri

C. rugosa

23. **Crataegus margaretta** Ashe

A tree up to 6–7 m. high with ascending or spreading branches, straight or slightly flexuous thorny branchlets and slightly scaly dark gray-brown bark, often fruiting as a large shrub; leaves extremely variable in shape, relatively small, seldom over 2.5–3 cm. wide except at ends of shoots where sometimes suborbicular and deeply incised near base. Var. **margaretta.** Flowers 1.2–1.5 cm. wide; anthers white or pale yellow; fruit 0.9–1.2 cm. thick, dull red or orange-red, or in forma **xanthocarpa** Sarg. bright yellow at maturity. Var. **brownei** (Britt.) Sarg. (*C. brownei* Britt.) has smaller flowers 0.8–1 cm. wide and fruit 6–8 mm. thick. Var. **meiophylla** (Sarg.) Palmer (*C. meiophylla* Sarg.) has smaller usually oval or obovate leaves and red anthers.

24. **Crataegus sicca** Sarg. var. **glabrifolia** (Sarg.) Palmer

An arborescent shrub or small tree up to 5–6 m. high with slender glabrous thorny branchlets.

Series VIII. BRAINERDIANAE Egglest.

Leaves ovate, oblong-ovate, or elliptic, finely serrate and usually with 4–6 pairs of small acuminate lateral lobes, or at ends of shoots irregularly dentate and more deeply lobed; flowers mostly 4–12 in glabrous corymbs; stamens about 20 (or less in var.); fruit subglobose or slightly oblong, 8–9 mm. thick, with small sessile calyx, thin flesh, and 2–3 (usually 3) nutlets
25. *C. brainerdi*
Leaves mostly oval or oblong-ovate, irregularly and deeply serrate nearly to base, usually with 3–4 pairs of small shallow lateral lobes above middle, or more deeply lobed nearly to base at ends of shoots; flowers mostly 6–10 in glabrous corymbs; fruit subglobose or short oblong, with rather thick firm flesh, slightly elevated calyx, and 3–5 nutlets...............26. *C. coleae*

25. **Crataegus brainerdi** Sarg.

An arborescent shrub or rarely small tree up to 6–7 m. high with slender flexuous or straight usually thorny branchlets and dark, close or slightly scaly bark.

Var. **brainerdi.** Leaves mostly elliptic or oblong-obovate, 3.5–6 cm. long, 2.5–4 cm. wide, thin but firm, dark bluish green and glabrous at maturity; stamens about 20; anthers pink; fruit oblong. Var. **scabrida** (Sarg.) Egglest. (*C. scabrida* Sarg.). Leaves mostly obovate or elliptic, slightly smaller, thicker and scabrate above at maturity; stamens 5–15; anthers pink or pale yellow; fruit short-oblong or subglobose.

Crataegus biltmoreana

Crataegus margaretta

Crataegus sicca var. glabrifolia

26. **Crataegus coleae** Sarg.

    C. inserta Sarg.

An arborescent shrub or small tree up to 8–10 m. high with stout spreading branches and slender glabrous thorny branchlets, yellow-green when they first appear, becoming red-brown and lustrous, finally brownish gray.

Series IX. TENUIFOLIAE Sarg.

Foliage and inflorescence glabrous except for short scabrate hairs on upper surface of young leaves; sepals entire or sometimes partly glandular-serrate.
    Leaves mostly ovate or oval, abruptly narrowed, rounded, truncate, or rarely subcordate at base, finely serrate and more or less lobed above middle with 3–5 pairs of triangular lateral lobes, often reflexed at tips; stamens 5–10; anthers pink . . . . .27. *C. macrosperma*
    Leaves ovate, oblong-ovate, or elliptic, deeply serrate, usually with 3–4 pairs of shallow triangular lateral lobes; stamens about 20 . . . . .28. *C. basilica*
Foliage and inflorescence more or less villous at least while young; sepals glandular-serrate or pectinate . . . . .29. *C. lucorum*

27. **Crataegus macrosperma** Ashe

    C. colorata Sarg., *C. cyanophylla* Sarg., *C. ignea* Sarg., *C. mineata* Ashe, *C. otiosa* Ashe, *C. prona* Ashe, *C. rubicunda* Sarg., *C. suavis* Sarg., *C. tenera* Ashe.

A tree up to 7–8 m. high with stout spreading or ascending branches and slender usually thorny branchlets; trunk and larger branches often fluted or buttressed; bark pale gray or brownish gray, somewhat scaly.

Var. **macrosperma.** Leaves mostly ovate or oval, sharply serrate, usually with 4–5 pairs of broad shallow lateral lobes; flowers 1.5–1.7 cm. wide; fruit oblong or obovoid, 0.8–1.5 cm. thick, with thin flesh becoming succulent when ripe in Sept. or Oct. Var. **acutiloba** (Sarg.) Egglest. (*C. acutiloba* Sarg.) has slightly larger ovate to deltoid leaves deeply divided with usually 5 pairs of acuminate spreading lateral lobes reflexed at tips; fruit usually less than 1 cm. thick. Var. **demissa** (Sarg.) Egglest. (*C. demissa* Sarg., *C. sextilis* Sarg.) is usually shrubby, 1–3 m. high; leaves mostly ovate or broadly ovate, rounded, truncate, or rarely subcordate at base; fruit subglobose or short-oblong, 6–9 mm. thick. Var. **matura** (Sarg.) Egglest. (*C. matura* Sarg., *C. acuminata* Sarg.) has leaves mostly broadly ovate or deltoid, often truncate or subcordate at base; fruit ripening in late Aug. or early Sept. Var. **pentandra** (Sarg.) Egglest. (*C. pentandra* Sarg., *C. exigua* Ashe) has leaves mostly ovate, abruptly narrowed or rounded at base; stamens 5–8, with

Crataegus brainerdi   Crataegus coleae   Crataegus macrosperma

large anthers; fruit with firm mellow flesh and slightly elevated calyx. Var. **roanensis** (Ashe) Palmer (*C. roanensis* Ashe, *C. bella* Sarg.) has leaves deeply indented with 4–5 pairs of acute spreading lateral lobes, acuminate but not reflexed at tips.

28. **Crataegus basilica** Beadle
    C. taetrica Sarg.
    A tree up to 6–7 m. high with wide-spreading or ascending branches, slender flexuous thorny branchlets, and dark scaly bark; often a stout irregular shrub.

29. **Crataegus lucorum** Sarg.
    C. decens Ashe
    An arborescent shrub or small tree up to 7–8 m. high with slender thorny branchlets, sometimes slightly villous when they first appear, soon glabrous.

Series X. SILVICOLAE Beadle

Leaves of flowering branchlets mostly 5–6 cm. long, 2.5–4 cm. wide, or about ⅓ larger at ends of shoots; fruit obovoid, oblong, or subglobose.
  Leaves relatively thin at maturity; flowers 1.2–1.5 cm. wide; fruit 0.8–1.2 cm. thick; nutlets 3–5.
    Leaves of flowering branchlets mostly broadly ovate or deltoid, finely serrate with narrow acuminate teeth; flowers mostly 3–7 in compact corymbs; stamens about 10; anthers pink or purple..30. *C. iracunda* var. *silvicola*
    Leaves of flowering branchlets ovate or oval, sometimes suborbicular at ends of shoots, more coarsely serrate with acute broad-based teeth; flowers mostly 6–12 in loose corymbs; stamens 10 or fewer; anthers pink or purple..........................31. *C. brumalis*
  Leaves firm or thick at maturity; flowers 1.5–2 cm. wide; fruit 0.9–1.6 cm. thick.
    Leaves of flowering branchlets mostly short-ovate or deltoid, often truncate or subcordate at base.
      Lobes of leaves broad and shallow; flowers mostly 5–10 in corymb; fruit 0.9–1.2 cm. thick..32. *C. stolonifera*
      Leaves of flowering branchlets with sharp spreading lobes; flowers mostly 6–12 in corymb; stamens 15–20; fruit 1.2–1.6 cm. thick; nutlets 5........33. *C. beata*
    Leaves of flowering branchlets mostly ovate or oval, abruptly narrowed, rounded, or rarely truncate at base; flowers 1.5–1.7 cm. wide; stamens 10 or fewer...34. *C. populnea*
  Leaves of flowering branchlets often 6–7 cm. long, 5–6 cm. wide, larger and broadly ovate to suborbicular at ends of shoots; flowers 1.5–1.8 cm. wide; stamens 10 or fewer; fruit obovoid or short-oblong, 0.8–1 cm. thick; nutlets 3–5.................................35. *C. gravis*

30. **Crataegus iracunda** Beadle var. **silvicola** (Beadle) Palmer
     C. silvicola Beadle, *C. drymophila* Sarg.
     An arborescent shrub or tree up to 5–6 m. high with thorny often flexuous branchlets and pale gray scaly bark.

Crataegus basilica | Crataegus lucorum | Crataegus iracunda

31. **Crataegus brumalis** Ashe

An arborescent shrub or small tree up to 6–8 m. high with spreading or ascending branches, stout flexuous thorny· branchlets at first reddish or olive, later gray; bark on trunk gray-brown, slightly furrowed.

32. **Crataegus stolonifera** Sarg.

Usually a shrub 2–3 m. high spreading into thickets; with crooked ascending branches and slender thorny branchlets.

33. **Crataegus beata** Sarg.

C. opulens Sarg.

An arborescent shrub (sometimes a small tree) with several crooked ascending stems and stout straight or flexuous thorny branchlets.

34. **Crataegus populnea** Ashe

C. blairensis Sarg., *C. marcida* Ashe, *C. propinqua* Ashe

An arborescent shrub or small tree up to 6–7 m. high with crooked ascending or spreading branches and slender thorny branchlets; bark on trunk brownish gray, slightly scaly.

35. **Crataegus gravis** Ashe

C. remota Sarg.

An arborescent shrub or small tree up to 6–7 m. high with ascending branches and stoutish thorny branchlets; often with compound thorns on trunk; bark gray and scaly on trunk, smooth and chestnut color on young branchlets.

Crataegus brumalis

Crataegus stolonifera

Crataegus beata

Crataegus populnea

Crataegus gravis

Series XI. PRUINOSAE Sarg.

a. Foliage and inflorescence glabrous or essentially so except for short appressed hairs on upper surface of young leaves of a few species.
   b. Fruit subglobose or short-oblong, as broad as long or broader.
      c. Leaves of flowering branchlets seldom over 4 cm. wide, bluish green or dark green at maturity.
         Leaves firm or thick at maturity; flowers 1.2–2.2 cm. wide.
            Leaves mostly ovate or deltoid, 3–4 cm. wide, slightly longer than wide except rarely at ends of shoots, usually more or less lobed.
               Leaves of flowering branchlets mostly abruptly narrowed or rounded at base, more than ½ grown when flowers open; bracts and bractlets inconspicuous, soon deciduous..36. *C. pruinosa*
               Leaves of flowering branchlets rounded, truncate, or rarely subcordate at base, less than ½ grown when flowers open; bracts and bractlets conspicuous and more persistent............................37. *C. mackenzii* var. *bracteata*
               Leaves mostly broadly ovate or deltoid, often as broad as long or sometimes broader than long at ends of shoots, with 3–5 pairs of acute spreading lateral lobes..38. *C. leiophylla*
            Leaves mostly ovate or oblong-ovate, 2.5–3.5 cm. wide, distinctly longer than wide, with shallow lateral lobes or unlobed.
               Leaves glabrous from the first, yellowish green at maturity; flowers mostly 4–6 in corymb, about 1.4 cm. wide; stamens 5–10; fruit subglobose, sometimes slightly narrowed at base, dark red.....................39. *C. franklinensis*
               Leaves slightly pilose above while young, becoming glabrous, thick, bluish green at maturity; flowers mostly 5–9 in corymb, 1.2–1.5 cm. wide; stamens about 20; fruit subglobose or oblong, sometimes slightly narrowed at base, orange-red with pale blotches..................................40. *C. compacta*
         Leaves comparatively thin, bluish green at maturity; flowers 1.4–1.7 cm. wide.
            Leaves ovate or deltoid, the terminal lobe often conspicuously elongate, glabrous from the first; flowers mostly 3–7 in corymb; stamens about 20..41. *C. gattingeri*
            Leaves mostly ovate, the terminal lobe not conspicuously elongated, sparsely short-villous above while young, becoming glabrous; flowers mostly 8–15 in corymb; stamens about 10...............................42. *C. milleri*
      cc. Leaves of flowering branchlets often 5–6 cm. wide, ovate, broadly ovate, or deltoid, sometimes broader than long at ends of shoots.
            Leaves of flowering branchlets with 2–4 pairs of broad, acute or rounded lateral lobes above middle; flowers 3–8 in corymb; fruit subglobose, 1–1.5 cm. thick, with prominent slightly elevated calyx.
               Leaves thick, yellowish green at maturity, with broad acute lateral lobes; flowers mostly 6–8 in corymb; stamens about 20; fruit subglobose or short-oblong; nutlets 2–3...43. *C. rugosa*
               Leaves comparatively thin, dull green at maturity, with shallow rounded or obscure lateral lobes; flowers mostly 3–6 in corymb; stamens 10 or rarely more; fruit subglobose, often slightly 5-angled; nutlets 2–5...........44. *C. disjuncta*
            Leaves of flowering branchlets with 3–4 pairs of narrow acute spreading lateral lobes; flowers mostly 7–10 in corymb; stamens 15–20; fruit subglobose or oblong, 0.9–1 cm. thick; nutlets usually 5.................................45. *C. formosa*
   bb. Fruit pyriform, obovoid or oblong, usually slightly longer than wide.
      c. Leaves more or less lobed but not deeply divided or laciniate even at ends of shoots.
         Leaves thick, bluish green at maturity; flowers mostly 3–7 in corymb; fruit bright red
46. *C. porteri*
         Leaves comparatively thin at maturity; flowers mostly 5–8 in corymb; fruit orange-red or green blotched with red.
            Leaves mostly ovate, glabrous from the first, yellowish green at maturity; flowers mostly 8–12 in corymb; stamens about 20.................47. *C. crawfordiana*
            Leaves ovate or oval, slightly hairy while young, becoming glabrous, dark green and slightly lustrous above at maturity; flowers mostly 6–7 in corymb; stamens 15–20...48. *C. gaudens*
      cc. Leaves sharply and deeply lobed, often laciniate at ends of shoots, relatively thin, yellowish green at maturity; stamens 10 or less......................49. *C. jesupi*
aa. Foliage and inflorescence pubescent while young, usually more or less pubescent throughout season.

b.   Leaves yellowish green, pubescent on both sides; flowers 2–2.4 cm. wide, mostly 7–10 in villous corymbs; stamens about 20; anthers pink or pale yellow...........50. *C. locuples*

bb. Leaves deep green, short pubescent above and slightly hairy beneath while young, becoming glabrous or nearly so; flowers 1.6–1.8 cm. wide, mostly 4–7 in slightly villous or glabrous corymbs; stamens 15–20; anthers pink........................51. *C. virella*

36.   **Crataegus pruinosa** (Wendl.) K. Koch

C. amoena Sarg., *C. ater* Ashe, *C. horridula* Sarg., *C. howeana* Sarg., *C. sitiens* Ashe.

A tree up to 7–8 m. high, or sometimes a stout shrub with spreading or ascending branches and slender thorny branchlets; bark dark gray, scaly; foliage and inflorescence glabrous.

Var. **pruinosa.**   Leaves usually more or less lobed; flowers 1.7–2 cm. wide, with about 20 pink or rarely creamy white anthers; sepals lanceolate or partly glandular-serrate; fruit subglobose or short-oblong, with prominent elevated calyx and thin dry flesh, remaining green or becoming dull crimson at maturity, pruinose; in forma **angulata** (Sarg.) Palmer (*C. angulata* Sarg., *C. placiva* Sarg.) the fruit is conspicuously 5-angled.   Var. **pachypoda** (Sarg.) Palmer (*C. pacyhpoda* Sarg.) has flowers 1.5–1.7 cm. wide, mostly 5–7 in nearly simple corymbs, pedicels 1–1.5 cm. long, 10 or rarely 20 stamens with white or pale yellow anthers, and fruit with nearly sessile calyx.   Var. **dissona** (Sarg.) Egglest. (*C. dissona* Sarg., *C. marriettensis* Sarg.) has flowers 1.2–1.5 cm. wide with 10 stamens or less, pink anthers, and depressed globose fruit with broad nearly sessile calyx.   Var. **latisepala** (Ashe) Egglest. (*C. latisepala* Ashe, *C. cognata* Sarg., *C. conjuncta* Sarg., *C. jejuna* Sarg.) has leaves obscurely lobed or sometimes unlobed, about 10 stamens with white or pale yellow anthers, and sepals attenuate from broad base.

37.   **Crataegus mackenzii** Sarg. var. **bracteata** (Sarg.) Palmer

C. bracteata Sarg.

A small tree up to 5–6 m. high or oftener a stout shrub with crooked spreading or ascending branches and stout often flexuous thorny branchlets; bark dark gray, slightly scaly on trunk and old branches.

38.   **Crataegus leiophylla** Sarg.

C. gracilis Sarg., *C. longipedunculata* Sarg.

An arborescent shrub or rarely tree up to 5–6 m. high with crooked ascending branches and stoutish very thorny branchlets, olive-green when they first appear, becoming gray.

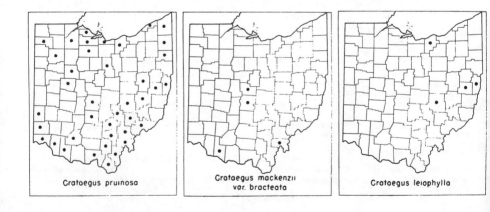

Crataegus pruinosa

Crataegus mackenzii var. bracteata

Crataegus leiophylla

Crataegus franklinensis

Crataegus compacta

Crataegus gattingeri

Crataegus milleri

Crataegus rugosa

Crataegus disjuncta

Crataegus formosa

Crataegus porteri

Crataegus crawfordiana

39. **Crataegus franklinensis** Sarg.

Usually an arborescent shrub with erect or ascending crooked branches and slender thorny branchlets, orange-brown when they first appear, becoming chestnut-brown and ultimately dull gray.

40. **Crataegus compacta** Sarg.

   C. ellipticifolia Sarg., *C. repentina* Sarg.

An arborescent shrub 3–4 m. high, forming thickets; branches crooked, ascending or erect; branchlets slender, somewhat thorny, orange-green when they first appear, becoming red-brown and ultimately dull gray-brown.

41. **Crataegus gattingeri** Ashe

   C. bedfordensis Sarg.

An arborescent shrub or small tree up to 6–7 m. high with stoutish ascending or spreading branches and slender usually very thorny branchlets, purplish brown while young, becoming dull gray; bark dark gray, slightly scaly on trunk.

42. **Crataegus milleri** Sarg.

A shrub 3–4 m. high with thick erect branches and stout often flexuous thorny branchlets, at first reddish brown, becoming gray. Perhaps a hybrid between *C. pruinosa* and a species of the *Silvicolae* series.

43. **Crataegus rugosa** Ashe

   C. onusta Ashe

An arborescent shrub or small tree up to 6–7 m. high with stoutish spreading or ascending branches and slender thorny branchlets.

44. **Crataegus disjuncta** Sarg.

An arborescent shrub or small tree up to 5–6 m. high with an irregular head of open branches and stout slightly flexuous thorny branchlets.

45. **Crataegus formosa** Sarg.

Usually an arborescent shrub up to 3–5 m. high with erect or ascending branches and slender thorny branchlets.

46. **Crataegus porteri** Britt.

An arborescent shrub or sometimes a small tree 3–5 m. high with erect or ascending branches and slender thorny branchlets.

47. **Crataegus crawfordiana** Sarg.

Usually an arborescent shrub up to 3–5 m. high with erect or ascending branches and slender thorny branchlets.

48. **Crataegus gaudens** Sarg.

Usually an arborescent shrub up to 4–5 m. high with ascending branches and stout thorny branchlets, light orange-green when they first appear, becoming red-brown and finally dark gray.

49. **Crataegus jesupi** Sarg.

   C. diversifolia Sarg.

An arborescent shrub or rarely a small tree up to 5–6 m. high with crooked ascending branches and slender slightly flexuous branchlets armed with stout thorns.

Crataegus gaudens

Crataegus jesupi

Crataegus locuples

Crataegus virella

Crataegus holmesiana

Crataegus hillii

Crataegus pringlei

50. Crataegus locuples Sarg.

A tree up to 7–8 m. high with stout ascending and spreading branches and slender slightly flexuous, more or less thorny branchlets. Possibly a hybrid between *C. pruinosa* or a related species and *C. mollis*.

51. Crataegus virella Ashe

An arborescent shrub or small tree with stout ascending branches and slender thorny branchlets. This species can be distinguished from *C. pruinosa* most readily by the slight pubescence of young leaves and sometimes pubescent inflorescence.

Series XII. COCCINEAE Loud.

a. Leaves oval, elliptic, or oblong-ovate, broadest about middle, abruptly narrowed or rounded at base, or at ends of shoots sometimes ovate and truncate at base; fruit oblong or obovoid.
  Leaves flat or plane; calyx-tube glabrous or nearly so; fruit bright red or crimson, glabrous.
  Leaves of flowering branchlets mostly elliptic or oblong-ovate, noticeably longer than wide, abruptly narrowed or rounded at base; fruit 1–1.2 cm. thick, crimson; nutlets usually 2. . . . . .52. *C. holmesiana*
  Leaves of flowering branchlets mostly ovate or oblong-ovate, often only slightly longer than wide, rounded at base, or at ends of shoots broadly ovate and sometimes truncate or subcordate at base. . . . . .53. *C. hillii*
  Leaves slightly cupped (concavo-convex); calyx-tube villous; fruit dark dull red, often slightly pubescent at ends at least while young. . . . . .54. *C. pringlei*
aa. Leaves of flowering branchlets mostly ovate, oblong-ovate, or deltoid, broadest at or below middle, rounded or truncate at base, or at ends of shoots sometimes broadly ovate and subcordate at base.
  Flowers 1.8–2.2 cm. wide; fruit oblong or obovoid (sometimes subglobose in a var.), usually longer than thick. . . . . .55. *C. pedicellata*
  Flowers 1.5–1.8 cm. wide; fruit subglobose or short-oblong, usually as thick as long or thicker than long.
    Leaves of flowering branchlets mostly ovate or deltoid, or at ends of shoots sometimes broadly ovate to suborbicular; foliage and inflorescence glabrous or essentially so; flowers few (4–7) in simple corymbs.
      Flowers 1.4–1.5 cm. wide; stamens about 10; fruit 1–1.3 cm. thick. . . . . . .56. *C. habereri*
      Flowers 1.6–1.8 cm. wide; stamens about 20; fruit 1.2–1.5 cm. thick. .57. *C. putnamiana*
    Leaves of flowering branchlets mostly oval or ovate, or at ends of shoots broadly ovate to deltoid; foliage and inflorescence slightly villous; flowers many (8–15) in compound corymbs. . . . . .58. *C. pennsylvanica*

52. Crataegus holmesiana Ashe

An arborescent shrub or tree up to 9–10 m. high with conical top of ascending branches and slender thorny branchlets; leaves relatively thin, yellow-green at maturity.

Var. **holmesiana.** Flowering corymbs glabrous; stamens about 10; nutlets usually 3. Var. **amicta** (Ashe) Palmer (*C. amicta* Ashe) has slightly villous flowering corymbs; stamens 5–8; nutlets 3–5.

53. Crataegus hillii Sarg.

A tree up to 8–9 m. high with stout ascending branches and slender thorny or nearly thornless branchlets, reddish and villous when they first appear; bark light gray, fissured on trunk.

54. Crataegus pringlei Sarg.

Tree up to 7–8 m. high with stout ascending or spreading branches, slender, often flexuous thorny branchlets, dark green and villous when they first appear, becoming glabrous and chestnut-brown and finally brownish gray.

55. Crataegus pedicellata Sarg.

C. sejuncta Sarg., *C. coccinea* of many auth.

Tree up to 6–8 m. high with ascending and spreading branches and slender thorny branchlets; sometimes fruiting as an arborescent shrub. Var. **pedicellata.** Leaves mostly ovate, abruptly narrowed or rounded at base; flowers 1.8–2 cm. wide in compound, many-flowered, slightly villous corymbs; fruit obovoid or oblong, 0.7–1 cm. thick. Var. **albicans** (Ashe) Palmer (*C. albicans* Ashe, *C. cristata* Ashe) has leaves relatively broader, glabrous except for short villous hairs on upper surface while young; flowers 1.7–1.8 cm. wide, mostly 5–10 in glabrous or nearly glabrous corymbs; fruit short-oblong or nearly globose. Var. **assurgens** (Sarg.) Palmer (*C. assurgens* Sarg.) has leaves mostly broadly ovate, thin but firm at maturity; flowers mostly 8–15 in villous corymbs; stamens 10 or 20, usually about 10; fruit 1–1.5 cm. thick. Var. **robesoniana** (Sarg.) Palmer (*C. robesoniana* Sarg.) has relatively large leaves, mostly ovate or oblong-ovate, deeply serrate and sharply lobed; flowers 1.6–1.8 cm. wide, in few-flowered slightly villous corymbs; fruit obovoid with small sessile calyx.

56. Crataegus habereri Sarg.

Usually an arborescent shrub 3–5 m. high with crooked spreading branches and slightly flexuous thorny branchlets light orange-green when they first appear, becoming reddish brown and ultimately brownish gray.

57. Crataegus putnamiana Sarg.

Tree up to 8–10 m. high, often fruiting as an arborescent shrub, with ascending or spreading branches and stout flexuous thorny branchlets.

58. Crataegus pennsylvanica Ashe

Tree up to 9–10 m. high with a broad head of ascending or spreading branches and slender thorny branchlets, villous when they first appear, later glabrous; bark brownish gray, scaly on trunk.

Series XIII. MOLLES Sarg.

Leaves ovate, oblong-ovate, or deltoid, abruptly narrowed, rounded or subcordate at base, copiously pubescent while young and somewhat pubescent throughout the season; stamens about 20; anthers creamy white or rarely pink; fruit subglobose or rarely obovoid, 1–1.8 cm. thick, ripening Aug. to Sept...59. *C. mollis*
Leaves mostly ovate or oval, rounded to truncate at base, less densely pubescent; stamens 10 or less; fruit pyriform or obovoid, 1.2–1.5 cm. thick, ripening in Sept.........60. *C. submollis*

Crataegus pedicellata   Crataegus habereri   Crataegus putnamiana

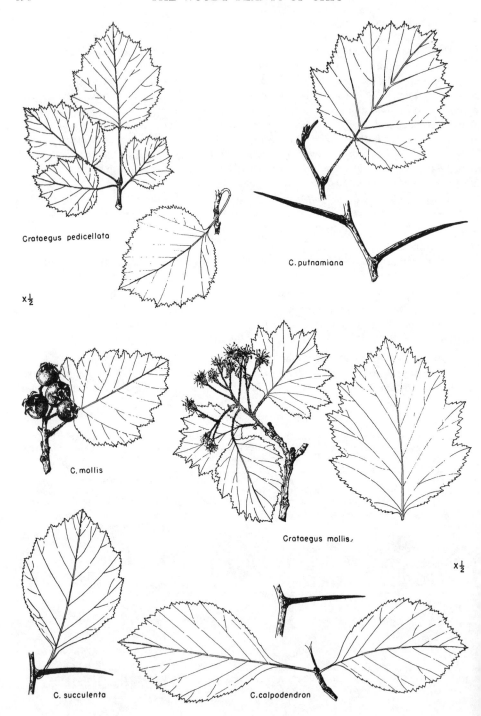

Crataegus pedicellata

C. putnamiana

$x\frac{1}{2}$

C. mollis

Crataegus mollis

$x\frac{1}{2}$

C. succulenta

C. calpodendron

59. **Crataegus mollis** (T. & G.) Scheele

C. redolens Ashe

A tree up to 10–12 m. high with broad rounded top of wide-spreading and ascending branches and stoutish sparingly thorny branchlets, villous when they first appear, and stout straight trunk sometimes 3 dm. in diameter; bark on trunk thick, brownish gray, deeply furrowed. The fruit is highly flavored and edible, but often infested with weevils.

Var. **mollis.** Leaves usually broadest below middle, often truncate or subcordate at base; fruit 1.3–1.8 cm. thick. Var. **sera** (Sarg.) Egglest. (*C. sera* Sarg., *C. mollipes* Sarg.) has leaves mostly ovate or oblong-ovate, broadest about middle, abruptly narrowed or rounded at base; fruit usually pyriform or obovoid, 1–1.2 cm. thick.

60. **Crataegus submollis** Sarg.

A tree up to 8–10 m. high, or sometimes an arborescent shrub, with stoutish ascending branches, flexuous thorny branchlets, and brownish gray bark.

Series 14. MACRACANTHAE Loud.

Foliage and inflorescence glabrous or sparsely pubescent (under surface of leaves villous in one var.); leaves subcoriaceous, with veins distinctly impressed above at maturity; stamens 10 or 20; young branchlets glabrous or rarely slightly villous.

Leaves of flowering branchlets mostly elliptic or oblong-ovate, 3.5–5 cm. wide, usually more or less lobed above middle, dark green; fruit 0.8–1.2 cm. thick........... 61. *C. succulenta*

Leaves of flowering branchlets mostly obovate or oblong-obovate, unlobed or rarely obscurely lobed, 3–4 cm. wide, yellow-green, lustrous above at maturity; fruit 1–1.3 cm. thick......
62. *C. laetifica*

Foliage and inflorescence distinctly pubescent; leaves rather thin, yellowish green at maturity; stamens about 20; young branchlets villous........................63. *C. calpodendron*

61. **Crataegus succulenta** Link.

C. gemmosa Sarg., *C. rutila* Sarg.

A tree up to 7–8 m. high with open ascending branches and slender often flexuous thorny branchlets; sometimes fruiting as a stout arborescent shrub.

Var. **succulenta.** Leaves elliptic, rhombic, or rarely ovate, usually with 4–5 pairs of short acute lateral lobes, glabrous or slightly villous along veins beneath; flowers 1.3–1.7 cm. wide, in many-flowered glabrous or slightly villous corymbs; stamens about 20, anthers pink or rarely white; fruit subglobose, 0.7–1.2 cm. thick, bright red, becoming succulent. Var. **macracantha** (Lodd.) Egglest. (*C. macracantha* Lodd.) has about 10 stamens with white, pale yellow, or rarely pink anthers, fruit about 1 cm. thick, and usually

Crataegus pennsylvanica

Crataegus mollis

Crataegus submollis

more numerous thorns up to 7-8 cm. long. Var. **michiganensis** (Ashe) Palmer (*C. michiganense* Ashe) has leaves mostly oval or oblong-obovate, rounded or abruptly pointed at apex, glabrous or nearly so, dark green, subcoriaceous at maturity, about 20 stamens, white or pink anthers, and subglobose fruit 7-8 mm. thick. Var. **neofluvialis** (Ashe) Palmer (*C. neofluvialis* Ashe, *C. tanuphylla* Sarg.) has mostly rhombic or oblong-elliptic leaves acute or short-acuminate at apex, usually sharply lobed, relatively thin, glabrous or nearly so; flowers 1.3-1.5 cm. wide, in many-flowered glabrous or slightly villous corymbs; stamens 10 or rarely more, anthers pink or white, and fruit 6-9 mm. thick, remaining hard and dry or becoming mellow late in the season. Var. **pertomentosa** (Ashe) Palmer (*C. pertomentosa* Ashe) has oval, rhombic, or rarely ovate leaves, dark green above, much paler and pubescent beneath, subcoriaceous at maturity; flowering corymbs more or less villous, stamens about 10; fruit subglobose, 6-9 mm. thick, usually pubescent at least while young.

62. **Crataegus laetifica** Sarg.

Small tree with dense spreading or ascending branches and stout slightly flexuous thorny branchlets.

63. **Crataegus calpodendron** (Ehrh.) Medic.

C. pubifolia Ashe, *C. structilis* Ashe, *C. tomentosa* of many authors.

Arborescent shrub or sometimes small tree up to 6-7 m. high with erect or ascending branches and straight slender branchlets, thorny or sometimes nearly thornless, villous or tomentose while young.

Var. **calpodendron.** Leaves ample, usually with 3-5 pairs of short acute lateral lobes above middle; fruit obovoid or rarely subglobose, becoming succulent and sweet at maturity. Var. **globosa** (Sarg.) Palmer (*C. globosa* Sarg.) has fruit subglobose or short-oblong, with thin flesh remaining hard and dry. Var. **microcarpa** (Chapm.) Palmer (*C. tomentosa* var. *microcarpa* Chapm., *C. tomentosa* var. *chapmani* Beadle, *C. chapmani* Ashe) has leaves smaller than in the typical variety with primary veins more crowded and deeply impressed; fruit subglobose, 3-8 mm. thick.

64. CRATAEGUS PRUNIFOLIA (Poir.) Pers.

Frequently cultivated, but rarely known as an escape; reported from Lake County. It was described from a tree of unknown origin found growing in a French garden. From its characters, it is believed to be a hybrid between two American species, *C. crus-galli* and *C. calpodendron* or a related species. The leaves are elliptic or slightly obovate,

Crataegus succulenta

Crataegus laetifica

Crataegus calpodendron

sharply serrate, unlobed or nearly so, and slightly pubescent while young; flowers mostly 3–5 in pubescent corymbs; fruit subglobose, red; nutlets usually 2, slightly pitted on ventral surfaces.

The following hybrids have also been found in Ohio:
C. *calpodendron* × *crus-galli*
C. *crus-galli* × *succulenta*
C. *margaretta* × *mollis*
C. *mollis* × *pedicellata* (?)
A few other hybrids or undescribed species or varieties are indicated by incomplete material.

COTONEASTER Ehrh.

1. COTONEASTER PYRACANTHA (L.) Spach FIRE-THORN
   Pyracantha coccinea Roem.
   An evergreen shrub of southern Europe and western Asia, with small white flowers and bright red fruit in dense cymes. Commonly planted and reported as an escape in Franklin and Lawrence counties.

POTENTILLA L.

A large genus of the northern hemisphere, mostly herbs; one woody species in our range.

1. **Potentilla fruticosa** L. SHRUBBY CINQUEFOIL
   Dasiphora fruticosa (L.) Rydb.
   Distinguished from all other rosaceous shrubs of our range by the yellow flowers with 5 roundish petals widely spreading, the flat calyx subtended by 5 bractlets alternate with the calyx-lobes, the more or less pubescent pinnate leaves with 5–7 narrow leaflets. A much-branched shrub, usually less than 1 m. (–1.5m) in height. In wet meadows and marly bogs, sometimes in dry situations. Essentially transcontinental in the North, and extending far southward in the western mountains.

Potentilla fruticosa

RUBUS L. BRAMBLES: RASPBERRIES, DEWBERRIES, BLACKBERRIES

A large genus of deciduous or evergreen shrubby, suffruticose, or herbaceous often prickly plants—some erect, some decumbent or trailing; most abundant in the north temperate zone of both hemispheres, and also present in the mountains of South America. The rose-like flowers often showy, usually white (sometimes pink or rose-purple), with 5-parted calyx (no bractlets), 5 deciduous petals, numerous stamens, many carpels (ripening into drupelets) on a spongy or fleshy receptacle. Shrubby species commonly send up biennial stems from the perennial roots. During the first year these are usually unbranched, do not flower, and are known as *primocanes*. During the second year, short lateral flowering and fruiting branches develop, but the canes do not increase in length; these are known as *floricanes*. Leaves on primocanes and floricanes are different, the former (with few exceptions) compound, some of the latter usually simple.

No estimate of the number of species can be given, because interpretation of "species" of *Rubus*, especially of the blackberries (subgenus *Eubatus*), differs greatly in different manuals. The genus is commonly subdivided into subgenera (and/or sections) which are more or less readily defined. Fernald (1950) recognizes five subgenera including 14 sections and 204 species in the Manual range. Gleason (1952) distinguishes five sections (the subgenera of Fernald) including 24 species, some of which he terms "collective species." In the same area, Bailey (in Gentes Herbarum) distinguishes about 400 "species." Interpretation of species of four of the subgenera (or sections) is essentially uniform. Ohio species in these can be determined readily. The fifth subgenus (or section), *Eubatus*, seems to consist of a relatively few wide-spread somewhat constant species and an indefinite number of local species which have developed in relatively recent time—since clearing and disturbance of natural conditions by the white man. Some of these are individual clones, some are the result of hybridization, many are perhaps best interpreted as incipient species. Gleason's "collective species" group many of these under one name, a few of which can be correlated with the sections of Fernald; others contain "species" from two or three of the sections recognized by Fernald.

Within the subgenus *Eubatus*, species, which from published range may be expected in Ohio, are listed and included in the keys. No attempt is made here to show the distribution of all of these in Ohio. Identification is almost impossible unless specimens include primocanes and floricanes, and notes as to habit of growth.

For further distinguishing features of the "species" included here, and of others which may occur, see Fernald (1950).

a. Principal leaves simple, 3–5 lobed, somewhat maple-like in shape; stems woody, unarmed, with exfoliating bark; young parts of stems, petioles, pedicels, and calyx with bristly-glandular hairs; flowers showy, rose-purple, 3–6 cm. broad; fruit red, flat-hemispheric, dryish. . . . .
                                                  2. *R. odoratus*

aa. Principal leaves compound (leaves of floricanes often simple).
  b. Stems slender, weak, 1–2 (–5) mm. in diam., trailing or low-arching, with erect flowering stems.
    c. Stems herbaceous or nearly so, unarmed (rarely with a few bristles), long-trailing or creeping, the erect herbaceous flowering stems 1–5 dm. tall, with long-petioled 3-foliate leaves and few flowers; leaves soft, dull, somewhat pubescent, coarsely serrate (central leaflet rhombic-ovate or rhombic-lanecolate, acute at both ends); fruit dark red with large juicy drupelets. . . . . .1. *R. pubescens*
    cc. Stems slender but woody, usually bristly, low-arching; leaves 3 (–5) foliate, coriaceous, evergreen, shining, blunt-toothed, obovate or broadly rhombic-ovate; inflorescence a lax corymb or irregular raceme; fruit finally blackening, seedy. . . . . .15. *R. hispidus*
  bb. Two-year old stems woody, less slender, trailing, arching, or nearly erect, usually armed; leaves of primocanes compound (rarely simple).
    c. Fruit readily separating from the dry receptacle, which remains attached to pedicel; flowers not showy, petals shorter than sepals. . . . . .Subg. *Idaeobatus*, raspberries (p. 203)
      d. Stems densely shaggy with long purple gland-tipped hairs; leaves 3-foliate, densely white-tomentose beneath, petioles with purple gland-tipped hairs; inflorescence a terminal panicle, glandular-villous; petals white, much shorter than the lance-attenuate sepals; fruit red. . . . . .3. *R. phoenicolasius*
      dd. Stems densely to sparsely armed with sharp prickles or bristles, not glandular-villous.
        Stems glaucous, erect to ascending, elongating and then rooting at tips; prickles with broad base; leaves 3-foliate, if 5-foliate then digitate, white beneath; fruit black. . . . . .5. *R. occidentalis*
        Stems prickly, bristly, or nearly smooth, not glaucous, not rooting at tips; leaves 3–5 foliate, if 5-foliate, then pinnate, softly white or grayish pubescent beneath; fruit red. . . . . .4. *R. idaeus*
    cc. Receptacle included in the fruit when it separates from stem; flowers showy, petals usually longer than sepals. . . . . .Subg. *Eubatus*, blackberries, dewberries (p. 205)
      d. Leaflets pinnately cut into small leaflets and segments; flowers pinkish-white, in villous prickly panicles. . . . . .6. *R. laciniatus*
      dd. Leaflets not deeply cut.

Leaves white-felted beneath, those of primocanes mostly with 5 leaflets; entire
plant (stems, petioles, principal veins beneath, inflorescence) armed with stout
curved and flattened prickles...............................7. *R. procerus*
Leaves not white-felted beneath.
  Stems prostrate, trailing, sometimes low-arching; prostrate primocanes rooting
  at tips; flowering branchlets upright from prostrate floricanes.....Dewberries
  Stems armed with strong scattered prickles; flowers large.
    Leaves coriaceous, somewhat evergreen; stems with glandular bristles
    among the hard curved prickles........................8. *R. trivialis*
    Leaves thinner, not evergreen; stems woody, the stronger ones angled,
    with prickles mostly along angles; fruit large, juicy.................
                  Section *Flagellares* (p. 205)
  Stems hispid or bristly, usually without strong prickles; leaflets 3; flowers
  small..............................Section *Hispidi:* 16. *R. hispidus*
  Stems erect or arched-ascending (sometimes becoming depressed in winter),
  not rooting at tips..Blackberries
    Stems armed with bristles or slender, not broad-based prickles...........
                   Section *Setosi* (p. 209)
    Stems armed with broad-based often hooked prickles, or without prickles;
    leaflets green or grayish beneath.
      Stems glabrous or nearly so, some straight prickles may be present; leaves
      glabrous on both surfaces, or but slightly pubescent beneath; rachis of
      inflorescence and pedicels smooth; northern and mountain species......
                 Section *Canadenses* (p. 209)
      Stems more or less pubescent, at least when young, with some strong
      hooked or broad-based prickles; leaves pubescent on one or both
      surfaces; inflorescence glandular or pubescent.
        Young growth and inflorescence stipitate-glandular; inflorescence an
        elongate or corymb-like raceme........Section *Alleghenienses* (p. 209)
        Young growth and inflorescence glandless (or with few inconspicuous
        glands); inflorescence short.................Section *Arguti* (p. 209)

WINTER KEY

Some of the species of *Rubus* are readily recognized in winter condition; others, the
dewberries and blackberries, are difficult to identify even in summer, and winter identification should not be attempted, other than to place them in their group or section.   All
have alternate buds placed singly or superposed in axil of persistent petiole-base.   This
petiole-base, which may or may not bear persistent stipules, is a diagnostic character of
the genus; it is shown in illustrations of *R. occidentalis* and *R. frondosus.*

a. Stems trailing or prostrate.
  Stems slender, wire-like, 1–2 (–5) mm. in diam., with weak prickles or unarmed.
    Leaves persistent, coriaceous, evergreen; stems bristly; short erect branches numerous....
                   16. *R. hispidus*
    Leaves deciduous; stems very slender, creeping on surface of ground; upright stems
    herbaceous, usually unarmed......................................1. *R. pubescens*
  Stems cord-like, usually over 2 mm. in diam., with strong scattered prickles.
    Leaves persistent, usually evergreen; stems terete, with hard curved prickles and glandular
    bristles..8. *R. trivialis*
    Leaves deciduous; stems usually angled, with prickles mostly along angles............
                Section *Flagellares*, dewberries
aa. Stems erect, ascending, or arching.
  Upright shrubs; stems unarmed, the bark light tan, shredding..............2. *R. odoratus*
  Stems variously ascending, upright, or arching; bark not shredding.
    Stems light purple, the youngest glaucous, armed with short hooked prickles, some arched
    and rooting at tip...5. *R. occidentalis*
    Stems not light purple *and* glaucous, not rooting at tip.
      Stems erect or ascending, bristly, often with a few curved slender-based prickles, terete,
      red, brown, or purple...4. *R. idaeus*
      Stems arched-ascending, usually armed with broad-based prickles, more or less angled,
      green to brown or dark red...........Sections *Alleghenienses* and *Arguti*, blackberries

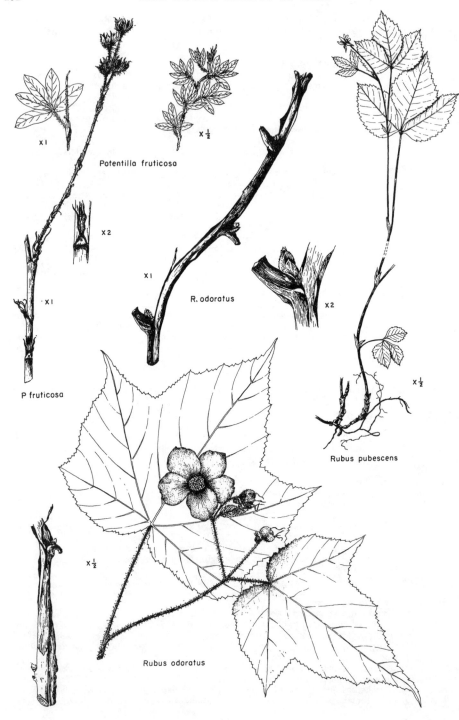

Potentilla fruticosa

X 1

X ½

X 2

X 1

X 1

P fruticosa

R. odoratus

X 1

X 2

X ½

Rubus pubescens

X ½

Rubus odoratus

1. **Rubus pubescens** Raf.

R. triflorus Richards, *R. americanus* Britt.

A slender almost herbaceous dwarf species with erect flowering stems and long trailing vegetative stems; fruit dark red, the drupelets large and juicy, finally separating from the receptacle. Northern in range, extending southward into our area in boggy places. At its most outlying station, Cedar Swamp in Champaign County, it grows in the arbor vitae forest.

2. **Rubus odoratus** L. FLOWERING RASPBERRY

A large and showy shrub of rich soil in mesic situations and rocky shaded slopes; easily identified by the large simple 3–5 lobed leaves, bristly-glandular dark hairs of younger stems, petioles, and inflorescence, large rose-purple flowers 3–4 cm. across, and red fruit. Blooms for a long period, hence flowers and ripe fruit occur at the same time. Essentially Appalachian in range, but extending west, locally, into Michigan and the Knobs of Indiana.

Subgenus IDAEOBATUS RASPBERRIES

3. RUBUS PHOENICOLASIUS Maxim. WINEBERRY

A shrubby species from eastern Asia, introduced in 1876 and now occasionally seen as an escape. Distinguished by its densely bristly pubescent stems, and by the leaves conspicuously white-pubescent beneath. Reported in Ohio from Adams, Ashtabula, Cuyahoga, Highland, Lake, and Meigs counties.

4. **Rubus idaeus** L. var. **strigosus** (Michx.) Maxim. RED RASPBERRY

R. strigosus Michx.

The circumboreal species, *R. idaeus*, is highly variable, and several varieties and forms have been recognized. The European raspberry, var. *idaeus* (to which the name *idaeus* is sometimes limited) is the source of our cultivated red raspberries; this has spread from cultivation in a few places. Our native red raspberry, var. *strigosus*, regarded as a distinct species by some authors, is a wide-ranging northern species, extending south into our area. This may be distinguished from var. *idaeus* by the glands and minute bristles of the inflorescence. Other varieties of more limited range are distinguished by Fernald (1950).

Rubus pubescens   Rubus odoratus   Rubus idaeus var strigosus

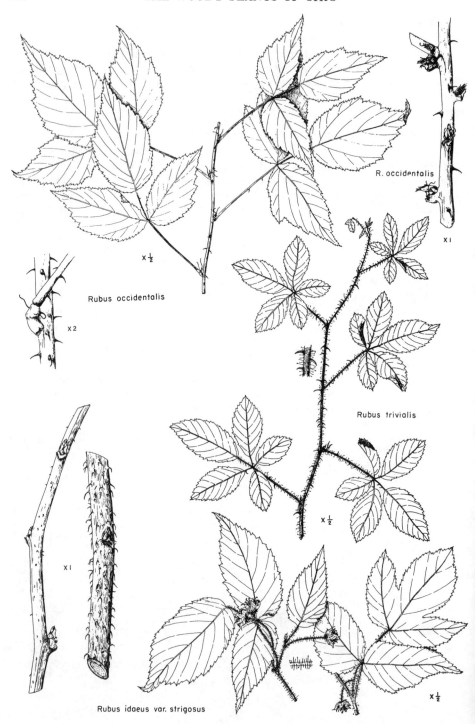

R. occidentalis

x 1

x ½

Rubus occidentalis

x 2

Rubus trivialis

x ½

x 1

Rubus idaeus var. strigosus

x ½

5. Rubus occidentalis L. Black Raspberry

A common species ranging almost throughout the Deciduous Forest, with closely related species southward into the Tropics. Generally in mesic situations in open woods of slopes and ravine flats. Widely cultivated. Easily recognized by its glaucous, strongly arching stems rooting at the tips, and ternate or quinate (digitate) leaves whitened beneath.

A hybrid between this species and our native red raspberry—× *R. neglectus* Peck—is recorded from a few northern Ohio counties (Ashtabula, Defiance, Trumbull). It is similar to *R. occidentalis* but has petioles and inflorescence more or less glandular, and inflorescence more compact than in *R. occidentalis*.

Rubus occidentalis

Subg. Eubatus Focke Blackberries, Dewberries

6. Rubus laciniatus Willd. Cut-leaf Blackberry

From the Old World; cultivated and sometimes escaped; reported from several counties. Flowers large, pinkish white, the obovate petals often lobed; sepal-lobes with long narrow foliaceous appendages.

7. Rubus procerus P. J. Muell. Himalaya-berry

A coarse blackberry, introduced from Europe, with leaves white felty beneath. Reported from the banks of the Ohio River in Brown and Gallia counties.

8. Rubus trivialis Michx. Southern Dewberry

Distinguished by its coriaceous somewhat evergreen leaves and glandular bristles. A southern species, known in Ohio only from Hamilton County.

Section *Flagellares* Dewberries

A section of some 44 species, most of which are grouped by Gleason under two "collective species," *R. flagellaris* and *R. Enslenii,* the latter, however, including a few species not in Fernald's section *Flagellares.* Seven of the species listed by Fernald occur locally in Ohio; these are *R. flagellaris, R. nefrens, R. tetricus, R. roribaccus, R. rosagnetus, R. enslenii,* and *R. baileyanus.* Of these, the "highly variable" *R. flagellaris* is most widespread; other very local species, known only from West Virginia, Kentucky, Indiana, or southern Michigan, may occur in Ohio, but are not included in the key.

The stems of all the dewberries are prostrate or nearly so, usually about ankle-high; flowering branchlets of the floricanes are short and erect.

a. Stems strongly woody; leaves (except shade leaves) firm; corymbs mostly 2–9 flowered.
　　Young stems, petioles, pedicels, and rachis glandless or nearly so.
　　　Mature leaflets glabrous or appressed-pubescent on veins beneath, not velvety to touch.
　　　　Petioles of bracteal leaves, and pedicels glabrous or appressed-pubescent, without spreading hairs; leaflets 3 or 5, terminal ovate to subelliptic, abruptly contracted to acuminate tip; floricanes often reddish..........................9. *R. flagellaris*
　　　　Petioles of bracteal leaves, and pedicels densely pubescent with long soft spreading hairs.
　　　　　Canes without prickles (or prickles few, 1–2 mm. long, subulate); petioles and pedicels unarmed (or with few bristles); leaflets coarsely toothed with simple or slightly notched teeth; corymbs 2–6 flowered, the pedicels long (to 7 cm.).....10. *R. nefrens*
　　　　　Canes with many strong prickles 2–4 mm. long; petioles, flowering branchlets, and pedicels prickly; leaflets sharply and often doubly serrate; corymbs 2–15 flowered..
　　　　　　　　　　　　　　　　　　　　　　　　　　11. *R. tetricus*

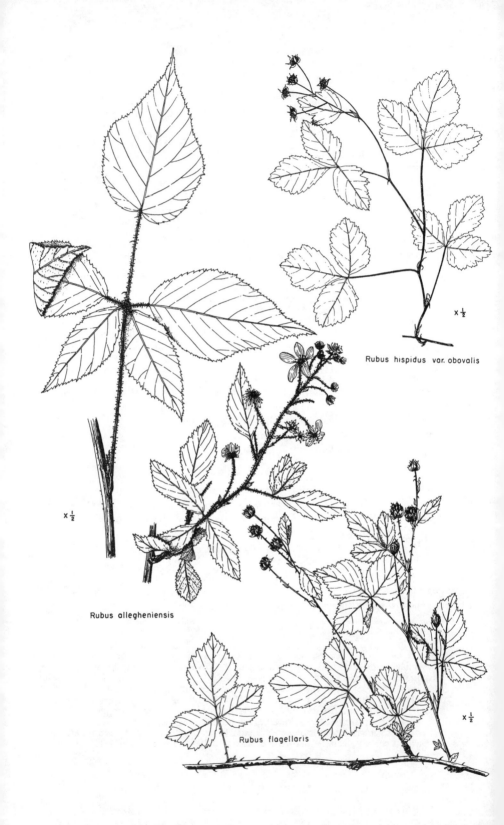

Rubus hispidus var. obovalis

x ½

x ½

Rubus allegheniensis

Rubus flagellaris

x ½

Mature leaflets pubescent beneath, velvety to touch, terminal leaflets much longer than broad; pedicels spreading-villous at flowering time, much longer than subtending bract
12. *R. roribaccus*
Young stems, petioles, pedicels and rachis glandular; lower leaf-surface velvety to touch; leaflets of primocane leaves 3, the terminal with long narrow acuminate tip.............
13. *R. rosagnetus*
aa. Stems slender, flexible, but slightly woody; leaves thin; petioles and pedicels appressed pilose, glandless; inflorescence usually 1-flowered.
Trailing stems cord-like, 1–3 mm. thick, without prickles, or with few short (0.5–2 mm.) bristle-like prickles; primocane leaves 3-foliate.........................14. *R. enslenii*
Trailing stems coarser, 2.5–5 mm. thick, with stout hooked prickles 1.5–3 mm. long; primocane leaves 3–5 foliate.......................................15. *R. baileyanus*

9. **Rubus flagellaris** Willd. DEWBERRY

R. villosus Ait. of ed. 7, *R. procumbens* of many auth.

This common dewberry with long woody prostrate stems, glabrous leaves (or veins beneath pilose), and essentially glabrous inflorescence is widely scattered in Ohio and doubtless more common than the map indicates.

Other species of the section (for the most part formerly included under the name *villosus*) are briefly characterized in the key. Each is doubtfully recorded from one or a few counties: 10. *R. nefrens* Bailey (Adams, Lucas, Scioto); 11. *R. tetricus* Bailey (Clermont); 12. *R. roribaccus* (Bailey) Rydb. (Erie); 13. *R. rosagnetus* Bailey (Erie, Hamilton, Knox, Montgomery); 14. *R. enslenii* Tratt. (Cuyahoga, Fairfield, Lawrence); 15. *R. baileyanus* Britt. (Lake, Warren).

Rubus, section Flagellares

On the accompanying map, the distribution of all species of the section is combined. A large proportion of specimens are too incomplete for more definite determination; about one-third are referred to *R. flagellaris*.

Section *Hispidi*

A section including 24 species, corresponding in large part with Gleason's "collective species" *R. hispidus*, which, however, includes some species in another of Fernald's sections. All of our specimens, with one exception, are referred to *R. hispidus*, represented by two varieties. A Jackson County specimen of the section, with larger less coriaceous leaves, strongly glandular and prickly pedicels, and inflorescences much exceeded by the bracteal leaves, may represent the West Virginian *R. vagulus* Bailey.

16. **Rubus hispidus** L.

Evergreen or nearly evergreen plants with slender trailing stems more or less heavily beset with bristles, and lustrous dark green leaves, the less bristly or smooth flowering branchlets erect, scarcely woody. Two varieties are recognized: var. *hispidus*, with the stronger stems 2–5 mm. thick and very bristly (30–200 bristles per cm.) is northeastern in range, and found in Ohio only in Ashtabula County; var. *obovalis* (Michx.) Fern., with the more slender stems (1–2 mm. thick) smooth or with only 1–20 bristles per cm., is more widespread. The terminal leaflets of var. *obovalis* are smaller (1.5–4.5 cm. long, 1–3.5 wide) than those of var. *hispidus* (2.5–7 cm.

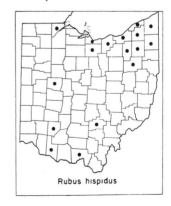

Rubus hispidus

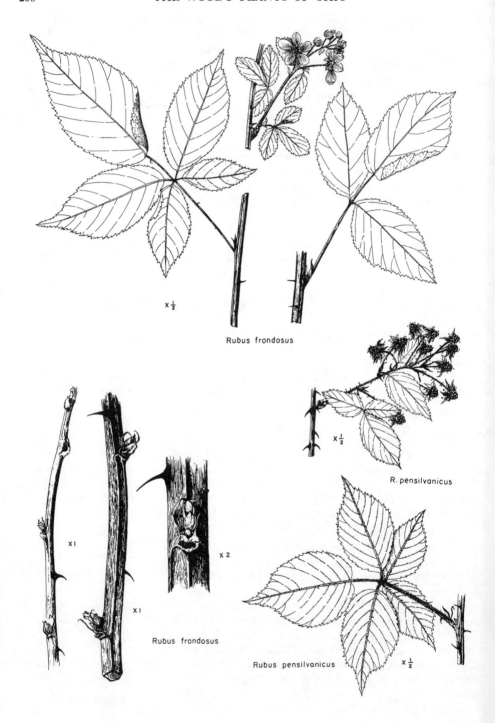

Rubus frondosus

R. pensilvanicus

Rubus frondosus

Rubus pensilvanicus

long, 2–5.5 wide). Intermediate forms occur in northeastern Ohio. The species is wide-ranging but in general northern in distribution, growing in wet soil, in swampy or boggy woods.

Section *Setosi*

A section including 25 species, few if any of which enter our area; most are northern or local in distribution. Several have been recorded from southern Michigan, West Virginia, New York, and Pennsylvania, and may occur in adjacent parts of Ohio. Members of the section, referred by Gleason to "collective species" *R. setosus*, are distinguished from other blackberries with erect or arch-ascending canes by the bristles or slender prickles, rather than broad-based or claw-like prickles, and the relatively low soft-stemmed or barely woody character.

Section *Canadenses*

A section of only 6 species, of which *R. canadensis* L. of the Northeast and the mountains of the East, may occur in Ohio. Plants essentially glabrous.

Section *Alleghenienses*

A section of 14 species, including, in large part, two "collective species" of Gleason— *R. allegheniensis* Porter and *R. orarius* Blanch. The branched and often large cylindric inflorescence, which is stipitate-glandular as is also the young growth, the pubescent stems and leaves distinguish this section. Most of the species are very local; only two, *R. allegheniensis* and *R. alumnus*, may be expected in Ohio. Both have corrugated or angled prickly canes, erect or high-arching, long-attenuate or caudate leaflets, and large succulent fruit up to 2 cm. long.

> Inflorescence an elongate raceme, cylindric, 2–4 times as long as broad; terminal leaflets of primocane leaves ovate to narrowly ovate, cordate or broadly rounded at base, usually less than ⅗ as wide as long . . . . .17. *R. allegheniensis*
> Inflorescence racemose-subcorymbose, broadest toward summit, ½ to as broad as long; terminal leaflets broadly ovate, ¾ to as wide as long . . . . .18. *R. alumnus*

17. **Rubus allegheniensis** Porter

A common and wide-ranging variable species; our most abundant blackberry in the Allegheny Plateau section of Ohio. The long-petiolate 5-foliate leaves of primocanes, soft-pilose or velvety beneath, with petioles stipitate-glandular, and the copiously stipitate-glandular pedicels and rachis of the elongate inflorescence aid in its identification.

18. **Rubus alumnus** Bailey

Not distinguished from *R. allegheniensis* in earlier Manuals, nor in any Ohio herbaria. Its published range should include Ohio, and one specimen from Monroe County is referred here. It is the source of several horticultural varieties of blackberry.

Rubus allegheniensis

Section *Arguti*

A large section with 38 species; includes Gleason's "collective species" *R. argutus*, *R. ostryifolius*, and *R. pensilvanicus*. The published ranges of several species of *Arguti* include Ohio: *R. argutus*, *R. pensilvanicus*, *R. frondosus*, *R. bushii*, and *R. recurvans*. The first of these is southern in distribution, and should be looked for in southern Ohio; the second and third are wide-ranging, and probably occur more or less throughout Ohio.

Upper 3 leaflets of mature primocane leaves 2–3 times as long as broad, acuminate; few-flowered racemes corymbiform, pedicels pilose, the axis with claw-like prickles; bracteal leaves narrow, coarsely and simply toothed; young primocanes angled and furrowed. .19. *R. argutus*
Upper leaflets of mature primocane leaves less than 2 times as long as broad, acuminate; sub-corymbose-racemose inflorescence with short axis.
  Leaf-margin serrate or doubly serrate.
    Foliaceous bracts in inflorescence few, most pedicels with stipule-like bracts; axis of inflorescence densely pilose or villous; stems finally purple. . . . . .20. *R. pensilvanicus*
    Corymbs leafy-bracted to above middle, bracts overtopping flowers, the larger coarsely toothed, 4–7 cm. long; pedicles short, densely pubescent; stems becoming terete. . . . .
                                                                         21. *R. frondosus*
  Leaf-margin irregularly lobulate, appearing jagged; bracteal leaves narrow, numerous, jagged, equaling or overtopping inflorescence. . . . . .22. *R. recurvans*

19. **Rubus argutus** Link.
    "Dry or moist thickets and borders of woods, Ga. and Ala. to Ark., n. to se. Mass., N. J., Pa., Ky and s. Ill." No specimens.

20. **Rubus pensilvanicus** Poir. BLACKBERRY
    A common blackberry with stout prickles, especially on the primocanes. Pedicels elongate, hairy, all but the lower subtended by narrow stipule-like bracts, the inflorescence extending above the lower leaf-like bracts and sometimes suggesting that of *R. alleghaniensis* from which it is easily distinguished by the absence of conspicuous stipitate glands. A widespread species including many phases (*andrewsianus* Blanch., *ostryifolius* Rydb., *abactus* Bailey, and many others).

21. **Rubus frondosus** Bigel BLACKBERRY
    A common and widespread species, perhaps often confused with *R. pensilvanicus*, in which "collective species" *R. frondosus* is included by Gleason. "Thickets and borders of woods, Mass. to Ind. and Va." Inflorescence more compact and more leafy than that of the preceding species, and primocanes with fewer and more slender prickles.

22. **Rubus recurvans** Blanc.
    The jagged-toothed leaves of this member of the *Arguti* give it a somewhat distinctive aspect. One specimen, Darke County, along railroad. "Dry to moist thickets, clearings or borders of woods, se. Que. to Minn., s. to N.S., N.E., Va., Ill. and Mo."

Rubus pensilvanicus

Rubus frondosus

RHODOTYPOS Sieb. & Zucc.

An Asiatic genus with one species.

1. RHODOTYPOS SCANDENS (Thunb.) Mak.

A shrub, occasionally appearing (from seed) in mesic woods or thickets. Easily distinguished by its simple, doubly-serrate, *opposite* leaves, solitary white, 4-merous flowers, and dry black drupes surrounded by persistent calyx. Often planted.

ROSA L. ROSE

A large genus of 100–200 species, all natives of the northern hemisphere. Shrubs, often prickly, some with long "climbing" stems; leaves compound (in our species), the stipules more or less adnate to the petiole; flowers large, the five petals and many stamens inserted near the opening of the globose to urn-shaped calyx tube (receptacle) which later develops into a fleshy fruit. Many species are variable and hybridization is frequent; some specimens cannot be identified with certainty, as they are intermediate in character. In addition to the native species, several Asiatic and European roses have become naturalized; others rarely escape, persist after cultivation, or spread from stem cuttings.

a. Styles united, forming a column exserted from throat of receptacle; stems long, trailing or climbing.
 b. Leaflets usually 3, rarely 5, stipules entire, glandular-ciliate; stems with stout somewhat decurved prickles, irregularly scattered (or wanting) and flattened at base; petioles, pedicels, receptacle and calyx glandular-hispid; flowers pink, 4–8 cm. across, few to many in corymb; native..1. *R. setigera*
 bb. Leaflets 7–9; stipules toothed or fimbriate; flowers 2–5 cm. across; introduced.
  c. Stipules fimbriate-pectinate and glandular-ciliate; stems ascending or arching; flowers white, 2–4 cm. across, in a many-flowered pyramidal inflorescence....2. *R. multiflora*
  cc. Stipules denticulate to jagged-dentate; stems trailing; flowers white or pink, 4–5 cm. across, few in corymb..3. *R. wichuriana*
aa. Styles not united, shorter, reaching about to throat of receptacle; stems ascending.
 b. Outer sepals pinnatifid, unlike inner; opening of receptacle about 1 mm. across, styles slightly exserted; introduced.
  c. Leaves stipitate-glandular on one or both surfaces, the teeth gland-tipped; achenes on wall as well as base of receptacle.
   Leaves glandular on both surfaces, aromatic; sepals long-persistent..4. *R. eglanteria*
   Leaves scarcely glandular above, not fragrant; sepals soon deciduous..5. *R. micrantha*
  cc. Leaves not glandular, not fragrant; teeth not gland-tipped; sepals soon deciduous....
                          6. *R. canina*
 bb. Outer and inner sepals similar; opening of receptacle 2–4 mm. across; styles included, brush of stigmas closing opening.
  c. Sepals reflexed or spreading (sometimes ascending) after flowering, deciduous in fruit; pedicels, receptacle, and fruit often bristly or stipitate-glandular; native species.
   Leaflets 7 (5–9), dull or slightly lustrous above, finely toothed (12–35 teeth above middle on each side); infrastipular prickles stout, decurved, internodal prickles few or absent; large shrub (to 2.5 m.) of wet soil and swamps......7. *R. palustris*
   Leaflets coarsely toothed (5–15 teeth above middle on each side).
   Leaflets 5–9 (–11), lustrous above, firm; stipules dilated above; infrastipular prickles usually present, stout, flattened toward base, base usually more than ½ length of prickle; flowers on branches from old wood; shrub (to 2 m.) of dry to moist soil...8. *R. virginiana*
   Leaflets 3–7, dull or slightly lustrous above; stipules trough-like, linear, with short diverging tips; infrastipular prickles straight, slender, base usually less than ½ length of prickle; flowers on 1-year old stems; low shrub (to 1 m.) of dry soil..
                         9. *R. carolina*
  cc. Sepals persistent in fruit, erect or somewhat connivent; pedicels, receptacle, and fruit smooth.
   Branches densely pubescent between crowded prickles; prickles pubescent; leaves strongly reticulate-veined; introduced............................10. *R. rugosa*
   Branches and prickles not pubescent.
   Infrastipular prickles present; exotic......................11. *R. cinnamomea*

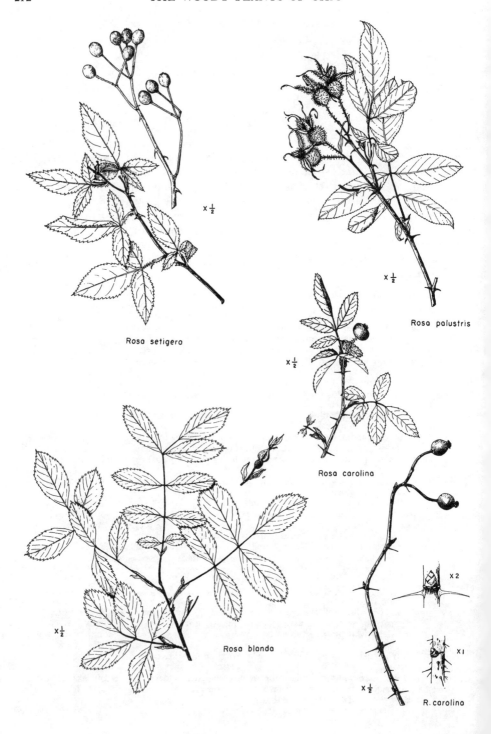

Rosa setigera

Rosa palustris

Rosa carolina

Rosa blanda

R. carolina

Infrastipular prickles absent, or not different from internodal prickles; pedicels and hypanthium glabrous; native.
Prickles absent, or only near base of stem; leaflets 5–7, obtuse or acute; petiole, rachis, and lower leaf-surface glabrous or finely pubescent; hypanthium often glaucous, calyx lobes caudate with dilated tips, often glandular hispid; shrub, to 2 m..12. *R. blanda*
Prickles usually present; leaflets 9–11, usually obtuse; petiole, rachis, and lower leaf-surface soft-pilose; low shrub (0.5 m.).......13. *R. arkansana* var. *suffulta*

WINTER KEY TO NATIVE SPECIES

Stems elongate, climbing or leaning, forming large circular mounds; prickles broad-based.....
  1. *R. setigera*
Stems ascending, not elongate; growing as scattered straggly bushes or extensive patches.
  Prickles absent, or only near base of stems.............................12. *R. blanda*
  Prickles usually present.
    Tall shrubs, 1–2 m., in patches in swamps and wet meadows; infrastipular prickles decurved, flattened at base which is about ½ as long as prickle; internodal prickles absent or few except near base...7. *R. palustris*
    Low shrubs, less than 1 m. tall, in dry situations; stems usually scattered; infrastipular and internodal prickles similar, straight, slender, terete, numerous or few or absent on branches..9. *R. carolina*

1. **Rosa setigera** Michx. Climbing Rose. Prairie Rose
  Our only native rose with elongate climbing or leaning stems; often forms large circular mounds in open habitats. Widely distributed, but in general southern and interior in range; found throughout Ohio.
  Two varieties may be distinguished, differing in pubescence of the lower leaf-surface.

Leaves lustrous above, glabrous beneath, or pilose only on veins.................var. *setigera*
Leaves dull above, tomentose beneath..............................var. *tomentosa* T. & G.

In our area, plants vary from one extreme to the other, and segregation of varieties seems unnecessary. In general, specimens from western Ohio, especially those from the Prairie Peninsula, tend to be more pubescent than those from eastern Ohio.

2. Rosa multiflora Thunb. Multiflora Rose
  From eastern Asia; commonly planted, and appearing spontaneously (from seed) on roadsides and in clearings. Rapidly spreading and reported from about 40 Ohio counties.

3. Rosa wichuriana Crépin
  A native of eastern Asia, sometimes planted on steep roadside banks. Persistent or spreading vegetatively. Reported from Ashtubula and Lake counties.

Rosa setigera

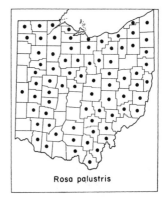

Rosa palustris

4. ROSA EGLANTERIA L. SWEETBRIER. EGLANTINE

A European rose, formerly widely planted, and escaped in clearings and on road-sides throughout Ohio. Some reports doubtless should be referred to the related *R. micrantha*.

5. ROSA MICRANTHA Sm. SWEETBRIER

R. rubiginosa var. *micrantha* Lindl.

Very similar to *R. eglanteria* and often confused with it. From Europe. Reported from Ashland and Hamilton counties.

6. ROSA CANINA L. DOG ROSE

From Europe. Occasionally persisting after cultivation, and reported (as an escape?) from several counties.

7. **Rosa palustris** Marsh. SWAMP ROSE

R. carolina of ed. 7.

Abundant in swampy habitats and along ditches and sluggish streams. Distinguished from our other native roses by habitat, minutely serrate leaves, and strongly decurved infrastipular prickles. Widely distributed throughout the Deciduous Forest; found in all parts of Ohio.

8. ROSA VIRGINIANA Mill.

R. lucida Ehrh.

A species which is not well understood, and is frequently confused with *R. carolina*. Deam (1924) has shown that lustrous leaves are not confined to this species, but occur in *R. carolina* also. The range as given by Fernald (1950) includes Ohio; Gleason (1952) states that it ranges from "Nfd. to Pa. and Va." and is "often reported from farther w. and s., probably always erroneously." Hybridizes with *R. carolina*. Following Gleason, no Ohio specimens are referred to this species.

9. **Rosa carolina** L. WILD ROSE

R. humilis Marsh., *R. virginiana* Mill. (Schaffner, 1932), *R. virginiana* var. *humilis* Schneid.

Our common wild rose of dry habitats—open woods, hillsides, prairies. Throughout eastern United States. A highly variable complex, and sometimes segregated into ill-defined varieties. A few specimens may be referred to var. *villosa* (Best) Rehd., which

Rosa carolina

Rosa blanda

has leaves soft-pubescent beneath, and two (Champaign and Preble cos.) to forma *glandulosa* (Crépin) Fern. with leaf-rachis very glandular. *R. carolina* is known to hybridize with *R. arkansana* var. *suffulta* in the Middle West, resulting in a variety of intermediate forms (see Gleason, 1952). Some specimens from southern Ohio prairies appear to be such intermediates. It also hybridizes with *R. palustris*.

10. Rosa rugosa Thunb.

A native of eastern Asia, commonly planted and rarely escaped. The tomentum between the prickles, the strongly reticulate lower leaf-surface, the large rose-purple (or white) flowers, and large (2.5 cm. diam.) brick-red fruit distinguish the species. Reported as an escape on the beach of Lake Erie in Ashtabula and Lake counties.

11. Rosa cinnamomea L. Cinnamon Rose

From Eurasia; sometimes planted, and reported as a common escape in Ashtabula County.

12. **Rosa blanda** Ait.

The generally smooth stems, and the persistent erect beak of sepals on the fruit are distinctive features; the glabrous and often glaucous hypanthium, and gradually widening stipules are additional characters of this species, which is more northern in range than our other roses, and local in occurrence.

13. **Rosa arkansana** var. **suffulta** (Greene) Cockerell

R. suffulta Greene, *R. pratincola* Greene

Known in Ohio only from Hamilton County where it is now probably extinct. Petiole, rachis, and lower leaf-surface soft-pilose. Some Adams County specimens appear to be intermediate between this and *R. carolina*.

PRUNUS L. Plum, Peach, and Cherry

A large genus, about 200 species, of the northern hemisphere and Andes of South America. Distinguished by floral structure—a cup-shaped to obconical calyx-tube, around the rim of which are the 5 calyx-lobes, 5 petals, and numerous stamens, and loosely surrounded by which is the simple pistil attached at the bottom of the cup; fleshy drupe; and simple alternate serrate leaves, with, in some species, petiolar glands. Branches and branchlets with prominent horizontal lenticels.

The genus may be divided into subgenera (sometimes elevated to generic rank) distinguished in large part by fruit characters. It is of great economic importance because of edible fruits, ornamental features, or timber. Many species, varieties, and horticultural forms are planted; a few of these have escaped, and others occasionally persist.

True terminal bud wanting; ovary and fruit covered by a bloom; fruit sulcate, the 2 opposite grooves longitudinal; stone somewhat flattened, longer than broad....................Subg. *Prunophora*
Terminal bud present.
    Ovary and fruit pubescent; fruit sulcate; stone sculptured and furrowed......Subg. *Amygdalus*
    Ovary and fruit glabrous, without bloom; fruit globose, not furrowed; stone subglobose.
        Flowers solitary, umbellate, corymbed, or in short 4–10 flowered racemes with long pedicels
                Subg. *Cerasus*
        Flowers in elongate racemes, pedicels much shorter than axis; racemes on branchlets of the
          current year..Subg. *Padus*

Subgenus *Prunophora* Plums

a. Leaf-margins sharply serrate with eglandular teeth; calyx glandless, or with few obscure glands; flowers opening before leaves, the lance-attenuate calyx-lobes soon reflexed; leaves oblong-ovate to obovate, acuminate, petioles usually without glands..............1. *P. americana*

aa. Leaf-margins with gland-tipped blunt teeth (these sometimes deciduous with age, then leaving a callous).
 Calyx-lobes without glands; flowers less than 1 cm. broad; leaves lanceolate to lance-oblong, very finely serrate, with gland near sinus, usually folded lengthwise or trough-like; petioles red, usually with 1–2 summit glands; southern.............2. *P. angustifolia*
 Calyx-lobes with marginal glands; flowers more than 1 cm. broad.
  Leaf-teeth relatively coarse, spreading, triangular to ovate; flowers 2–3 cm. broad, calyx-lobes 3.5–5 mm. long, abruptly reflexed; leaves obovate to oval, abruptly acuminate, less than twice as long as wide; petiole with 2 apical glands; northern...
                       3. *P. nigra*
  Leaf-teeth relatively fine, appressed; flowers 1–1.5 cm. broad, calyx-lobes 1.5–3 mm. long, tardily reflexed; leaves tapering to tip; southwest interior in range.
   Flowers on short lateral branches, opening before or with the leaves; leaves longitudinally rolled in bud and somewhat folded at maturity, very finely serrate (teeth project less than 0.5 mm.), with red gland adjacent to sinus; stones plump, rugose, obliquely truncate at base....................................4. *P. munsoniana*
   Flowers chiefly on prolonged slender branches of preceding year, opening with the leaves or later; leaves folded in bud, flat at maturity, finely serrate, the teeth larger, triangular, with terminal gland; stones pointed at both ends, rugose.......
                     5. *P. hortulana*

Subgenus *Amygdalus* PEACH

Leaves lanceolate to oblong-lanceolate; flowers pink; one introduced species in our area......
                     6. *P. persica*

Subgenus *Cerasus* CHERRIES

a. Coarse shrubs or trees; leaves toothed from base to apex.
 Flowers large, 2.5–3.5 cm. broad, on pedicels 2–4 cm. long, in few-flowered umbels subtended by enlarged inner scales of winter-bud; leaves coarsely toothed, ovate-oblong to obovate; fruit 1.5–2.5 cm. in diam.
  Leaves pubescent on veins beneath, large (to 15 cm. long); enlarged inner bud-scales reflexed; fruit sweet..7. *P. avium*
  Leaves glabrous, smaller (to 10 cm. long); enlarged inner bud-scales upright; fruit sour
                     8. *P. cerasus*
 Flowers smaller, 1–1.5 cm. broad, on pedicels 1–2 cm. long, in umbel-like or racemose clusters, without subtending involucre; leaves finely toothed; fruit 5–10 mm. in diam.
  Leaves orbicular to round-ovate, abruptly short-pointed or obtuse; flowers in short racemes; fruit nearly black; bark aromatic-fragrant..................9. *P. mahaleb*
  Leaves narowly elliptic to oblong-lanceolate, acute to acuminate; flowers in umbel-like clusters; fruit red..10. *P. pensylvanica*
aa. Low shrubs, 1 (–3) m. high; leaves with firm or cartilaginous margin, entire or nearly so toward base, toothed above middle with low teeth; inflorescence umbellate; fruit subglobose, 1–1.5 cm. in diam.
 Leaves narrowly oblanceolate, apex acute to acuminate, base narrowly cuneate, lustrous dark green above, paler beneath; stone 6–8 mm. broad, subglobose........11. *P. pumila*
 Leaves elliptic to obovate, obtuse to subacute at apex, base acute to rounded, pale green above, glaucous beneath; stone 5–6 mm. broad, ellipsoid.............12. *P. susquehanae*

Subgenus *Padus* RACEMED CHERRIES

a. Leaves lanceolate to oblong or oblanceolate, the fine teeth blunt, incurved and callous-tipped; calyx persistent under fruit, its lobes narrow, acute; fruit dark red to purple-black, bitter or sweetish, 7–10 mm. in diam.; twigs with odor of bitter almond.............13. *P. serotina*
aa. Leaves oblong to obovate, sharply serrate with slender ascending teeth; calyx deciduous, its lobes broadly triangular, blunt; fruit dark red, astringent, 8 mm. in diam.; twigs strongly and disagreeably scented...14. *P. virginiana*

WINTER KEY

a. Terminal bud absent. (Plums and Apricot)
 b. Uppermost bud distinctly lateral, twig prolonged to a point ½–⅔ as long as bud; twigs shining bright brown; bud-scales dark, finely ciliate......................*P. armeniaca*
 bb. Uppermost buds appearing terminal; tip-scar at base of bud.
  c. Buds finely puberulent, light brown, ovoid; planted.
   Twigs pubescent..*P. domestica*
   Twigs glabrous...*P. insititia*

cc. Buds glabrous or bud-scales ciliate.
Buds round-ovoid, little longer than thick; twigs and buds red-brown, lustrous.
Twigs slender, less than 2 mm. thick; flower-buds broad ovoid, short-acute,
mostly collateral on elongate twigs, or crowded on short laterals; leaf-buds short-
conical, ½–⅔ as long as flower-buds; upper edge of leaf-scar ciliate..........
2. *P. angustifolia*
Twigs moderate, more than 2 mm. thick; flower-buds almost spherical, collateral
on leading twigs, crowded on short spur-like twigs; leaf-buds between flower-buds
very small, those on leading twigs triangular, 3–4 mm. long, the scales black-
apiculate...5. *P. hortulana*
Buds ovoid to conical, distinctly longer than thick; twigs gray-brown to gray, not
lustrous.
Twigs pubescent.................................1. *P. americana* var. *lanata*
Twigs glabrous.
Buds pubescent, acute; twigs slender; cultivated for flowers and foliage.......
P. cerasifera
Buds glabrous; native species.
Twigs zigzag; buds black to grayish brown, 4–6 mm. long, subconical;
northern..3. *P. nigra*
Twigs scarcely zigzag, spine-tipped twigs numerous; buds dark reddish brown
to gray, 3–4 mm. long, lance-ovoid, long-acute, flower-buds slightly larger
and thicker than leaf-buds, mostly crowded on short lateral twigs; widespread
1. *P. americana*
aa. Terminal bud present. (Peach and Cherries)
b. Lenticels minute, numerous; buds densely short-pubescent at least toward tip; twigs
bright colored, green to red; collateral buds usually present...............6. *P. persica*
bb. Lenticels neither minute nor numerous; buds not densely downy; collateral buds usually
absent; twigs not green to red.
c. Twigs light coppery yellow; buds red, 4–6 mm. long, ovoid...............*P. serrulata*
cc. Twigs not light coppery yellow.
Terminal buds clustered; twigs slender, usually under 2 mm. in diam., red, with
easily removable white coating...............................10. *P. pensylvanica*
Buds not clustered at tips of twigs except on fruiting spurs.
Buds clustered on short fruiting spurs, ovoid or round-ovoid, obtuse; collateral
buds often present; exotic species.
Twigs stout, 4–6 mm. in diam.; buds large, often 10 mm. long......7. *P. avium*
Twigs more slender, buds smaller; tree with more spreading habit..8. *P. cerasus*
Fruiting spurs absent; native or naturalized species.
Twigs fragrant, light brown, older ones silvery; buds ovoid or round-ovoid,
obtuse...9. *P. mahaleb*
Twigs not fragrant, but aromatic or rank, reddish-brown, covered with an
easily removable gray coating; buds ovoid to conical, acute or obtuse.
Large tree; buds bright brown, small, 4 mm. long or less, narrowly acute;
bud-scales smooth, slightly keeled, apiculate; twigs with odor of bitter al-
monds or aromatic...13. *P. serotina*
Small tree or large shrub; buds dull brown or the margins of scales straw-
color, mostly 6–10 mm. long; bud-scales inconspicuously roughened, appearing as if dotted, rounded at apex; twigs with strong rank odor...........
14. *P. virginiana*

Subg. PRUNOPHORA Focke PLUMS

All of our native species of plum are variable; hybrids are known between *P. americana*
and *P. angustifolia*, *P. hortulana*, *P. munsoniana*, and between *P. hortulana* and *P. munsoniana*; hence many specimens will be found with intermediate characters. Satisfactory
determination is not always possible. In addition to the native plums, occasional trees
of exotic or horticultural forms may be found persisting after cultivation. Many more
are to be found in ornamental plantings; among these, the cherry plum, *P. cerasifera*
Ehrh., with many horticultural forms. The apricot, *P. armeniaca* L., is a member of this
subgenus. *P. domestica* L., garden plum, and *P. insititia* L., damson plum, are planted
for their fruit; the green-gage plum is a form of the latter.

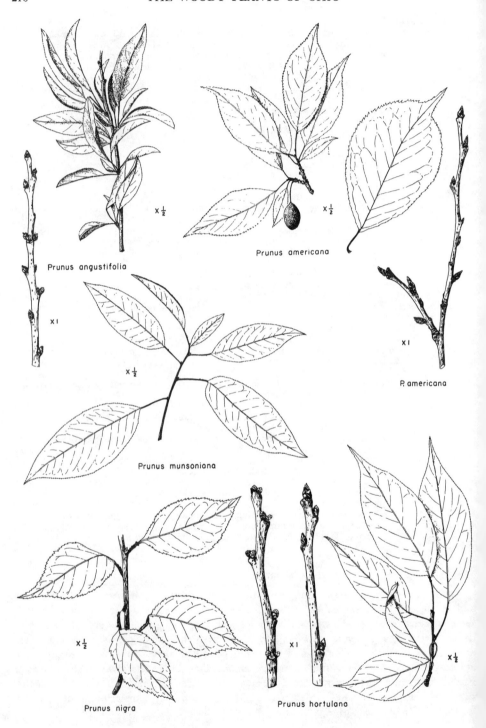

Prunus angustifolia

x ½

x I

Prunus americana

x ½

Prunus munsoniana

x ½

P. americana

x I

Prunus nigra

x ½

Prunus hortulana

x I

x ½

1. **Prunus americana** Marsh. WILD PLUM

Our most abundant and most widespread species, occurring throughout Ohio which is near the center of its extensive range reaching westward almost to the Rocky Mountains. A shrub or small tree developing root-shoots and thus forming large thickets; usually with spine-tipped branchlets. The source of many pomological varieties. Variable in pubescence of leaves and stems:

Lower surface of leaves glabrous or nearly so except when young; branchlets glabrous........
                                                                        var. *americana*
Lower surface of leaves densely soft-pubescent; branchlets pubescent.........var. *lanata* Sudw.

The latter variety, often regarded as a distinct species, is more western in range; known in Ohio only in an open post oak-blackjack oak woodland in Adams County, and from Table Rock, Lawrence County.

2. PRUNUS ANGUSTIFOLIA Marsh. CHICKASAW PLUM

A southern species, originally very local in range—central Texas and Oklahoma (Sargent, 1922)—and now naturalized northward as far as southern Ohio. Specimens from south-central Adams Co. and from Hamilton Co. The narrow, usually folded leaves, and red petiole aid in identification.

3. **Prunus nigra** Ait. CANADA PLUM

A northern species ranging southward into Ohio and Indiana. Similar to *P. americana*; differing in its larger flowers, coarse blunt gland-tipped teeth of the leaves, and glandular-serrate calyx-lobes.

4. **Prunus munsoniana** Wight & Hedrick WILD GOOSE PLUM

This and the next species (*P. hortulana*) are variable and difficult to distinguish from one another. Range, from Ohio westward and southwestward to Kans., Okla., and Texas.

5. **Prunus hortulana** Bailey WILD GOOSE PLUM

According to Rehder (1940), this is a "tree to 10 m., not sprouting from the root." As *P. munsoniana* is said to produce abundant suckers from the roots, this character may help in the recognition of the two species in the field. Because of the difficulty of distinguishing *P. munsoniana* and *P. hortulana* from incomplete herbarium material, our records are incomplete. Range: "s. Ind. to Ia., s. to n. Ala., w. Tenn., Ark. and Okla."

Prunus americana   Prunus nigra   Prunus munsoniana

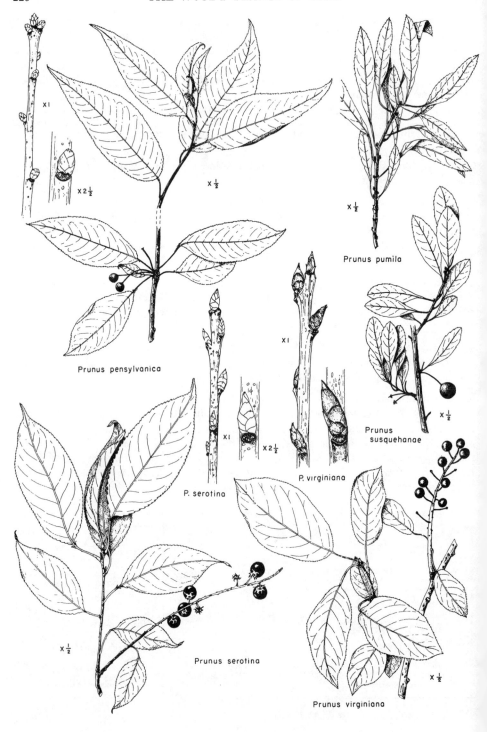

Prunus pumila

Prunus pensylvanica

Prunus
susquehanae

P. serotina

P. virginiana

Prunus serotina

Prunus virginiana

Subg. Amygdalus (L.) Focke Peach

No member of this subgenus is American; only one, peach, has spread slightly from cultivation.

6. Prunus persica (L.) Batsch. Peach

Native of China, and cultivated since ancient times. Occasionally adventive from discarded stones, but not naturalized.

Subg. Cerasus Pers. Cherries

Both native and exotic species occur in our flora. All are rather well-defined and easily recognizable. In addition to those mentioned below, a number of ornamental species are often planted, among them *P. serrulata* Lindl., the oriental flowering cherry, with numerous horticultural forms.

7. Prunus avium L. Sweet Cherry

Occasionally found as an escape along roadsides, in thickets and second-growth woodlands. From Europe and western Asia, cultivated since ancient times, and introduced into North America about 1625. Represented by many horticultural forms.

8. Prunus cerasus L. Sour Cherry

Rarely found as an escape. Like the sweet cherry, cultivated since ancient times, and represented by many horticultural forms.

9. Prunus mahaleb L. Mahaleb Cherry. Perfumed Cherry

A small tree naturalized from Eurasia. Reported from a dozen widely scattered counties. Readily distinguished by its fragrance and its roundish leaves. It is used as a stock for grafting other cherries.

10. **Prunus pensylvanica** L.f. Pin Cherry. Fire Cherry

A tree or large shrub of the North and the Appalachians; often very abundant on cut-over and burned-over forest land. Found in Ohio only in the northern and northeastern counties. Showy in spring when white with flowers, and in late summer in fruit.

11. **Prunus pumila** L. Sand Cherry

Often interpreted as a polymorphic species made up of four varieties, more or less distinct geographically. Following Fernald (1950), who elevated each of these to specific rank, the name *P. pumila* here is used in a restricted sense, synonymous with Gleason's

Prunus hortulana

Prunus pensylvanica

Prunus pumila

P. pumila var. *pumila*. This is restricted to the Great Lakes area, where it is confined chiefly to dunes and shores. Found in Ohio at Cedar Point, Erie County, and adventive in Hamilton County where it grew in gravelly soil of an abandoned gravel pit near a railroad.

12. **Prunus susquehanae** Willd.   SAND CHERRY

P. pumila var. *susquehehanae* Jaeger, *P. cuneata* Raf., *P. pumila* var. *cuneata* (Raf.) Bailey

Differing from typical *P. pumila* principally in leaf-shape; also, slightly longer petioles, more glaucous lower leaf-surface, shorter stipules (5–8 mm. as compared with 12–15 mm.) and less globose more pointed stone. General range more extensive than that of *P. pumila*, extending from Me. and Que. to Man. and Minn., and in the East, southward locally to L. I., Va., and N. C. Rare and local in Ohio.

Subg. PADUS (Moench) Koehne   RACEMED CHERRIES

All members of the subgenus are characterized by the many-flowered racemes on leafy branchlets of the current season. Flowers small (less than 1 cm. across) in our species, larger (to 1.5 cm.) in the European bird cherry, *P. padus* L., which is sometimes planted.

13. **Prunus serotina** Ehrh.   WILD BLACK CHERRY

P. virginiana of auth., not L. (in Schaffner, 1932), *Padus virginiana* (L.) Mill.

Attaining much larger size than any of our other species of *Prunus*; rarely (in Great Smoky Mountains) 4 ft., d.b.h., and 100 ft. in height; and (at Rocky Fork Lake in Highland County, Ohio) 5 ft. 4.5 inches d.b.h., and 80 ft. in height (Ohio Forestry Ass'n). Most large specimens have long since been removed from our forests, for wild black cherry is a valuable timber tree whose wood resembles mahogany. Ranges throughout Ohio in mesic woods, second-growth woodlands, and scattered along fence-rows; reaches its best development in mesic sites.

14. **Prunus virginiana** L.   CHOKE CHERRY

P. nana DuRoi (in Schaffner, 1932), *Padus nana* (DuRoi) Roem.

A tall shrub or small tree, more northern in range than *P. serotina*, and, as distinct varieties, extending farther west. In Ohio, more frequent northward, and on the Allegheny Plateau, in a variety of habitats. The common name refers to the astringent fruit.

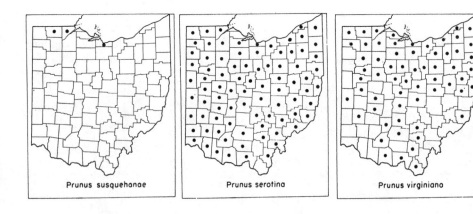

Prunus susquehonae   Prunus serotina   Prunus virginiana

LEGUMINOSAE

A very large family, almost 500 genera and 15,000 species, cosmopolitan in distribution, but best developed in tropical and subtropical lands; few woody species in temperate latitudes. Commonly divided into three subfamilies, these treated as families by some authors. The familiar papilionaceous corolla (well illustrated by black locust) is almost confined to one of the subfamilies; the fruit of all is a legume, short and few-seeded in some genera, longer and many seeded in others. Leaves are alternate, mostly compound, simple in a few genera. A family of great economic importance because of food (peas, beans, etc.), forage (alfalfa, clover, etc.), and ornamental plants.

SUBFAMILY I. MIMOSOIDEAE

Flowers regular, not papilionaceous, small, the stamens exceeding the perianth. Distinguished by the usually bipinnate leaves and small flowers in heads, dense spikes, or racemes, the showiness due mostly to the stamens. About 35 genera and 2000 species, mostly trees and shrubs of warm climates. Represented in our area only by the herbaceous species, *Desmanthus illinoensis* (Michx.) MacM. The so-called "Mimosa" or "Acacia," a widely spreading low tree, sometimes planted in southern Ohio, and very commonly in the South, is *Albizzia julibrissin* Durazz., a tree with pink flowers in tassel-like globular clusters, and large broad flat legumes; this is native from Persia to central China.

SUBFAMILY II. CAESALPINIOIDEAE

Flowers regular or irregular, or in *Cercis*, imperfectly papilionaceous, the upper or odd petal enclosed by the others. Leaves simple, pinnate, or bipinnate. Almost 60 genera and 2200 species, mostly trees and shrubs. In addition to the three genera of woody plants below, a fourth genus, *Cassia*, is represented in our area by 4 or 5 herbaceous species.

GYMNOCLADUS Lam.

A genus of only two species, one in eastern America, the other in China. Name from the Greek *gymnos*, naked, and *klados*, branch, referring to the stout branchlets without smaller twigs.

1. **Gymnocladus dioica** (L.) K. Koch KENTUCKY COFFEE-TREE
Medium size or rarely large tree of open woods, with gray fissured bark, the plates curling at the sides; one Ohio tree in Franklin Co. is reported to be 12 ft. 4 in. in circumference at 4½ ft. above the ground (Ohio Forestry Ass'n.). Easily recognized in summer by its large (up to 1 m. long) twice-pinnate leaves, the lower pair of pinnae usually reduced to single leaflets, the others with 7–13 ovate, petiolulate entire leaflets, each 3–4 cm. long; in winter, by its coarse brownish twigs with whitish coating and salmon-pink pith. Staminate and pistillate flowers on different trees, hence the specific name *dioica*; the large heavy pods which persist through the winter are as a result produced only on some trees. Used by early settlers in Bluegrass Kentucky as a coffee substitute; roasted seeds eaten by some Indians. A tree of interior range, most frequent on limestone soils; in Ohio, most abundant in the southwestern quarter of the state.

GLEDITSIA L. HONEY-LOCUST

A genus of about 12 species, in North and South America, central and eastern Asia, and tropical Africa. In addition to our species, another, *G. aquatica* Marsh., occurs in

x 1

x 2

x ½

Gymnocladus dioica

x ½

alluvial swamps of the Mississippi embayment and southern Coastal Plain. Generic name in honor of a German botanist, contemporaneous with Linnaeus, and director of the Botanic Garden in Berlin.

1. **Gleditsia triacanthos** L. HONEY-LOCUST

A tree often heavily armed with stout branched thorns—the thorns often more than the 3-times branched suggested by the specific name; or, in forma *inermis* (Pursh) Schneid., unarmed, and then a desirable and beatuiful shade-tree, because of its lacy foliage. Leaves pinnate or bipinnate, the leaflets small (1–2 cm. long on bipinnate leaves, 2–4 cm. on pinnate leaves). The large flat, somewhat twisted, pods contain a sweet edible pulp around the hard seeds; greedily eaten by cattle. A tree of interior range, reaching its greatest size in the Wabash bottoms of Indiana; the largest recorded Ohio tree (in Trumbull Co.) measures 14 ft. 4 in. in circumference 4½ ft. above the ground (Ohio Forestry Ass'n.). Locally naturalized east of the Appalachians; widely distributed in Ohio but most abundant in the southwestern part of the state.

CERCIS L. REDBUD

A genus of seven (or eight) species, two American, ours and *C. occidentalis* Torr. of the Southwest (or three, if *C. reniformis* Engelm. of northeastern Mexico, Texas, and Oklahoma is regarded as a distinct species rather than as *C. canadensis* var. *texensis* (S. Wats.) Hopkins); the others in southern Europe and Asia. *Cercis* is the classical name of *C. siliquastrum* L., the Judas-tree of southern Europe and western Asia, which has been cultivated since ancient times. Flowers imperfectly papilionaceous, in lateral clusters from the old wood—a habit common to many tropical trees, and known as cauliflory (stem flowers). Leaves simple, cordate, with pulvinus at tip of petiole.

1. **Cercis canadensis** L. REDBUD

A small tree, our only woody legume with simple leaves; wide-ranging, from Florida Texas, and northeastern Mexico northward to the southern Great Lakes area. Flower-buds (which are evident in the winter) are killed by extreme winter temperature, a limiting factor in its northward distribution. In Ohio, infrequent or local in the northern third of the state, more common southward; abundant on calcareous soils in southwestern Ohio, where it is sometimes the dominant tree of young second-growth stands, especially on slopes of southerly exposure. A white-flowered form (forma *alba* Rehd.) is sometimes seen

Gymnocladus dioica

Gleditsia triacanthos

Cercis canadensis

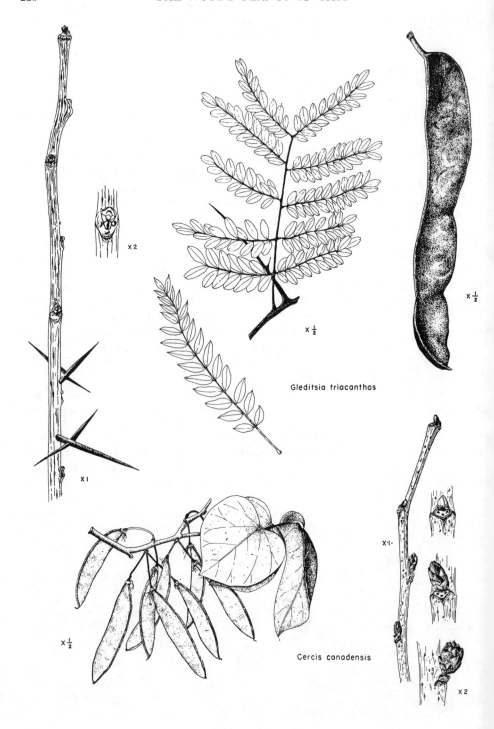

X 2

X ½

Gleditsia triacanthos

X ½

X I

X·I·

X ½

Cercis canadensis

X 2

in cultivation; a glabrous-leaved form (forma *glabrifolia* Fern.) is sometimes distinguished from the typical, which has leaves more or less pubescent beneath at maturity. When it is desirable to distinguish our eastern redbud from varieties in Mexico or Texas, it should be called *C. canadensis* var. *canadensis*.

SUBFAMILY III. PAPILIONOIDEAE

Flowers typically papilionaceous (zygomorphic); the upper or odd petal (the standard) enclosing the others in bud, later spreading; the two lateral petals (wings) exterior to the two partly coherent lower petals (keel) which enclose the stamens and style. This subfamily (or family FABACEAE) includes by far the greater number of the members of the inclusive family Leguminosae—about 400 genera and 10,000 species of herbs, shrubs, or trees. Many are ornamental, some are valuable food and forage plants. A few exotic woody species are frequently seen in cultivation: *Sophora japonica* L., Japanese Pagoda-tree with odd-pinnate leaves with opposite, subopposite, or occasionally alternate leaflets, green branchlets, yellowish-white flowers in large terminal panicles, and stipitate pods constricted between the seeds, native to China and Korea; *Laburnum anagyroides* Med., Golden-chain, a tall shrub or small tree with bright yellow flowers in long pendant racemes, from southern Europe; *Cytisus scoparius* (L.) Link, Scotch Broom, a bushy shrub from Europe, with stiff green stems, small leaves with 3 obovate leaflets (or only one) and yellow flowers placed singly or in pairs in the upper axils, sometimes persisting and spreading after cultivation; *Colutea arborescens* L., Bladder-senna; *Pueraria lobata* (Willd.) Ohwi, the Kudzu-vine; and several species of *Caragana*, of which *C. arborescens* Lam. has been used in shelter-belt planting in the Great Plains.

CLADRASTIS Raf. YELLOWWOOD

Trees with odd pinnate leaves with alternate rather large leaflets, and white flowers in panicled racemes. Five species, one American and four Asian (two in China, two in Japan); *C. lutea* is very closely related to one of the Japanese species, which in turn is close to a Chinese species. The common name refers to the color of the wood, which yields a dye.

1. CLADRASTIS LUTEA (Michx. f.) K. Koch YELLOWWOOD

A handsome tree with smooth gray bark (somewhat like that of beech) and large loose panicles of white flowers (about the size of those of black locust); pods narrow, 6–9 cm. long, few-seeded. Confined, in the wild, to ravine slopes and limestone bluffs in the Western Mesophytic Forest region, except for a few local occurrences in the Southern Appalachians. Often planted as an ornamental; trees in Adena Woods near Chillicothe, Ross Co., Ohio, are evenly spaced along an old roadway, and were evidently planted.

AMORPHA L.

Shrubs (rarely herbs), principally in warm-temperate and tropical North America. Flowers small, blue to dark purple, in dense axillary and terminal spike-like racemes; leaves pinnate, leaflets small; pods small, 6–10 mm. long, 1–2 seeded.

1. **Amorpha fruticosa** L. FALSE INDIGO

A shrub with locust-like foliage, and flowers so dark that they are inconspciuous; pods 6–8 mm. long, with large blister-like dots. Generally restricted to alluvial bottoms and moist or rocky river banks; not definitely known to be native in Ohio, although adventive in several counties. A variable species, which has been segregated into several varieties; ours belong to the typical variety.

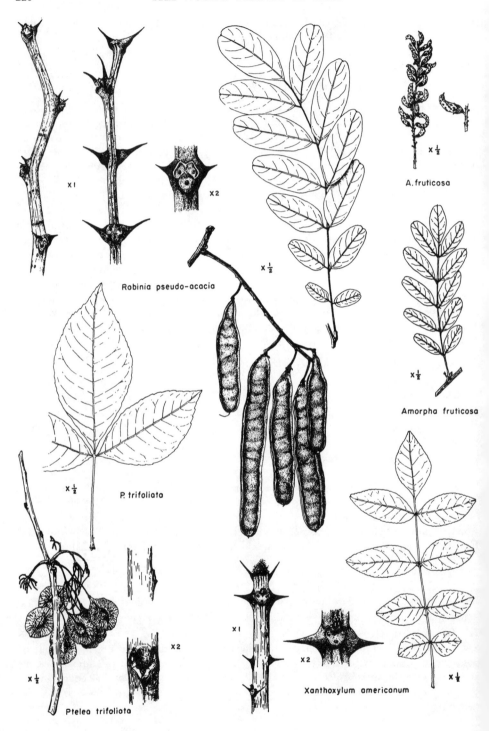

x 1

x2

A. fruticosa

Robinia pseudo-acacia

x ½

Amorpha fruticosa

x ½

P. trifoliata

x ½

x ½

x2

Ptelea trifoliata

x 1

x2

Xanthoxylum americanum

x ½

ROBINIA L. Locust

About 20 species, trees or shrubs, of the United States and Mexico. Leaves pinnate, the stipules in some species modified into spines; flowers large, white, pink, or pale purple; the seeds are poisonous if eaten.

Branchlets glabrous or thinly pubescent when young, usually with paired spines at nodes; leaflets glabrous; flowers white, in drooping racemes 1–2 dm. long; ovary and pod glabrous
1. *R. pseudo-acacia*
Branchlets bristly or viscid-glandular; flowers pink; ovary and pod hispid.
Branchlets and peduncles with long (2–5 mm.) stiff brown spreading hairs; flowers rose-colored to purplish, in open racemes; pods bristly......................2. *R. hispida*
Branchlets and peduncles viscid with large sessile or subsessile glands; flowers pink with yellow spot on standard, in short crowded racemes; pods hispid with stalked glandular hairs..
3. *R. viscosa*

1. **Robinia pseudo-acacia** L. BLACK LOCUST
Well known thorny tree, most abundant in waste places and on dry hillsides. The original range is doubtful, as this species has spread widely following settlement. It probably grew throughout the southern Appalachians, and ranged northward to the Ohio River and westward to Oklahoma; now naturalized northward into Quebec, Ontario, and Iowa; planted and now naturalized in many countries. Found throughout Ohio although once limited to the southern part (our map shows present range). Daniel Drake, in his *Picture of Cincinnati in 1815*, states that it was seldom seen more than 30 miles north of the Ohio River. Valuable for reforestation and erosion-control; as a timber, for fence-posts.

2. ROBINIA HISPIDA L. Bristly Locust; Rose-acacia
This showy stoloniferous shrub (0.5–3 m. high) of the southern Appalachian area is frequently planted, and may persist or spread after cultivation; sometimes found on roadsides. Reported from several counties, but without evidence that it is indigenous to Ohio.

3. ROBINIA VISCOSA Vent. CLAMMY LOCUST
A tall shrub or small tree (to 10 or 12 m.) of mountain woods; a frequent escape, and reported from several Ohio counties, sometimes occurring in ravines away from roadsides. It is not indigenous to Ohio. Although this species and *R. hispida* are sometimes reported as if native, evidence indicates that both are native only in limited areas in the Southern Appalachians, *R. viscosa* "only in a small area in western North Carolina" (quotation from Carroll E. Wood, Jr., in Clarkson, 1958).

Amorpha fruticosa

Robinia pseudo-acacia

WISTERIA Nutt. Wisteria

Twining woody vines of eastern America and eastern Asia. The two American species, *W. frutescens* (L.) Poir. and *W. macrostachya* Nutt., inhabit alluvial or wet woods and river banks in the South; the latter ranges northward in the Mississippi embayment to southwestern Indiana; both are occasionally planted, and the latter has been reported from two Ohio counties (Licking and Meigs). Two of the Asiatic species, *W. floribunda* (Willd.) DC. and *W. sinensis* (Sims) Sweet, are often planted and may persist in old home-sites. The American species have ovary and pod glabrous, the Asiatic have ovary and pod pubescent or velvety. The herbaceous *Apios americana* Medic. has similar leaves.

COLUTEA L. Bladder-senna

Shrubs of southern Europe, western Asia, and northern Africa, with yellow flowers in axillary racemes, and inflated or bladdery pods. *C. arborescens* L. is occasionally planted, and is reported as an escape "on hillsides" in Ashtabula County.

PUERARIA DC.

A genus of high-climbing woody or half-woody vines of eastern and tropical Asia. *P. lobata* (Willd.) Ohwi (*P. thunbergiana* (Sieb. & Zucc.) Benth.), the Kudzu-vine, with 3 large leaflets, has been planted on high railroad embankments and road-cuts for erosion control, and may spread vegetatively to surrounding areas. Much used in the South.

RUTACEAE

A large family of trees, shrubs, or herbs; about 140 genera and 1500 species, chiefly at low latitudes and best developed in Australia and South Africa; a few species in cool-temperate latitudes of America and Eurasia. Of greatest economic importance is the genus *Citrus*, whose wild members are natives of tropical and subtropical Asia and the Malayan Archipelago. In addition to the plants mentioned below, the Trifoliate Orange, *Poncirus trifoliata* (L.) Raf., and Cork-tree, *Phellodendron* spp., are often planted. Neither is known in Ohio as an escape. *Poncirus*, with green twigs and stout spines, 3-foliate leaves, fragrant white flowers (like those of orange), and small hard and inedible orange-like fruit, is used as a grafting stock for oranges. The leaves of all members of the family have clear or translucent spots (glands) and emit a pungent or aromatic odor when crushed.

XANTHOXYLUM Gmel. Prickly-Ash

The generic name, derived from the Greek *xanthos*, yellow, and *xylon*, wood, is ortho-graphically correct when spelled *xanthoxylum*, as in Fernald (1950). However, Linnaeus' original spelling was *Zanthoxylum*, and the International Code recommends the retention of his name. The genus, of about 200 species, is mostly tropical in distribution, with about 50 species in North America, five of which reach the United States, one northward into Ohio.

1. **Xanthoxylum americanum** Mill. Prickly-Ash
    Shrub; leaves resembling those of ash, but pellucid-dotted; prickly or thorny stems resembling those of black locust, except for the rusty-red buds. (The prickles, although resembling the thorns of black locust, are detachable, and believed to be prickles rather than thorns.) The somewhat fleshy, berry-like, 1–2 seeded follicles are aromatic, as is the foliage. Wide-ranging, from the Gulf slope northward to Quebec, Minnesota, and

South Dakota. Widely distributed in Ohio, in open woods, and on steep, often calcareous banks; sometimes forming dense thickets. Leaflets of mature leaves vary in length from 3 to 10 cm.

PTELEA L. Hop-tree

A small genus of temperate North America. The generic name is the Greek name for elm, and alludes to the resemblance of the samaras to those of elm.

1. **Ptelea trifoliata** L. Hop-tree. Wafer-Ash

Tall shrub, sometimes tree-like; recognized by its glossy, pellucid-dotted, 3-foliate leaves, the leaflets subsessile, red-brown twigs with small superposed buds, and nearly orbicular samaras in terminal compound cymes. Variable in amount and persistence of pubescence:

Branchlets glabrous or nearly so...var. *trifoliata*
Leaflets glabrous or glabrate...var. *trifoliata*
Leaflets permanently pilose beneath......................forma *pubescens* (Pursh) Voss
Branchlets densely pubescent.......................................var. *mollis* T. & G.

P. trifoliata ranges almost throughout the eastern forest area, and, as var. *mollis*, far westward. Ohio specimens are referable to the typical variety, and a few to its forma *pubescens*; widely scattered in Ohio, but apparently absent from many of the eastern counties.

Xanthoxylum americanum

Ptelea trifoliata

SIMAROUBACEAE

A family of chiefly tropical trees or shrubs, a few of which reach southern Florida; none native in temperate latitudes of America; one, *Ailanthus*, naturalized.

AILANTHUS DESF. Ailanthus. Tree-of-heaven

Tall-growing trees of eastern Asia and Australia. The generic name is said to be derived from a Moluccan name meaning tree-of-heaven, referring to the great height reached in the native habitat. One species naturalized in our area.

1. Ailanthus altissima (Mill.) Swingle Tree-of-heaven
   A. glandulosa Desf.

An ill-scented tree, native of eastern Asia; commonly planted and spreading from root-shoots and seeds, now often naturalized. Occurs throughout Ohio; reported from

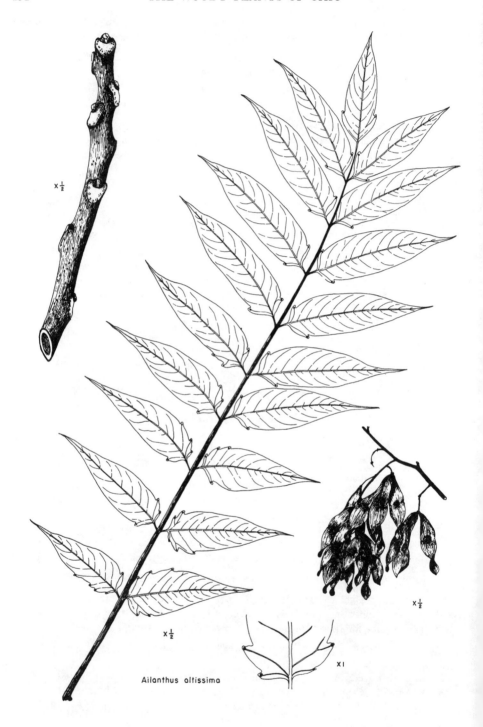

Ailanthus altissima

about 50 counties. Leaves large (3–6 dm. long), compound, with 11–25 entire or few-toothed acuminate leaflets, a gland in each tooth just within margin; 1–5 elongate samaras with centrally placed seed develop from the 2–5 parted ovary.

ANACARDIACEAE

Trees, shrubs, or woody climbers; about 60 genera and 600 species, mostly in tropical and subtropical latitudes, relatively few in temperate latitudes. Flowers small; fruit a drupe, in ours small and berry-like. Some species are poisonous, some are economically important: mango, *Mangifera indica* L., cultivated for its edible fruit in tropical lands; *Spondias*, also cultivated for fruit; *Anacardium occidentale* L., from the American tropics, yields cashew nuts; *Pistacia vera* L. from the Mediterranean region, pistachio nuts; *Schinus molle* L. the "California pepper-tree" from Peru, much planted as a shade and ornamental tree in warm climates.

COTINUS Duham. SMOKE-TREE

Of the two or three species, the European *C. coggygria* Scop. is often planted; the American *C. obovatus* Raf. (*C. americanus* Nutt.) is local in distribution, in the Ozark region and in an Alabama-Tennessee area. The smoky appearance of the bush is due to the greatly elongated plumose sterile pedicels in the panicles.

1. COTINUS COGGYGRIA Scop. SMOKE-TREE
   Reported as an escape in Ashtabula, Erie, Highland, Jefferson, and Trumbull counties.

RHUS L.

A genus of some 150 species of wide distribution, ours divisible into three natural groups, sometimes elevated to generic rank: the red-berried sumacs, with erect terminal panicles and pinnate leaves; the fragrant sumac, with trifoliate leaves and red berries; and the poisonous *Toxicodendron* species, with pinnate or trifoliate leaves, and white or whitish berries in axillary panicles. The varnish-tree or lacquer-tree, *Rhus verniciflua* Stokes, belongs to the third group; it is the source of Japanese lacquer. All our species color brilliantly in autumn. Hairs on the surface of sumac berries containing malic acid, give them a pleasant acid taste.

Key A.

a. Fruit red, the inflorescence a terminal more or less pyramidal panicle or cluster of short spikes.
  Leaves pinnate, with more than 3 leaflets; tall or tree-like shrubs with coarse twigs and erect
    terminal panicles...SUMACS
    Petiole and rachis wing-margined; leaflets entire or with few teeth, lustrous above; fruit
    pubescent with minute hairs.......................................1. *R. copallina*
    Petiole and rachis not wing-margined; leaflets serrate.
    Branches and petioles densely long-pubescent with soft hairs; fruit long-villous........
                                                                            2. *R. typhina*
    Branches and petioles glabrous or nearly so; hairs of fruit short, appressed..3. *R. glabra*
  Leaves 3-foliate, leaflets all sessile or nearly so; straggling shrub with slender twigs, and loose
    terminal clusters of soft-pubescent red fruit; flowers yellow, early..........4. *R. aromatica*
aa. Fruit white or whitish, in spreading or drooping axillary panicles; POISONOUS.............
                                                                          section *Toxicodendron*
  Leaves pinnate, with 7–13 entire, glabrous leaflets; coarse shrub or small tree; POISONOUS
                                                                          5. *R. vernix*
  Leaves 3-foliate, the leaflets entire or toothed, glabrous or nearly so, the terminal leaflet long-
    stalked; high-climbing by aerial roots, bushy, or (in var. *arenaria*) low; POISONOUS.....
                                                                          6. *R. radicans*

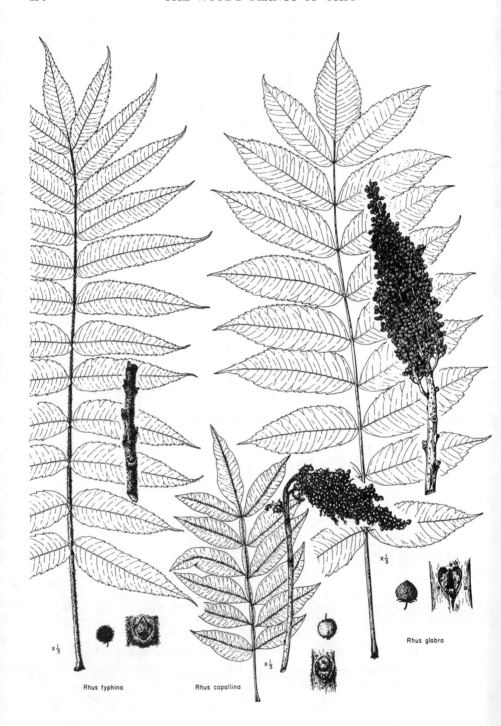

x⅓

Rhus typhina

x⅓

Rhus copallina

x⅓

Rhus glabra

Key B. For use with non-fruiting specimens; an aid to field determination of poisonous species. Key A should be consulted for additional characters.
a. Leaves 3-foliate.
    Terminal leaflet long-stalked; high-climbing, bushy, or low; POISONOUS.....6. *R. radicans*
    Terminal leaflet sessile or nearly so; straggling shrub with slender twigs; not poisonous.....
                                          4. *R. aromatica*
aa. Leaves pinnate, leaflets usually more than 5.
    Leaflets 7–21, entire or nearly so; tall or tree-like shrubs.
    Leaf-rachis winged, the wings 1–5 mm. wide and interrupted at place of attachment of
      leaflets..1. *R. copallina*
    Leaf-rachis not winged; POISONOUS.................................5. *R. vernix*
    Leaflets 7–31, serrate.
    Twigs densely soft-pubescent with long hairs..........................2. *R. typhina*
    Twigs glabrous or nearly so..3. *R. glabra*

Key C. For winter use; includes *Cotinus*.
a. Terminal bud present.
    Drooping clusters of white or whitish berries often persistent through the winter; terminal
    bud 5–15 mm. long; VERY POISONOUS.
    Shrub or small tree; buds purplish, downy...............................5. *R. vernix*
    Low straggling shrub or vine, sometimes high-climbing; buds light brown...6. *R. radicans*
    No fruit present; finely branched erect empty panicles sometimes seen; terminal buds very
    small, 2–3 mm. long...COTINUS
    Twigs and buds reddish brown...........................*C. obovatus (americanus)*
    Twigs and buds yellowish brown..................................*C. coggygria*
aa. Terminal bud absent; low trees or shrubs with erect or arching panicles of hairy red berries.
    Lateral buds hidden in a transverse fold in axil of leaf-scar; catkin-like flower-buds usually
    present; low shrub..4. *R. aromatica*
    Lateral buds not hidden.
    Twigs densely long-hairy...2. *R. typhina*
    Twigs glabrous or finely pubescent.
    Leaf-scars almost encircling the bud; lenticels small....................3. *R. glabra*
    Leaf-scars about half surrounding the bud.
    Lenticels very prominent and crowded, frequently touching one another; planted....
                                        R. chinensis
    Lenticels less prominent and crowded; native.......................1. *R. copallina*

1. **Rhus copallina** L. SHINING SUMAC. WINGED SUMAC

Frequently forms extensive patches of rather uniform height; brilliant in fall coloration; fruit darker and less showy than that of other sumacs. In Ohio, largely confined to the Allegheny Plateau, and acid soils of the flats of southwestern Ohio (Illinoian Till Plain) and Lake area of northern Ohio. Variable in number and width of leaflets, characters used by Fernald (1950) to distinguish varieties.

Rhus copallina   Rhus typhina   Rhus glabra

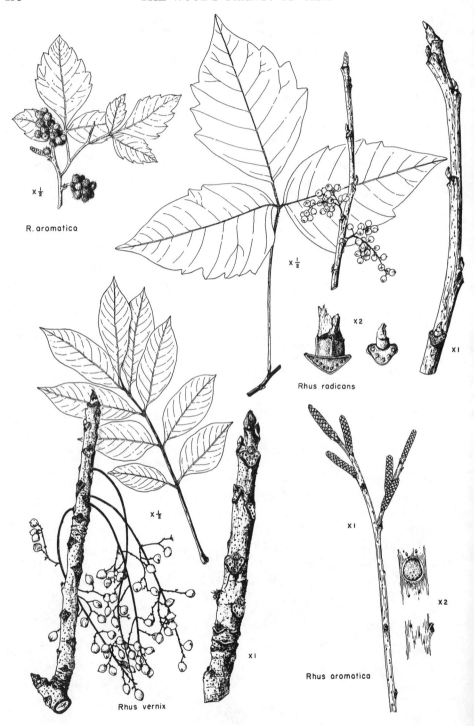

R. aromatica

x ½

Rhus radicans

x 2

x 1

Rhus vernix

x ½

x 1

Rhus aromatica

x 1

x 2

Leaflets 11–23, lance-oblong or linear oblong; southeastern (Fla.—s.e. N. Y.).....var. *copallina*
Leaflets 5–13, broadly oblong to narrowly ovate; wide-ranging............var. *latifolia* Engler
These varieties grade into one another, and are not recognized by Gleason (1952) or Little (1953.) *R. chinensis* Mill. (*R. javanica* Thunb.), also with rachis winged, is sometimes planted; it may be distinguished by its glabrous yellowish branchlets, crowded lenticels, and coarsely crenate-serrate leaflets.

2. **Rhus typhina** L. Staghorn Sumac
   R. hirta (L.) Sudw.
   Growing in groups, or individually; attains larger size than the other sumacs, reaching a height of 12–14 m. and trunk diameter of 2–3 dm. The largest of a group of seven trees near Lake Pike, Pike County, measured 2.8 dm. in diameter (d.b.h.). Lenticels of trunk horizontal, cherry-like, and show a rich reddish-brown color. More northern in range than other species of sumac; in Ohio, widely scattered and generally absent from the west-central counties (the Till Plains). Easily recognized by its long-pubescent branchlets and fruit.

3. **Rhus glabra** L. Smooth Sumac
   Range more extensive than that of any other sumac—from Maine and Quebec westward to British Columbia and southward to Florida, Texas, and Mexico. Our most abundant sumac, occurring throughout Ohio. A showy species in summer or fall, its fruit brighter red than that of the staghorn sumac. The northern var. *borealis* Britt., which may be a hybrid of *R. glabra* and *R. typhina*, is recorded from Ashtabula County; its branches are puberulent to short-pilose.

4. **Rhus aromatica** Ait. Fragrant Sumac
   R. canadensis Marsh., not Mill., *Schmaltzia crenata* (Mill.) Greene
   An attractive shrub of dry gravelly or rocky banks (calcareous) and sand dunes. Although a 3-leaflet *Rhus*, it is easily distinguished from poison ivy by its smaller sessile leaflets, bright red, long-hairy fruit in small irregular clusters, and short naked aments formed in summer for flowering the following spring. Showy in flower (flowers yellow, before the leaves in typical variety), in fruit, and in autumn color. Variable in amount of pubescence, shape of terminal leaflet, and time of flowering, characters used to distinguish varieties; of these, var. *aromatica* and var. *arenaria* are in our range.

Rhus aromatica

Rhus vernix

Rhus radicans

Terminal leaflet rhombic-ovate to elliptic, 3–9 cm. long, base subcuneate, apex acute or acutish; leaves more or less pubescent when young, later glabrous or nearly so; flowers opening before leaves. . . . . .var. *aromatica*
Terminal leaflet broadest above middle, 1.5–3.5 cm. long, base cuneate, apex rounded; leaves velvety beneath; flowers opening with leaves; low, depressed shrub of sand dunes. . . . .
var. *arenaria* (Greene) Fern.

The typical variety is widespread; in Ohio, it is most frequent along the north-south band of limestone outcrops extending from Richland County south-southwest to Adams County; many specimens have leaves persistently pubescent beneath; var. *arenaria* (*R. trilobata* Nutt. var. *arenaria* (Greene) Barkley) is confined to dunes of northern Ohio, northern Indiana, and northeastern Illinois.

5. **Rhus vernix** L.   Poison Sumac
    Toxicodendron vernix (L.) Kuntze
    A coarse shrub or small tree (2–7 m.) of bogs, swamps, and swampy woods, with smooth gray bark and glaucous and glabrous branchlets.   Leaves pinnate, the 7–13 leaflets acute or short-acuminate.   Northern and Coastal Plain in range, absent in the interior from the Ohio Valley southward.   In Ohio, most frequent toward the northeast, very local southward in relic bogs; long-extinct in Hamilton County, the record based on a specimen collected in 1837.   A beautiful but poisonous shrub, in autumn orange-scarlet, in winter with clusters of grayish-white berry-like drupes.

6. **Rhus radicans** L.   Poison Ivy
    R. toxicodendron of ed. 7, in large part, not L., *Toxicodendron radicans* (L.) Kuntze
    Extremely variable in habit, trailing, shrubby, or high-climbing; when climbing, stems attached to support by aerial roots.   Best recognized in the field by its alternate 3-foliate leaves with terminal leaflet long-stalked, silky-pubescent gray winter-buds, and axillary clusters of whitish berry-like drupes.   Seedlings in lawns or flower-beds may be distinguished from seedlings of Virginia creeper (which often have only 3 leaflets) by the narrow-oblong cotyledons; cotyledons of Virginia creeper are broad-ovate or cordate.   Very wide-ranging, almost throughout temperate North America except the extreme west; doubtless in every county in Ohio, but not always collected because of danger of poisoning.   A number of varieties and forms have been named (see Fernald, 1950); most of these are connected by intermediates, and difficult to distinguish.   Ohio specimens vary greatly in amount of pubescence.

AQUIFOLIACEAE

Trees or shrubs with simple alternate leaves and small white, greenish, or yellowish axillary flowers.   Three genera—two in our range—and about 300 species in temperate and tropical regions of both hemispheres.

ILEX L.   Holly

Deciduous or evergreen trees or shrubs, about 300 species, most numerous in the tropics.   Pistillate flowers few or solitary in the leaf-axils, staminate usually in clusters. Fruit a berry-like drupe containing 2–8 bony nutlets, red (or rarely yellow) or black. *Ilex* is the classical name of the holm oak or holly oak, *Quercus ilex* L.   Several species are in cultivation; English holly, *Ilex aquifolium* L., less hardy than American holly, is brought in from the Pacific Northwest (where grown) for the Christmas trade.

Leaves thick, evergreen, spine-toothed; tree with smooth light gray bark. . . . . .1. *Ilex opaca*
Leaves thin, deciduous, lanceolate to round-obovate, usually acuminate, serrate; shrub of swamps or wet woodlands. . . . . .2. *I. verticillata*

1. **Ilex opaca** Ait. AMERICAN HOLLY

A southern species, ranging north in the interior to northern Kentucky (Lewis Co.) and southern Missouri, and in the East near the coast to New Jersey and eastern Massachusetts. Each of our Ohio records is based on a single tree: one in Lawrence County which is said to be native; one in Scioto County, a "tree 18 feet tall and 5 inches in diameter" which has been known to have been there for at least 40 years (C. A. Eulett) and is far removed from any planted holly; and one in Fairfield County "about 10 to 12 feet in height . . . along an old fence-row" and "about a quarter of a mile away from a one-room school" (C. R. Goslin). The two southern stations doubtless represent natural extensions of the more continuous range not far to the south.

2. **Ilex verticillata** (L.) Gray WINTERBERRY

Shrub (rarely tree-like) of swamps, wet woods, and sandy stream margins; in southwestern Ohio, limited to wet flats of the Illinoian Till Plain. Generally northern in distribution, but extending south locally to Tennessee and Georgia. The red fruit, persisting after the leaves have fallen, is used in Christmas decorations. Variable in foliage characters, on which several intergrading varieties are based; the following, not separated on our map, occur in our range.

Leaves without transparent dots, pubescent at least on veins beneath, lanceolate to round-obovate; bark of 2-year branchlets dark or brownish.
Leaves with principal veins beneath appressed-pilose or downy; widespread....var. *verticillata*
Leaves downy beneath on tissue as well as veins; widespread...var. *padifolia* (Willd.) T. & G.
Leaves dotted with minute transparent spots, glabrous or sparsely downy on principal veins beneath, oblanceolate to elliptic-obovate; bark of 2-year branchlets whitish-gray; northern..
var. *tenuifolia* (Rorr.) S. Wats.

NEMOPANTHUS Raf.

A monotypic genus distinguished from *Ilex* by its entire leaves, separate petals, and minute deciduous sepals. Name from the Greek *nema*, a thread, *pous*, foot, and *anthos*, flower, referring to the long slender axillary pedicels.

1. **Nemophanthus mucronata** (L.) Trel. MOUNTAIN-HOLLY

A much-branched shrub to 3 m. in height, with ashy-gray bark, smooth entire (or with few remote teeth) leaves, slender often reddish petioles, fascicled staminate and solitary pistillate axillary flowers on slender pedicels. Northern in range; in Ohio confined to bogs and wet woods of the northeastern and northwestern counties.

Ilex opaca

Ilex verticillata

Nemopanthus mucronata

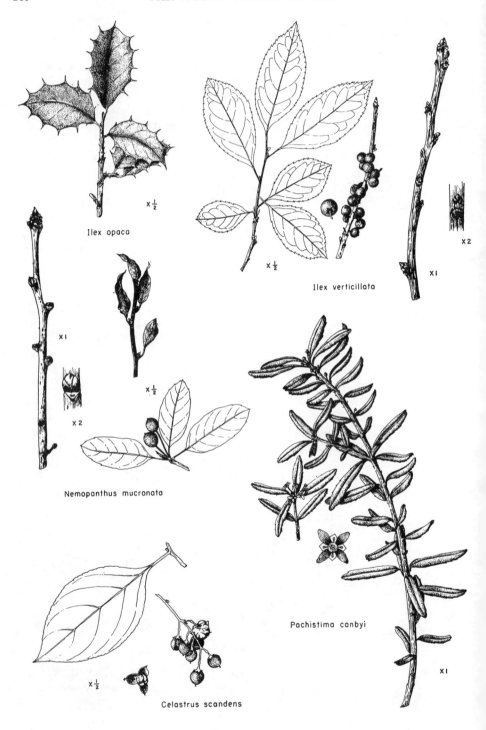

Ilex opaca

$x\frac{1}{2}$

Ilex verticillata

$x\frac{1}{2}$

X2

X1

X1

X2

$x\frac{1}{2}$

Nemopanthus mucronata

Pachistima canbyi

X1

$x\frac{1}{2}$

Celastrus scandens

CELASTRACEAE

A family of woody plants of about 40 genera (three in our range) and 400–500 species, widely distributed through both hemispheres, except in arctic latitudes. The small flowers have a prominent fleshy disk on the margin of which the stamens are inserted; the seeds bear arils, appendages at or about the scar or point of attachment, and sometimes surrounding the seed.

EUONYMUS L. (*Evonymus*)

A large genus (about 120 species) of bushy, climbing, or tree-like species, most abundant in eastern Asia; only five occur in the United States, three in the East, two in the West. Leaves opposite; stems 4-sided, green in our species; flowers in few- to several-flowered cymes; capsules 3–5 lobed, when splitting exposing the "fleshy" seeds (seeds enclosed in a fleshy red aril). Many are planted for their ornamental fruit or foliage; a few of these appear sporadically as escapes, especially *E. europaeus* L., similar in habit, foliage, and fruit to *E. atropurpureus*; *E. alatus* (Thunb.) Sieb., a tree-like shrub characterized by the thin but broad corky wings on branches and branchlets; and *E. fortunei* (Turez.) Hand-Mazz., a shrubby, climbing (by aerial roots), or trailing species (many forms) with half-evergreen leaves and smooth, subglobose 4-parted fruit.

Tree-like shrub; leaves elliptic to ovate-lanceolate, on petioles 1–2 cm. long; flowers brown-purple, 4-merous; fruit red, smooth, 4-lobed, the aril bright red.........1. *E. atropurpureus*
Not tree-like; petioles short, 1–5 mm. long; flowers greenish, 5-merous; fruit pink-red, tuberculate, 3–5 lobed, the aril red.
Bushy, to 2 m., with slender divergent branches; leaves lanceolate to narrow-ovate, 2–4 times as long as wide, acute, almost sessile (petioles 1–3 mm. long).......2. *E. americanus*
Prostrate, with erect or leaning branches 2–5 dm. long; leaves oblong to obovate, often less than twice as long as wide, obtuse, on petioles 2–5 mm. long...............3. *E. obovatus*

1. **Euonymus atropurpureus** Jacq. WAHOO. BURNING-BUSH

Wide-ranging, extending west to Montana and Oklahoma; in woods and on sunny banks (here fruiting most prolifically). A tree-like shrub, showy in autumn after leaf-fall, when the brilliantly colored angular 4-lobed fruits are most conspicuous. The bright red flesh-covered seeds are poisonous.

2. **Euonymus americanus** L. STRAWBERRY-BUSH. HEARTS-BURSTING-OPEN-WITH-LOVE

A southern species occurring in only a few counties of southern Ohio; reports from more northern counties should be referred to *E. obovatus*, many years ago known as *E.*

Euonymus atropurpureus

Euonymus americanus

Euonymus obovatus

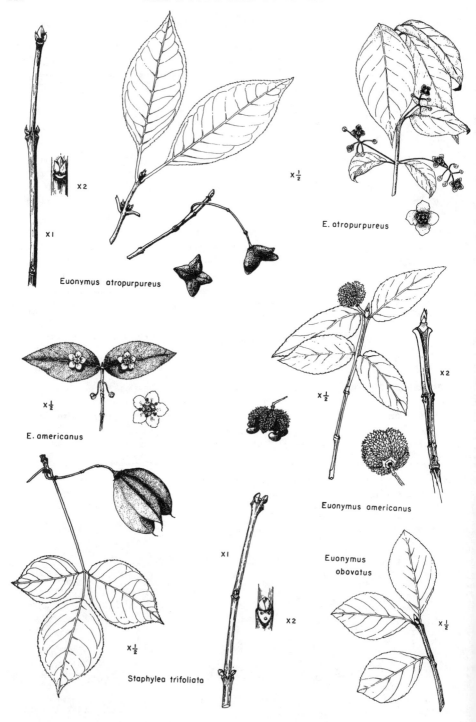

X2

X1

Euonymus atropurpureus

X½

E. atropurpureus

E. americanus

X½

X½

X2

Euonymus americanus

X1

Euonymus
obovatus

X2

X½

Staphylea trifoliata

X½

americanus var. *obovatus*. Shrubby, sometimes forming large straggly clumps 1–2 m. high, the 5-lobed tuberculate fruit very ornamental.

3. **Euonymus obovatus** Nutt. Running Strawberry-bush
A procumbent shrub, sometimes forming extensive patches. More northern in range than *E. americanus*; in rich woods on alluvial flats, till plains, and occasionally on rocky slopes; widespread in Ohio, but local in the Unglaciated Allegheny Plateau section.

PACHISTIMA Raf. (*Pachystima*)
Two species, ours and *P. myrsinites* Raf. of the Rocky Mountains and Northwest. Low evergreen shrubs with opposite serrulate to entire leaves, and small axillary flowers.

1. **Pachistima canbyi** Gray. Cliff-green. Canby's Mountain-lover
Known in Ohio from only a few patches, each perhaps a single clone of great age; on rocky (dolomite) knolls near Ohio Brush Creek. This plant is local throughout its range; a large proportion of the known localities are in the New-Kanawha-Ohio drainage, i.e. in the drainage basin of the preglacial Teays River whose old valleys are prominent topographic features of southern Ohio. The type locality is near Eggleston, Giles Co., Va., where it grows at the top of limestone bluffs of the New River. It was first discovered in 1858 by William Canby, but not described until 1873 (Massey, 1940). A low shrub, 1–3 dm. tall, patches spreading centrifugally by root-stalks and rooting of decumbent stems; flowers very small, in March or early April, on slender pedicels; petals rich dark red, calyx lobes red or tinged with red on outer surface (our manuals state that flowers are green); fruit leathery, 2-valved, about 4 mm. long.

CELASTRUS L. Bittersweet
A genus of ornamental twining shrubs, with alternate leaves and (usually) yellow to orange 3-valved globose capsules splitting at maturity and disclosing the crimson seeds (seeds enclosed in a fleshy aril); about 50 species, the majority in Asia. The oriental bittersweet, *C. orbiculata* Thunb., is often planted; distinguished from our native species by its broader crenate leaves, and axillary inflorescence.

1. **Celastrus scandens** L. Bittersweet. Waxwork
Climbing on fences, over shrubs, or high into trees; most conspicuous and best known in fruit. Inflorescence a terminal panicle; leaves alternate, oblong to ovate-oblong, finely serrate, acuminate. Wide-ranging, from Quebec and Ontario to Manitoba and Wyoming, southward to the Gulf slope.

Pachistima canbyi

Celastrus scandens

Staphylea trifolia

STAPHYLEACEAE

A small family of trees or shrubs; five genera with about 25 species; one genus represented in the United States.

STAPHYLEA L. BLADDERNUT

Shrubs or small trees of temperate North America and Eurasia. Flowers white, 5-merous, in drooping panicles terminating lateral branchlets; leaves pinnate, with 3 leaflets (in ours); fruit an inflated balloon-like capsule containing a few globular bony seeds. One eastern species, another in California.

1. **Staphylea trifolia** L. BLADDERNUT

An erect shrub spreading by root-shoots and forming large patches. Bark striped; branches forking dichotomously because of absence of terminal bud; leaves opposite, serrate, long-petioled, 3-foliate, the terminal leaflet long-stalked. The large inflated pendant capsules, 3-lobed at apex, are conspicuous all summer. A widespread shrub, in Ohio most abundant on shaded banks.

ACERACEAE

A family of about 120 species in two genera: *Dipteronia* with two species in China, and *Acer*, in North America, Asia, Europe, and North Africa. All are trees or shrubs with opposite leaves, usually 5-merous flowers, and flat winged fruit splitting into two samaras.

ACER L. MAPLE

Maples are confined to temperate latitudes of the northern hemisphere. Thirteen species (some of which include several varieties) in the United States reach tree size, of which nine are in the East. A number of species from Europe and Asia are in cultivation, especially Norway maple, *A. platanoides* L., frequently planted as a street tree, sycamore maple, *A. pseudoplatanus* L., hedge maple, *A. campestre* L., a low rounded tree—all three introduced from Europe—and the shrub-like *A. ginnala* Maxim. and Japanese maple, *A. palmatum* Thunb. (with a bright red form) introduced from Asia. The range of floral structure in species of *Acer* leads to recognition of sections of the genus. Staminate, pistillate, and perfect flowers occur in most species; the staminate and pistillate usually in different racemes on the same plant in *A. pensylvanicum*, in the same raceme in *A. spicatum* (the pistilate toward base), in the same or in different clusters on the same tree or on different trees in *A. saccharum* and *A. nigrum*, in separate clusters on the same or on different trees in *A. rubrum* and *A. saccharinum*, on different trees in *A. negundo*. Leaves in all our species except *A. negundo* simple and palmately lobed or cleft; buds, leaves, and branchlets opposite.

a. Leaves simple, palmately lobed.
   b. Inflorescence a panicle or raceme; leaves serrate, the lobes acute to acuminate, the sinuses between the lobes angular.
      Inflorescence a slender drooping glabrous raceme; leaves 3-lobed, glabrous or nearly so at maturity, finely and doubly serrate; samaras 2.5–3 cm. long, without noticeable veins over seed; bark of young trunk and branches striped.............1. *A. pensylvanicum*
      Inflorescence paniculate; leaves coarsely toothed.
         Panicles erect or ascending; leaves 3 (–5)–lobed, pubescent, coarsely glandular serrate; samaras 1.5–2.5 cm. long, often red, reticulate-veined over seed; bushy tree........
                                          2. *A. spicatum*
         Panicle drooping; leaves 5-lobed, glabrous, coarsely crenate-serrate; samaras 3–4 cm. long, not reticulate veined over seed; tree, from Europe...........*A. pseudoplatanus*

bb. Inflorescence not racemose or paniculate; tall trees.
    Leaves 3–5-lobed, with open V-shaped angular sinuses between lobes; the lobes usually less than half the leaf length, coarsely toothed, pubescent beneath when young, at maturity glabrous above, below white, glaucous, or pubescent; petioles red; flowers very early, long before leaves, scarlet, sessile or short pedicelled in rounded lateral clusters; samaras red, on slender pedicels, ripening in spring, 1.5–2.5 cm. long; branchlets red..3. *A. rubrum*
    Leaves with rounded sinuses between lobes (sinuses narrow or subacute in no. 4).
        Leaves 5-lobed, with narrow to subacute sinuses, the bordering leaf-margin slightly concave, lobed to below middle, irregularly and remotely dentate, silvery-white beneath; flowers early, long before leaves, greenish- or yellowish-red, in rounded lateral clusters; samaras green to straw-colored, on slender pedicels, ripening in spring, 4–8 cm. long; branchlets greenish-red, curved upwards toward tip.........
                                           4. *A. saccharinum*
        Leaves 3–5 (–7)-lobed; flowers appearing with the leaves; fruit ripening in late summer; branchlets gray or brown.
            Petioles with milky juice; flowers in erect corymbs; samaras spreading horizontally (diverging at angle of 180°); exotic species.
                Leaves large, 1–2 dm. across, acuminate, sinuses broad; flowers yellow (reddish in a horticultural form), in many-flowered erect open glabrous corymbs; samaras 3.5–4.5 cm. long...*A. platanoides*
                Leaves smaller, 5–10 cm. across, obtuse, sinuses narrow, sometimes angular; flowers yellowish-green, in few-flowered erect pubescent corymbs; samaras 2.5–3.5 cm. long...*A. campestre*
           Petioles without milky juice; flowers light yellow, on filiform pendulous pedicels, from terminal leaf-buds and lateral leafless buds; samaras divergent but not spreading horizontally; native.
                Leaf-blades flat, often pale beneath, basal sinus broad, open; petioles slender, gradually thickening toward base; year-old branchlets light brown, lustrous like varnished wood...5. *A. saccharum*
                Leaf-blades curved downwards at sides, green beneath, basal sinus closed or nearly so; petioles stout, abruptly thickening near base and enclosing axillary buds, often with foliar appendages; year-old branches light brown but with whitish overcast..6. *A. nigrum*
aa. Leaves pinnately compound, leaflets 3–7, irregularly coarsely serrate or slightly lobed; staminate flowers on filiform pedicels, fascicled; pistillate in racemes; samaras 3–4 cm. long, the wings ascending or converging; year-old branchlets green, or purplish-red where exposed to sun....
                                        7. *A. negundo*

WINTER KEY

a. Terminal buds usually large, 7–15 mm. long.
  b. Buds broadly ovoid or spherical; planted species.
    Buds reddish; opposite leaf-scars meeting............................*A. platanoides*
    Buds green; outermost scales sometimes brownish: opposite leaf-scars not meeting.....
                                            A. pseudoplatanus
  bb. Buds narrowly ovoid or oval.
    Buds bright red, stalked; bud-scales keeled.......................1. *A. pensylvanicum*
    Buds brown, not stalked; lower scales long-ciliate; low tree-like exotic shrub..*A. palmatum*
aa. Terminal buds smaller, 7 mm. or less in length.
  b. Lower bud-scales long-ciliate; lower margin of leaf-scar much elevated, forming a flaring collar about the bud; tree-like exotic shrub.............................*A. palmatum*
  bb. Lower bud-scales not ciliate (inner ciliate in some species); leaf-scars without prominently elevated margins.
    Buds white-tomentose, some nearly spherical; opposite leaf-scars meeting; twigs usually bright green (purple-green where exposed to sun); pistillate trees with pendant racemes of samaras..7. *A. negundo*
    Buds not white-tomentose.
      Terminal buds acutely pointed at apex.
        Twigs red, appressed hairy near tip; buds red.....................2. *A. spicatum*
        Twigs brown or grayish-brown; buds brown or almost black.
          Twigs brown with brown buds; hairs at upper edge of leaf scar brown.........
                                        5. *A. saccharum*
         Twigs grayish or dull brown with dark brown or almost black buds; hairs at upper edge of leaf-scar pale.......................................6. *A. nigrum*

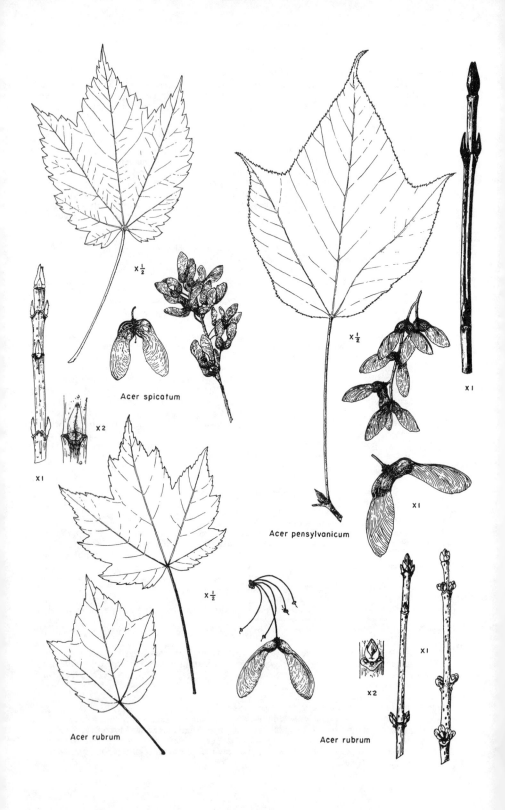

Acer spicatum

Acer pensylvanicum

Acer rubrum

Acer rubrum

$\times \frac{1}{2}$

$\times \frac{1}{2}$

$\times \frac{1}{2}$

$\times \frac{1}{2}$

X2

X1

X1

X1

X1

X2

Terminal buds rounded or blunt pointed at apex.
  Bud scales glabrous.
    Terminal bud about 4–6 mm. in length; spherical flower buds usually present, clustered at nodes.
      Twigs usually turned upward at tip; bud-scales apiculate, inner scales of swelling flower-buds plainly white-ciliate; twigs with rank odor when broken
.... 4. *A. saccharinum*
      Twigs not curving upward at tip; bud-scales not apiculate, inner scales of swelling flower-buds short-ciliate; twigs without rank odor......3. *A. rubrum*
    Terminal bud small, about 2–3 mm. long; no flower-buds present; tip of scales often bright orange-red; exotic shrub.............................*A. ginnala*
  Bud-scales pubescent, at least toward tip; terminal bud about 4 mm. long; no flower-buds present; small round-headed exotic tree.................*A. campestre*

1. **Acer pensylvanicum** L.   Striped Maple

A small slender tree of the Northeast, ranging southward in the mountains to Georgia. Known in Ohio only from Ashtabula Co. Easily recognized by its striped green bark, leaves with three long-tapering lobes, and pendulous racemes.

2. **Acer spicatum** Lam.   Mountain Maple

Small bushy tree or tall shrub of northern range extending from Newfoundland to eastern Saskatchewan and southward into Iowa, Michigan, Ohio, Pennsylvania, and New Jersey, and in the mountains to northern Georgia. In Ohio, more or less frequent in the northeast, local southward; in gorges and on shaded cliffs, often with hemlock.

3. **Acer rubrum** L.   Red Maple

A large forest tree ranging throughout the Deciduous Forest. In Ohio, common and widely distributed in the Allegheny Plateau, the "flats" of southwestern Ohio, and the Lake area. A tree of mesophytic forest in hilly country, of swamp forest in the flats, in both situations in acid to neutral soils. Conspicuous in early spring because of its brilliant red flowers and young samaras, in autumn because of its brilliant scarlet to orange coloration. Leaves vary greatly in lobing and coarseness of teeth, some (Adams and Perry counties) approaching var. *trilobum* K. Koch, with 3 shorter small-toothed lobes, some (Clermont County) approaching var. *drummondii* (H. & A.) Sarg., with leaves more deeply cleft (more than half-way) and pubescent on veins beneath.

4. **Acer saccharinum** L.   Silver Maple

A large tree of stream banks and alluvial bottoms, usually where subjected to flooding; fast-growing, and commonly planted as a street tree. The largest Ohio tree of this

Acer pensylvonicum

Acer spicatum

Acer rubrum

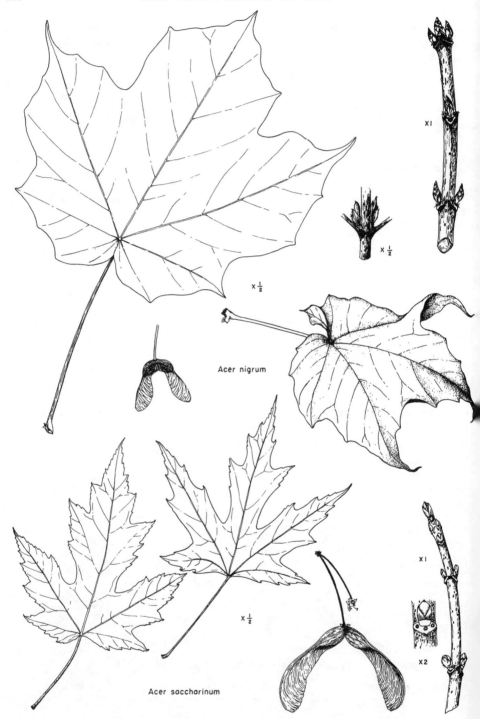

Acer nigrum

Acer saccharinum

species, 24 ft. 2 in. in circumference 4½ ft. above the ground, was found by Karl Maslowski near Utopia, Clermont Co., close to the Ohio River. More yellow in flower than red maple, and with large green samaras becoming straw-colored; leaves turning dull yellow in autumn, rarely with some red coloration. In winter condition, the upward curve of twigs is distinctive. This species and red maple are sometimes called soft maple.

5. **Acer saccharum** Marsh. SUGAR MAPLE

A large forest tree, and one of the dominants of the Beech-Maple Forest region, and of the Hemlock-White Pine-Northern Hardwoods region; in moist, rich, but well drained soil. Widely distributed in Ohio, but less common in the area of the Prairie Peninsula and on the Illinoian Till Plain of southwestern Ohio. Conspicuous in spring because of the numerous yellow flowers in tassel-like clusters; leaves turning yellow to orange and deep red in fall. A valuable timber tree, and the chief source of maple syrup. Variable in leaf shape, and in pubescence and color of the lower leaf-surface, on the basis of which characters several forms and varieties have been named. Of those distinguished by Fernald (1950), forma *rugelii* (Pax) Palmer & Steyerm., with leaves of fertile branches 3-lobed due to suppression of lateral lobes, occurs in our area (recorded from Adams, Hamilton, Highland, Ross, Scioto cos.); var. *schneckii* Rehd., with leaves less deeply lobed, pubescent on petioles and veins of lower leaf-surface, should be looked for. See note about variation under *A. nigrum*.

6. **Acer nigrum** Michx. f. BLACK MAPLE. BLACK SUGAR MAPLE

Similar to *A. saccharum*, but with darker bark, larger leaves drooping at the sides and with narrow or closed basal sinus, heavier petioles much enlarged at base, sometimes with leaf-like outgrowths, velvety pubescent to pilose in some specimens. Like sugar maple, a valuable timber tree and source of maple syrup.

Typical *Acer saccharum* and typical *Acer nigrum* are easily distinguished from one another. However, many trees are seen which have some characters of one, some of the other. This appears to be the result of introgressive hybridization. Studies of the sugar maples over a wide geographic area (22,718 trees in 465 different localities) indicate the desirability of considering *Acer saccharum* to be an inclusive species made up of six subspecies, of which subsp. *saccharum* and subsp. *nigrum* are "two populations, each having its own characteristics, united by a large intermediate population representing a variety of recombinations of the characteristics of the two subspecies. These intermediate indi-

Acer saccharinum

Acer saccharum

Acer nigrum

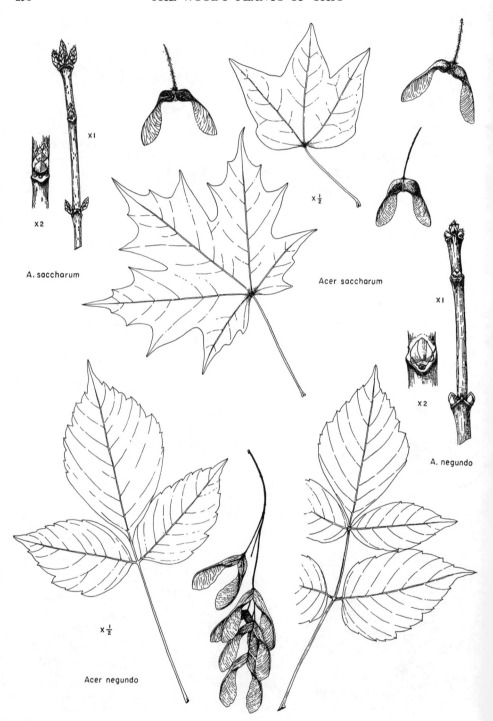

A. saccharum

Acer saccharum

A. negundo

Acer negundo

viduals are grouped together in the cline *saccharum-nigrum*." (Desmarais, 1952). Although 'ranges overlap, *saccharum* extends farther north than *nigrum*, and *nigrum* extends farther west than *saccharum*. Subsp. *schneckii* "is intermediate between *saccharum*, *floridanum* [southern], and *grandidentatum* [western]; it is found along a zone where the ranges of *saccharum* and *floridanum* meet." The forma *rugelii* is included in subsp. *schneckii*. If this modern interpretation of our sugar maples be followed, the nomenclature is:

> *Acer saccharum* Marshall subsp. *saccharum*
>
> subsp. *nigrum* (Michx. f.) Desmarais
>
> subsp. *schneckii* (Rehd.) Desmarais

For further details, see Dansereau and Desmarais, 1947; Desmarais, 1952. A study by Kriebel (1957) based upon experimental plantings supports the contention that the sugar maples, *A. saccharum* and *A. nigrum*, are extreme variants of a single species.

Acer negundo

7. **Acer negundo** L. Box-elder

Our only maple with compound leaves; pistillate trees readily recognized in winter from a distance by the dangling clusters of straw-colored samaras. The species, including its many varieties the races, ranges essentially across the United States and (as var. *interius* (Britt.) Sarg.) into the prairie region of Canada from southwestern Ontario to Alberta; in Canada, it is known as Manitoba maple. Found throughout Ohio, usually in alluvial soil, and sometimes on dry banks. Within our range, three varieties may be recognized:

Twigs green, glabrous; wide-ranging...var. *negundo*
Twigs glaucous and purple beneath the "bloom;" western.........var. *violaceum* (Kirsch.) Jaeg.
Twigs velvety pubescent; southern...var. *texanum* Pax.

The leaves of the typical variety usually glabrous beneath at maturity; those of var. *violaceum* usually with tufts of hairs in the vein-axils; those of var. *texanum* pubescent beneath. Ohio specimens vary in amount of pubescence.

HIPPOCASTANACEAE

A small family of three genera and about 30 species, most of which are in the genus *Aesculus*.

AESCULUS L. Buckeye. Horse-chestnut

Trees or shrubs of North America and Eurasia, with opposite palmate leaves, the lateral veins of the leaflets straight; flowers zygomorphic; fruit a somewhat globose capsule, the thick leathery walls splitting into three parts and exposing the 1–3 large red-brown seeds, each with a light brown circular scar. In addition to our two native species, several others are commonly seen in cultivation: *A. hippocastanum* L., horse-chestnut; *A. parviflora* Walt., a shrub of the Southeastern States with white flowers in upright panicles; and the red-flowered *A. pavia* L. (including *A. discolor* Pursh) native in the South. Neither of our native buckeyes occurs east of the mountains, both are interior in range.

Variation in species of *Aesculus*, although in part geographic, is largely the result of introgression resulting from hybridization.   *A. glabra* and *A. octandra* have hybridized in the past, and occasionally first-generation hybrids are found; many intermediates occur, which are the result of back-crosses with one of the parent species.   (For detailed study of hybridization in *Aesculus*, see Hardin, 1957 a, b.)

Aesculus glabra

The ranges of our two species of buckeyes display interesting correlations with the glacial boundary.   *A. glabra*, a calcareous soil species, is in every county of glaciated Ohio except in the northeast.   In unglaciated southeastern Ohio, a spot-map does not show the true distribution pattern; here, it seems to occur only where there is calcareous glacial outwash or exposures of calcareous rock (Beatley, 1959), thus is limited to certain drainages. On the other hand, *A. octandra* is limited to unglaciated southeastern Ohio and the Ohio valley at the margin of Illinoian glaciation. Reports north of this area cannot be substantiated.

Winter-buds glutinous; leaflets usually 7, cuneate-obovate, irregularly or doubly serrate; flowers showy, white marked with red, the petals cordate at base and clawed; fruit about 5 cm. in diam., echinate; planted tree...1.   *A. hippocastanum*
Winter-buds not glutinous; leaflets usually 5 (sometimes 6–7), narrow-elliptic to oblong-obovate, finely serrate; flowers greenish-yellow, yellow, or yellow and reddish, the petals not cordate.
  Fruit echinate, prickly when young, 3–4 cm. in diam.; flowers greenish-yellow, petals 4, about equal; stamens exserted; pedicels and calyx glabrous or finely tomentulose; year-old twigs branchlets with strong odor..2.   *A. glabra*
  Fruit not echinate, 5–6 cm. in diam.; flowers yellow or in part reddish to reddish-purple, petals 4, very unequal; stamens usually shorter than petals; pedicels and calyx glandular-villous...3.   *A. octandra*

WINTER KEY

Winter-buds glutinous..1.   *A. hippocastanum*
Winter-buds not glutinous.
  Bud-scales keeled (at least the outer 2–3 pair), longitudinally striate, apiculate; median leaf-scars triangular, their sides but slightly convex, gray-tan; year-old twigs brownish-gray, with skunk-odor when freshly cut...2.   *A. glabra*
  Bud-scales not keeled (or smaller ones with suggestion of keel), scarcely striate, apexes rounded, the heavy central point slightly apiculate or emarginate; median leaf-scars with more convex sides, thus rounded, orange-tan; year-old twigs yellow-brown, without fetid odor...3.   *A. octandra*

1. AESCULUS HIPPOCASTANUM L. HORSE-CHESTNUT
  Frequently planted for shade and ornament, rarely self-seeding; a dense round-crowned tree, native of the Balkan Peninsula and adjacent Asia. Fruit well-armed with stout sharp prickles up to 1 cm. long.

2. **Aesculus glabra** Willd. OHIO BUCKEYE
  A medium size forest tree, flowering and fruiting when very young; most abundant in second-growth woodlands and thickets; bark of older trees dark, furrowed. Range interior, from western Pennsylvania to southern Michigan and eastern Nebraska, south to Alabama, Arkansas, and Oklahoma. In Ohio, widely distributed, but less frequent eastward; largely confined to calcareous soil. To residents of the "Buckeye State," the origin of the common name, Ohio buckeye, may be of interest. Writing of early botanical explorations, F. Michaux (1818) in North American Sylva, states, "It is unknown in the Atlantic parts; I have found it only beyond the mountains and particularly on the banks of

Aesculus glabra

the Ohio . . . where it is common. It is called "Buckeye" by the inhabitants, but as that name has been given to the Yellow Buckeye (*lutea*) I have called it 'Ohio Buckeye'." Thus the ill-smelling or fetid buckeye became "Ohio buckeye."
  Individuals of **A. glabra** having the capsules scarcely echinate or with prickles confined to one section of capsule, or having pedicels with stipitate glands (a character of *A. octandra*) give evidence of introgression from *A. octandra*. As interpreted by Hardin, only one variety, *glabra*, occurs in Ohio; var. *sargentii* (credited to Ohio by Fernald, 1950) and reported from Clinton and Lorain counties, is not recognized as distinct; forms com-

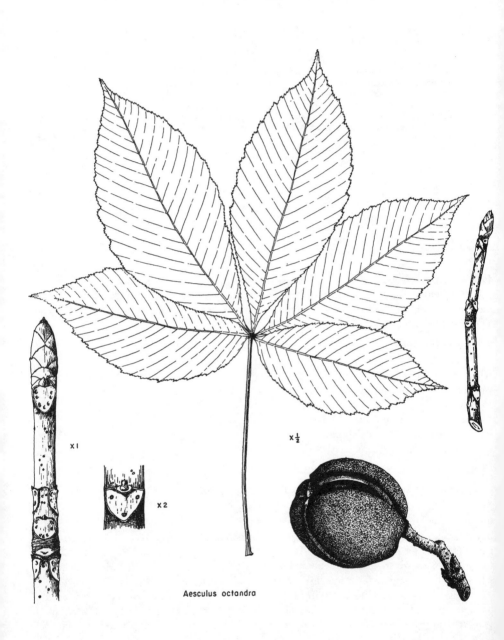

x1

x2

x½

Aesculus octandra

monly referred to this variety appear somewhat intermediate between var. *glabra* and the western var. *arguta* of Kansas, Oklahoma, and Texas.

3. Aesculus octandra Marsh. SWEET BUCKEYE.

YELLOW BUCKEYE

Aesculus octandra

One of our largest forest trees, a constituent of the Mixed Mesophytic Forest climax. The bark of old trees is light gray, splitting into more or less circular, oval, or quadrate plates somewhat concentrically marked. Range circumscribed, largely confined to the upper Ohio Valley and southern Appalachian area; most abundant in the Mixed Mesophytic Forest region and in the cove hardwoods and higher altitude hardwood forests of the Southern Appalachians (Oak-Chestnut Forest region); local west of the Appalachian Plateau; known in one county of southern Illinois as a shrub. In Ohio, in unglaciated southeastern Ohio, and westward near the Ohio River just within the limits of Illinoian glaciation. Daniel Drake, in his *Picture of Cincinnati and the Miami Country* (1815) wrote: "This species delights in rich hills, and is seldom seen far from the Ohio River. It frequently arrives at the height of 100 feet and the diameter of four feet." The largest known Ohio tree (in Washington Co.) measures 14 ft. 8 in. in circumference at 4½ ft. above the ground (Ohio Forestry Assn.) Not readily distinguished by foliage alone from *A. glabra*; flowers and fruit are distinctive, as is also the curiously marked bark; not ill-scented.

Because of the occurrence of hybrids between *A. octandra* and *A. glabra*, some of which have been incorrectly identified as *A. octandra*, the range of this species has been given as more extensive than it actually is (Sargent, 1922; Little, 1949; Fernald, 1950); Richland County specimens are *A. glabra* (× *octandra*) according to Hardin; that is, are best determined as *A. glabra* but with some genes of *A. octandra*.

SAPINDACEAE

A large family of over 100 genera and 1000 species, chiefly confined to the tropics. One Asiatic species, *Koelreuteria paniculata* Laxm., Golden Raintree, with bipinnate coarsely toothed leaves, broad panicles of yellow flowers, and inflated papery ovoid capsules about 5 cm. long, is planted as an ornamental tree.

RHAMNACEAE

Trees, shrubs, or woody climbers with simple, alternate or opposite leaves and small flowers. About 45 genera and 600 species, cosmopolitan in distribution.

RHAMNUS L. BUCKTHORN

Shrubs or small trees with small greenish flowers solitary or in clusters in the axils of lower leaves of the current season's growth. Fruit a drupe containing 2–4 seed-like nutlets. About 100 species, chiefly native to the north-temperate zone, a few in South Africa and Brazil. Many species have spinescent branchlets.

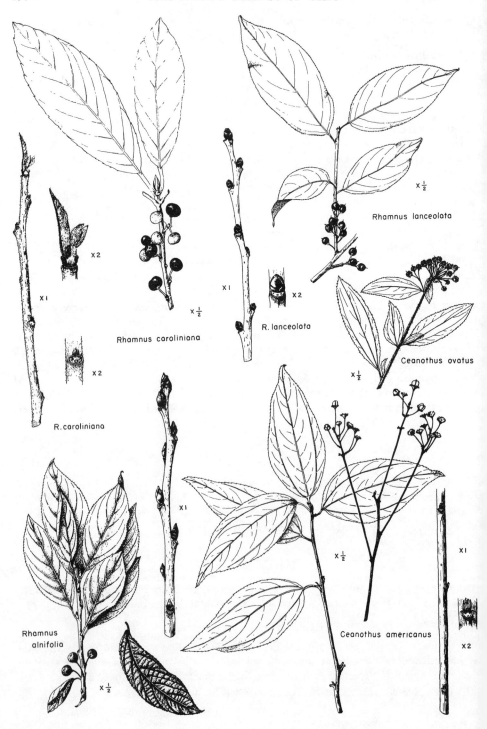

Rhamnus lanceolata

x½

R. lanceolata

x2

x1

Rhamnus caroliniana

x½

x2

R. caroliniana

Ceanothus ovatus

x½

x1

x½

Rhamnus
alnifolia

x½

Ceanothus americanus

x1

x2

Leaves opposite, broad-ovate to elliptic, usually with 3 lateral veins on each side of midrib; flowers mostly dioecious, 4-merous; winter buds scaly.....................1.   *R. cathartica*
Leaves alternate, with more than 3 pairs of lateral veins.
  Winter-buds naked; flowers perfect, appearing after the leaves; leaves lustrous green above, with 8–10 pairs of lateral veins.
    Leaves obscurely serrulate, about 3 times as long as wide; petioles 6–16 mm. long; southern
                                                             2.   *R. caroliniana*
    Leaves entire, usually less than twice as long as wide; petioles 6–12 mm. long; introduced
                                                             3.   *R. frangula*
  Winter-buds scaly; flowers perfect or dioecious, opening with the leaves; leaves with 6–9 pairs of lateral veins; petioles 3–8 (–12) mm. long.
    Leaves finely serrulate, ovate- to oblong-lanceolate, acute to acuminate (or obtuse on flowering branchlets); flowers 4-merous; drupe 2-seeded................4.   *R. lanceolata*
    Leaves unequally crenate-serrate, more or less oval; flowers 5-merous; drupe 3-seeded...
                                                             5.   *R. alnifolia*

WINTER KEY

Buds opposite or subopposite; some twigs spine-tipped.....................1.   *R. cathartica*
Buds alternate; twigs not spine-tipped.
  Buds naked, silky-hairy, tip-bud 5–10 mm. long; veins visible on "leaves" of terminal bud.
    Buds pale grayish ocherous, about 10 mm. long, loose; youngest twigs light brown, soon gray; lenticels inconspicuous; pedicel-scars on short axillary peduncle...2.   *R. caroliniana*
    Buds ocherous-brown to sienna, 5–6 mm. long; youngest twigs dull dark brown, becoming red-brown; lenticels conspicuous; pedicel-scars on scarcely elevated base...3.   *R. frangula*
  Buds with definite overlapping bud-scales, glabrous; tip-bud 2–5 mm. long.
    Buds bright brown, tip-bud about 2 mm. long; twigs dull brown..........4.   *R. lanceolata*
    Buds dark red-brown, tip-bud 4–5 mm. long; twigs red-brown, or gray before epidermis scuffs off....   ....5.   *R. alnifolia*

1. RHAMNUS CATHARTICA L.   EUROPEAN BUCKTHORN
    A coarse shrub with spine-tipped branchlets.   Introduced from Europe; reported from three counties: Greene, Hamilton, Lucas.   Fruit black when ripe.

2. **Rhamnus caroliniana** Walt.   CAROLINA BUCKTHORN.   INDIAN-CHERRY
    A tall shrub or small tree of the South, extending north into southern Ohio and southern Indiana.   Handsome in September with glossy-green foliage and bright red (unripe) fruit resembling holly berries; ripe fruit black.   Although the habitat of this species is given as "rich woods and sheltered slopes" and "moist woods and alluvial soil", we find it on dolomite ledges, dry rocky south slopes, and borders of prairie patches.
    Leaves vary in amount of pubescence on lower surface:

Leaves sometimes slightly pubescent when young, soon glabrous.................var. *caroliniana*
Leaves permanently velvety pubescent beneath..............................var. *mollis* Fern.

Rhamnus caroliniana

Rhamnus lanceolata

Intermediates occur; we have only one record of var. *mollis*, from a post oak—blackjack oak woodland in Adams County.

3. RHAMNUS FRANGULA L.   EUROPEAN ALDER BUCKTHORN
Introduced from Europe, and rapidly becoming naturalized in northern Ohio; reported from nearly a score of northern counties, and locally well established.   Fruit red, changing to purple-black.

4. **Rhamnus lanceolata** PURSH   LANCE-LEAF BUCKTHORN
A tall shrub of gravelly banks, woodland borders, and ravines, usually in calcareous soil.   The abundant small yellow-green flowers of two forms: one with short included style and short clustered pedicels, the other with long exserted style and few or solitary in the axils; fruit black.   Southern in range, but extending somewhat farther north than *R. caroliniana*.   Two varieties, based on differences in pubescence, are recognized:

Young leaves and branchlets pubescent; leaves permanently pubescent beneath...var. *lanceolata*
Young leaves and branchlets glabrous or nearly so; leaves glabrous or glabrate beneath......
              var. *glabrata* Gleason

5. **Rhamnus alnifolia** L'Her.   ALDER BUCKTHORN
A low shrub of swamps and bogs, ranging across the continent in the North.   In the bogs of northeastern Ohio, and in Cedar Swamp, Champaign County.

CEANOTHUS L.

Shrubby North American plants, about 500 species, chiefly in the Pacific Coast region southward to Mexico.   Flowers small, white (in ours), blue or pinkish in many western species, perfect, 5-merous, the petals clawed; in terminal or axillary panicles made up of small umbel-like clusters; fruit 3-lobed, falling at maturity from the cup-like persistent hypanthium; leaves alternate, 3-nerved, serrate.

Leaves ovate to ovate-oblong, acute or acuminate (rarely obtuse); panicles on axillary peduncles
    progressively longer downward...1.   *C. americanus*
Leaves elliptic to elliptic-lanecolate, obtuse or acutish; peduncles short, terminating leafy shoots
    of the season...2.   *C. ovatus*

1. **Ceanothus americanus** L.   NEW JERSEY-TEA
An attractive summer-flowering shrub up to 1 m. tall, easily recognized by its clusters of small white flowers on long axillary peduncles (these occasionally bearing one or two

Rhamnus alnifolia   Ceanothus americanus   Ceanothus ovatus leaves) and 3-nerved leaves; flowering stems new each year. A wide-ranging species (Que. to Minn., south to Fla. and Tex.) in which several varieties have been distinguished:

Leaves mostly 5–10 cm. long.
Leaves glabrous or nearly so beneath, except on veins.....................var. *americanus*
Leaves velvety pubescent on the lower leaf-surface....................var. *pitcheri* T. & G.
Leaves mostly less than 5 (2–4, –6) cm. long, glabrous; plant much branched..............
var. *intermedius* (Pursh) K. Koch

The typical variety is best developed east of the mountains; var. *pitcheri* in the prairies from Illinois westward and southwestward. Between the two extremes (and within our range), these two varieties intergrade; some specimens have leaves nearly glabrous, others are referable to var. *pitcheri*. The third variety is southern and probably does not reach our area. New Jersey-tea was used as a substitute for oriental tea during the American Revolution. As this plant does not contain caffeine, it does not have the stimulating effect of tea.

2. **Ceanothus ovatus** Desf.
A low shrub of generally western range, occuring locally in four northern Ohio counties. The amount of pubescence on the lower leaf-surface varies; in forma *pubescens* (S. Wats.) Soper the leaves are sordid-villous below; this occurs in Lake Co.

VITACEAE

A family of about 600 species, many tropical or subtropical in distribution; of the ten genera, three are represented in the Ohio flora. Woody plants, mostly climbing by tendrils, the tendrils and usually the inflorescences opposite the leaves; leaves simple and palmately veined or compound; flowers 4-5-merous, small, greenish; fruit a berry.

AMPELOPSIS Michx.

Tendril-bearing vines of North America and central and eastern Asia; distinguished from *Vitis* by the close bark with numerous lenticels, white pith, and flowers with 5 petals expanding and free at anthesis, not connate at tips. In addition to the one species native to southern Ohio, the Asiatic *A. brevipedunculata* (Maxim.) Trautv. (*A. heterophylla* Sieb. & Zucc.), with 3-lobed leaves, is reported as an escape on railroad embankments in Lake Co. *A. arborea* (L.) Koehne, a southern species with leaves twice pinnate or ternate, is often planned. Name from the Greek *ampelos*, grape, and *opsis*, likeness or appearance, referring to the similarity to grape.

Ampelopsis cordata

1. **Ampelopsis cordata** Michx.
Cissus Ampelopsis Pers.
A high-climbing vine of alluvial woodlands from South Carolina, Florida, and the Gulf Coast northward in the valley of the Mississippi River and its tributaries to southeastern Kansas, and to southern Illinois, Indiana (especially lower Wabash valley) and Ohio (where it is known from a few Ohio River counties). Leaves simple, heart-shaped, and coarsely toothed; berries bluish or greenish-blue.

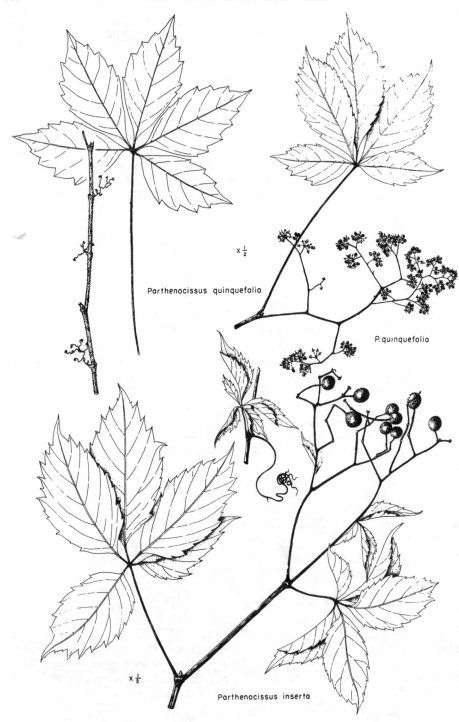

x ½

Parthenocissus quinquefolia

P. quinquefolia

x ½

Parthenocissus inserta

PARTHENOCISSUS Planch.   (*Psedera* Neck.)   Virginia Creeper.   Woodbine

The American species of this small genus (10 species) of North America and eastern Asia have the leaves palmately compound, typically with 5 sharply serrate leaflets, hence the name "five-leaved ivy" sometimes used for these climbers.   The commonly planted Boston-ivy, *P. tricuspidata* (Sieb. & Zucc.) Planch., has the leaves of basal sprouts 3-foliate, the others 3-lobed.   Tendrils of most species end in adhesive disks; flowers in compound cymes, these sometimes panicled.

Leaves dull above, leaflets subsessile or with petiolules to 10 mm. long; much-branched tendrils ending in adhesive disks; inflorescence many-flowered (25–200), with definite central axis; berries 5–7 mm. in diam., with 1–3 seeds..1.   *P. quinquefolia*
Leaves lustrous above, leaflets on petiolules to 3 cm. long; tendrils with only 3–5 branches, without adhesive disks; inflorescence dichotomously branched, fewer flowered (10–60); berries 8–10 mm. in diam. with 3–4 seeds...2.   *P. inserta*

1.   **Parthenocissus quinquefolia** (L.) Planch.   Virginia Creeper
Psedera quinquefolia (L.) Greene
Widely distributed throughout the Deciduous Forest in a variety of habitats.   Most conspicuous in early autumn because of the brilliant coloration of the leaves; branches of inflorescence red in late fall, contrasting with the blue-black berries.   Sometimes confused with poison-ivy, from which it is readily distinguished by growth habit: stems of that species cling by aerial roots, no tendrils are produced; leaflets are 3 instead of 5, few-toothed (dentate) or entire, not sharply serrate, berries whitish instead of blue-black.   Variation in pubescence, leaf-shape and serration, and branching of inflorescence of *P. quinquefolia* has lead to designation of several forms and varieties of which forma *hirsuta* (Donn) Fern. (pubescent) occurs in our area.

2.   **Parthenocissus inserta** (Kerner) K. Fritsch
P. vitacea (Knerr) Hitchc.
More northern and western than our common Virginia creeper.   Apparently rare in Ohio, but to be expected in the western and northwestern part of the state, as it is "rather

Parthenocissus quinquefolia

Parthenocissus inserta

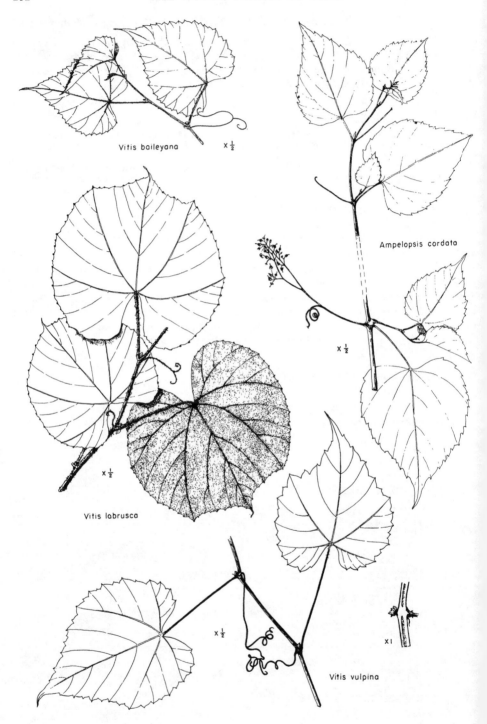

Vitis baileyana $\times \frac{1}{2}$

Ampelopsis cordata

$\times \frac{1}{2}$

Vitis labrusca $\times \frac{1}{2}$

Vitis vulpina $\times \frac{1}{2}$ $\times 1$

frequent in the open throughout the lake area" in Indiana. Rarely high-climbing, more often growing as a ground-cover or sprawling over bushes. The dichotomous inflorescence and larger often 4-seeded berries are distinctive.

VITIS L. GRAPE

A genus of some 60 species, mostly in the temperate regions of the northern hemisphere. The Mediterranean *V. vinifera* L. is widely cultivated in California. Eastern cultivated grapes, derived chiefly from our native *V. labrusca*, sometimes with an admixture of *V. vinifera*, are known collectively as *V. labruscana* Bailey. Although hybrid grapes are known, they appear to be uncommon in nature; however, some anomalous forms may be descendants of hybrids, back-crossed to one of the parents but carrying genes of both. As the number of successive nodes at which tendrils and/or panicles are produced is sometimes diagnostic, specimens should be collected with stems long enough to show this feature. Flowers in panicles, very fragrant (5 nectar-bearing glands alternate with the stamens), functionally unisexual, their petals cohering at the tips, falling without expanding. The brown pith is interrupted at the nodes (except in *V. rotundifolia* Michx., south of our area) by diaphragms which differ in thickness in different species, hence specimens cut longitudinally through one of the nodes are desirable. The bark of all our species separates in long strips and fibers. Leaves various, those of long vegetative shoots frequently more lobed than others.

a. Leaves permanently felted beneath with dense rusty (occasionally grayish) tomentum concealing the leaf-surface; tendrils or flower-clusters at 3 or more successive nodes; fruit large, 1.5–2.5 cm. in diam..1. *V. labrusca*
aa. Leaves glabrous beneath, or with thin or cobwebby flocculent pubescence, not concealing the leaf-surface except on young leaves; tendrils or flower-clusters intermittent, produced at 2 successive nodes, the third with none; fruit smaller, usually less than 1.0 cm. in diam.
  b. Leaves green and glabrous beneath at maturity except the vein-axils and sometimes veins; pubescent when young with short erect hairs, not cobwebby.
    Leaves not lobed, or with short shoulder-like lobes (if lobed, the lobes point outward, forming a wide angle with terminal lobe); teeth irregular, their sides generally convex, converging at more than 90°; basal sinus narrow; diaphragms at stem-nodes 2–6 mm. thick; berries black, without bloom, about 8 (4–10) mm. in diam........2. *V. vulpina*
    Leaves with two lateral acute lobes, the lobes forming an acute angle with the terminal, the teeth sharp, more acute, the margin ciliolate; basal sinus wide; diaphragms 0.5–1 (–2) mm. thick; berries black, with heavy bloom, about 9 (6–12) mm. in diam.......
                                                     3. *V. riparia*
  bb. Leaves glaucous and strongly whitened or pubescent beneath at maturity, at least on veins.
    Young branches and stem-tips terete, these and petioles glabrous, or with persistent reddish tomentum (whitened beneath in var. *argentifolia*); leaves roundish in general outline, shallowly to deeply 3–5–lobed; diaphragms at nodes 3–4 mm. thick; berries black or purple-black, more or less glaucous, about 8 (5–11) mm. in diam...........
                                                  4. *V. aestivalis*
    Young branches and stem-tips angled, these and petioles permanently pubescent with grayish hairs; berries small.
    Leaves broad-ovate to roundish in general outline, unlobed or with 2 short lateral lobes, terminal lobe somewhat prolonged; diaphragms 3–4 mm. thick; berries black, without or with slight bloom, about 7 (4–9) mm. in diam., inflorescence open......
                                                  5. *V. cinerea*
    Leaves ovate to orbicular-ovate in general outline, unlobed or 3-lobed near apex, the lobes short; berries black, about 6 (4–7) mm. in diam., inflorescence compact......
                                                  6. *V. baileyana*

Certain characters from the above key—especially number of nodes with tendrils, thickness of nodal diaphragms, and angularity of stems—can be used for winter identification, but will not in all cases be conclusive.

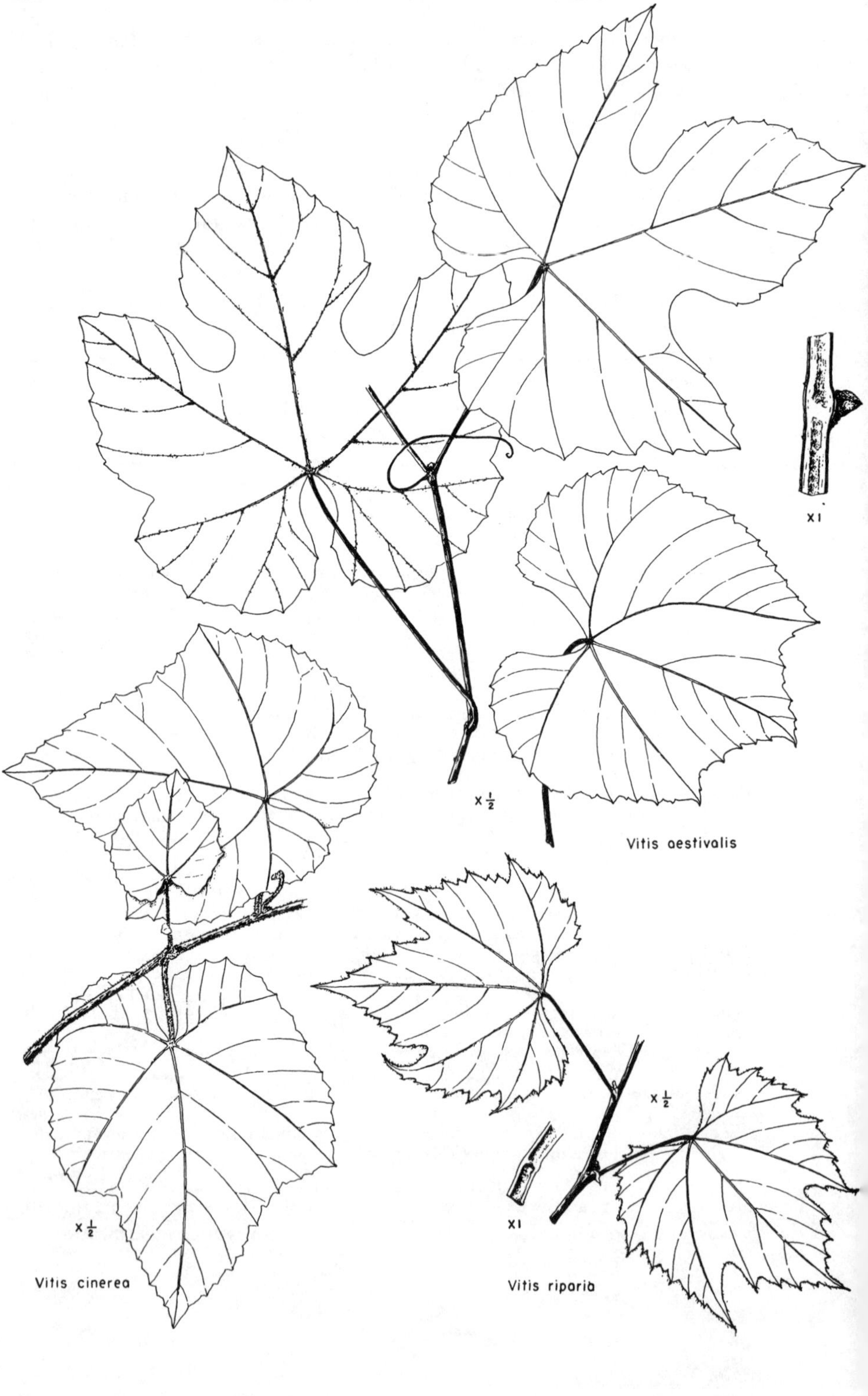

x1

$x \frac{1}{2}$

Vitis aestivalis

$x \frac{1}{2}$

$x \frac{1}{2}$

x1

Vitis cinerea

Vitis riparia

1. Vitis labrusca L. Fox Grape

More northern in range than our other species of grape, occurring more or less locally through the northern half of the Deciduous Forest; in Ohio, found in the northeastern and southwestern parts of the state—chiefly in the Lake and northern Allegheny Plateau areas and wet flats of the Illinoian Till Plain—a distribution comparable to that displayed in Indiana where it is restricted to the northwestern (Lake area) and southeastern (Illinoian Till Plain) parts of the state. Easily recognized by the heavy felted whitish to rusty tomentum of the lower leaf-surface, and large fruit much like that of Concord grapes (which are derived from this species) but with tougher skins and more fibrous pulp. Fruit commonly dark purple when ripe, occasionally red. Economically a very important species as it is the source (by selection and crossing) of the cultivated grapes (known as *V. labruscana*) of eastern United States.

2. Vitis vulpina L. Frost Grape

V. cordifolia Michx.

A common and widespread species, high-climbing, forming dense canopies over the crowns of small trees; trunks sometimes 50–60 cm. in diameter. Usually distinguished by the coarsely toothed unlobed or obscurely lobed leaves, green and glabrous beneath except for tufts of hairs in the vein-axils, margin not ciliolate, and black fruit without bloom. Almost throughout the Deciduous Forest, except its northern part.

3. Vitis riparia Michx. Riverbank Grape

V. vulpina of previous manuals, not L.

Vigorous high-climbing vine, similar in habit to *V. vulpina*; somewhat more northern in range than the frost grape, and more frequent than that species in northern Ohio. Lobes of leaves usually prolonged and acuminate, margin ciliolate, basal sinus wide, nodal diaphragm thinner than that of other species, and black fruit heavily glaucous. Both scientific and common names suggest the usual habitat of this species.

4. Vitis aestivalis Michx. Summer Grape

Less frequent in our range than the two preceding species, except in the Allegheny Plateau area. The deeply 3–5 lobed leaves of vigorous shoots are distinctive. The amount of pubescence and hence color of the lower leaf-surface vary greatly; two intergrading varieties are recognized:

Vitis labrusca

Vitis vulpina

Vitis riparia

Pale color of lower leaf-surface obscured by rusty tomentum; young branchlets and petioles
pubescent with reddish persistent to deciduous woolly hairs.....................var. *aestivalis*
Lower leaf-surface pale, distinctly blue-green or silvery; young branchlets and petioles glabrous
or glabrate, often glaucous..............................var. *argentifolia* (Munson) Fern.

In general, the latter variety, formerly known as *V. bicolor* LeConte, is more northern,
and in northern Ohio we have more records of it than of var. *aestivalis*; an Erie County
specimen bears a note "vine . . . about 80 ft. high and 28¼ in. in circumference." Both
varieties are frequent southward, where some intermediate forms occur.

5. **Vitis cinerea** Engelm. SWEET WINTER GRAPE
    More southern in range than our other species, extending northward into Ohio. Dis-
tinguished by the more or less angled branchlets and persistent dense, short, gray pubescence, at first with some cobwebby hairs.

6. **Vitis baileyana** Munson POSSUM GRAPE
    Rather similar in appearance to *V. vulpina*, but with the petioles and veins beneath
densely and permanently pubescent, young twigs angled or striate, leaves small (5–10 cm.
long), and teeth of leaf-margin small. High-climbing vine with short internodes, and
often with many short branchlets. Southern in range.

Vitis aestivalis
↓ var aestivalis ← var argentifolia

Vitis cinerea

Vitis baileyana

TILIACEAE

About 400 species in 40 genera, only one of which is represented in north-temperate
latitudes. The Tiliaceae is one of seven families of the Malvales, an order containing a
number of plants from which well known, economically important products are derived,
among them kapok, cola (from the kola-nut of Africa), cacao (chocolate), cotton, jute
(from annual species of the Tiliaceae), and the very light-weight wood known as balsa.
The family contains trees, shrubs, and herbs.

TILIA L. BASSWOOD. LINDEN

Trees of temperate eastern North America (south into the mountains of Mexico),
Europe, and Asia. Wood soft and white, easily worked, and much used. Inner bark
tough and fibrous, early used by the Indians and later by the pioneers for cordage, and
woven into mats, cords, fish-nets, etc. Flowers fragrant and nectar-bearing, eagerly
sought by bees; oil distilled from the flowers is used in perfume. European and American
species are often planted as shade trees; of the former, *T. platyphyllos* Scop., *T. europea* L.
and *T. cordata* Mill. are most used.

Trees in our area usually produce one to several basal shoots, a characteristic aiding in field recognition; leaves of these basal shoots generally yield unsatisfactory herbarium specimens. Leaves long-petioled, simple, palmately veined, serrate with gland-tipped teeth, usually prominently asymmetric at base, and either cordate or truncate; flowers cream-colored, perfect, 5-merous, in cymes, the axillary peduncle adnate to a strap-shaped short-stalked or almost sessile leaf-like bract (bracts variable in shape within a species); fruit nut-like, 6–8 mm. in diam., tomentose.

The taxonomy of the American species of *Tilia* is much confused. No two manuals agree, either in number of species recognized, or in characters and limits of species. Until detailed field studies, over the entire range of the genus in America, are undertaken to determine probable hybridization and subsequent introgression resulting from Pleistocene migrations and commingling of older species, no satisfactory treatment is possible.

Three well-marked and fairly constant species (as interpreted here) occur in Ohio: *americana* (northern) and *T. heterophylla* and *T. floridana* (southern). In Ohio, *T. americana* is almost confined to the area of Wisconsin glaciation. Yet within this area, many of our specimens depart, by one character or another, from more typical *T. americana*; some, although superficially resembling *americana*, have few characters in common with it. *T. heterophylla* is confined to the unglaciated part of the state and the southern border of the area of Illinoian glaciation. Its most distinctive character is the dense or felt-like mat of fine stellate pubescence on the lower leaf-surface., Some specimens, with thin or sparse stellate pubescence, and intermingling simple hairs cannot be referred to *T. heterophylla*. Evidence afforded by the variable characters, and combinations of these characters, suggests extensive hybridization in the past between *T. americana* and *T. heterophylla*, followed by back-crossing to one or the other parent, and giving rise to a variable complex, the fairly distinct forms of which may be called *T. neglecta*. *T. neglecta*, in this more restricted sense, is apparently most frequent within or near the southern border of glaciation, but is represented far northward in Ohio. Fernald (1950) calls *T. neglecta* "an inconstant and nondescript series;" how much to include in it is questionable. The distribution of variants of this complex suggests that hybridization between *T. americana* and *T. heterophylla* took place in late-glacial time; that extensive back-crossing with *T. americana* during late-glacial and post-glacial migrations has carried introgression into all parts of Ohio. *T. floridana*, which generally resembles *T. americana* in leaf-margin, is distinguished (in part) by the velvety feel of the lower surface of leaves. This character is sometimes so diluted as to suggest hybridization (in the past) with *T. americana*; geographic occurrence supports this interpretation. Rarely, some characters of *T. heterophylla* and *T. floridana* are combined, further complicating the range of variation in *Tilia* in Ohio. (For more complete discussion, see Braun, 1960).

The demarcation of species here is tentative, and arranged to take care of Ohio material. The key cannot make allowance for all variation within each species.

Leaves coarsely toothed, about 4 (3–6) teeth per 2 cm. of leaf-margin; margin jagged in appearance, teeth 4(3–6) mm. long on upper edge, abruptly contracted into linear tip about ½ as long as tooth.

    Leaves glabrous, not velvety beneath, often with inconspicuous tufts in axils of lateral veins; bracts glabrous on both surfaces...1. *T. americana*

    Leaves beneath velvety to the touch, appearing glabrous, but with minute imperceptible pubescence; small tufts usually present in axils of lateral and cross-veins; bracts often minutely pubescent on back; adnate part of peduncle often pubescent........2. *T. floridana*

Leaves less coarsely toothed; about 5–8 teeth per 2 cm. of leaf-margin; margin not jagged, generally serrate to crenate-dentate; teeth 1–4 mm. long on upper edge, contracted to short mucronate or apiculate tip.

Leaves thinly pubescent beneath with branched and simple hairs; axillary tufts usually
prominent along lateral and cross-veins; teeth 5–6 per 2 cm. of leaf-margin, about 2 (1.5–3)
mm. long on upper edge; bracts glabrous on both surfaces..................3. *T. neglecta*
Leaves densely pubescent beneath with felt-like mat of white (sometimes brownish)
branched hairs; axillary tufts present, often inconspicuous; teeth 5–8 per 2 cm. of leaf-
margin, about 2 (1–3) mm. long on upper edge; bracts glabrous on side to which peduncle
is adnate, pubescent on reverse side.................................4. *T. heterophylla*

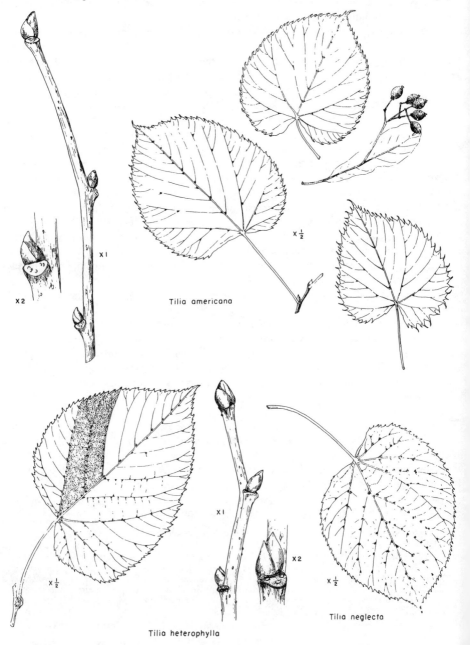

Tilia americana

Tilia heterophylla

Tilia neglecta

1. **Tilia americana** L. Basswood

    T. glabra Vent.

    A forest tree of northern range, most abundant toward the northwest, where it is a dominant tree of the Maple-Basswood Forest region. Frequent in Ohio in the glaciated area. Best recognized by characters of the leaf-margin: the large long-pointed teeth, often very irregular in size, giving a ragged or jagged appearance to leaf; leaves essentially glabrous beneath, although some specimens have slightly pubescent veins. The map shows, with dots, the distribution of specimens considered to be typical for the species, and, with circles, those which are considered to be *T. americana* in large part.

2. **Tilia floridana** (V. Engler) Small

    A tree of southern range, occurring in a few counties in the Unglaciated Allegheny Plateau. Leaves similar in appearance to those of *T. americana*, but distinct in the velvety (to the touch) under surface of leaves, very different from the soft but not velvety feel of *T. heterophylla*.

3. **Tilia neglecta** Spach Basswood

    A wide-ranging species or species-complex, perhaps intermediate between *T. americana* and *T. heterophylla*, but not synonymous with either. Fernald suggests that its forms might better be treated as variations of *T. americana*; Gleason and Little include it in *T. americana*. However, the smaller and more numerous teeth of the leaf-margin, and

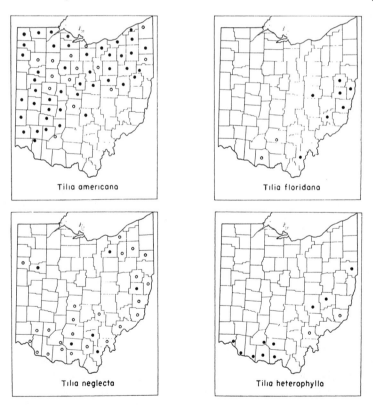

Tilia americana

Tilia floridana

Tilia neglecta

Tilia heterophylla

branched hairs of lower leaf-surface or veins indicate some relationship to *T. heterophylla*. Dots on the distribution map show occurrence of the more characteristic representatives of the complex, circles those of doubtful affinity.

4. **Tilia heterophylla** Vent. WHITE BASSWOOD

One of the two most characteristic trees of mixed mesophytic climax forests; confined to the unglaciated area and dissected parts of the area of Illinoian glaciation. Easily recognized by the dense mat of stellate hairs, and white or whitish (sometimes tawny) color of the lower leaf-surface, a feature conspicuous even at a distance, especially when leaves are upturned in a breeze. A few specimens have less dense pubescence, i.e., are intermediate with *T. neglecta*; a few combine a very fine stellate pubescence with the jagged margin of *T. floridana*, suggesting introgression into that species.

MALVACEAE

A large family—about 1200 species—of tropical and temperate regions over the whole world; many are herbaceous plants, among which okra and cotton are well known economic plants. No woody species is native in Ohio, nor in the "manual" range; one is an occasional escape.

HIBISCUS L.

About 200 species, mostly tropical, with large showy, usually solitary, axillary flowers. Petals 5, stamens numerous, the united filaments forming a column; the 5-cleft calyx surrounded by a circle of numerous bractlets is persistent into winter around the 5-valved capsule.

1. HIBISCUS SYRIACUS L. ROSE-OF-SHARON

Often spontaneous in the vicinity of planted bushes, but not naturalized. The 2–3 cm. long capsules, surrounded by calyx and involucel conspicuous in winter.

HYPERICACEAE

A family of only three genera and about 300 species, often included in the much larger polymorphic family GUTTIFERAE which is mostly tropical in distribution, and made up chiefly of trees and shrubs. The HYPERICACEAE is a family of herbs or shrubs with simple, opposite, entire leaves dotted with translucent spots (internal glands) and often also black-spotted, and in most of our species with yellow flowers.

ASCYRUM L.

A small genus of southeastern United States, Mexico, Central America, and the West Indies, with one endemic species in the Himalayas of Asia. Low shrubs, the young stems 2-edged; calyx of two large outer leaf-like sepals and two small inner sepals; 4 petals diverging 2 and 2; capsule flattened, enclosed by the persistent calyx.

1. **Ascyrum hypericoides** L. var. **multicaule** (Michx.) Fern. ST. ANDREW'S CROSS

The complex species, *A. hypericoides*, is made up of three more or less distinct but intergrading varieties, only one of which occurs in Ohio. This variety, more northern than the other two, is locally abundant in southern Ohio, mostly in or adjacent to the Allegheny Plateau. It is a low, decumbent or diffuse sub-shrubby plant of dry, usually non-calcareous soil, in summer easily mistaken for an herbaceous plant; in late fall and winter, the woody stems, with semi-persistent but thin leaves, are more apparent.

HYPERICUM L. St. John's-wort

A genus chiefly of the northern hemisphere and containing the majority of species of the family. Flowers yellow (except in subg. *Elodea*, usually separated as the genus *Triadenum*), sepals and petals 5, stamens numerous (or 5–10 in some small herbaceous species), often united at base into groups; styles united or separate; capsule ovoid or conical. Most of our common species herbaceous; two shrubby species in Ohio; others (as *H. sphaerocarpum* Michx.) with somewhat woody bases.

Shrubs; inflorescence interrupted, in part axillary, in part a few-flowered loose terminal cyme; flowers 2–3 cm. across.

Styles 3, capsules 3-celled, 8–14 mm. long; leaves oblong to narrowly elliptic, 3–7 cm. long, usually with several small leaves in the axils, these often persisting through the winter......
 1. *H. spathulatum*
Styles 5, capsules 5-celled, 7–10 mm. long; leaves linear to oblanceolate, thick, often revolute, 2–4 (–5) cm. long..2. *H. kalmianum*

1. **Hypericum spathulatum** (Spach) Steud. Shrubby St. John's-wort
   H. prolificum sensu Amer. auth., not L.
   A common and widespread shrub, often forming large patches; various habitats. Flowers showy, opening in July. The slender conic-cylindric capsules subtended by the rotate calyx are green until ripening in the fall, then brown and persistent through the winter.

2. **Hypericum kalmianum** L.
   A shrub of limited range, most frequent near the Great Lakes; in Ohio, local, and confined to a few northern counties; on rocks, sand-dunes, and sandy shores, and rarely in woodlands.

Ascyrum hypericoides
var multicaule

Hypericum spathulatum

Hypericum kalmianum

CISTACEAE

A family of seven genera and about 175 species, mostly in the Mediterranean region and North America, a few in South America and eastern Asia; three genera represented in Ohio, *Helianthemum* and *Lechea* (herbaceous), and *Hudsonia* (shrubby).

HUDSONIA L.

Low, dense, much-branched evergreen (or ever-gray) heath-like shrubs with alternate, sessile, scale-like or subulate imbricate leaves. Flowers yellow, numerous, each tipping a

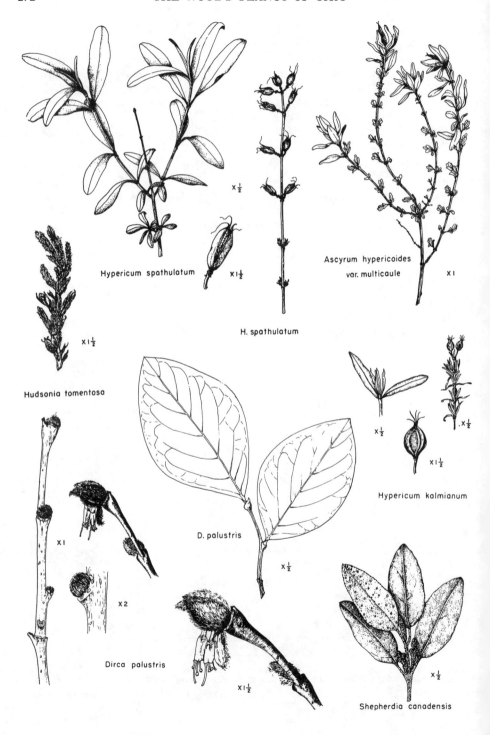

Hypericum spathulatum

$\times \frac{1}{2}$

$\times 1\frac{1}{2}$

Ascyrum hypericoides
var. multicaule $\times 1$

H. spathulatum

$\times 1\frac{1}{2}$

Hudsonia tomentosa

$\times \frac{1}{2}$

$\times 1\frac{1}{2}$

Hypericum kalmianum

$\times 1$

$\times 2$

D. palustris

$\times \frac{1}{2}$

Dirca palustris

$\times 1\frac{1}{2}$

$\times \frac{1}{2}$

Shepherdia canadensis

short lateral branch, the petals lasting but a short time. Three species, all North American, one very local in northern Ohio.

1. **Hudsonia tomentosa** Nutt. WOOLLY HUDSONIA

Gray, densely pubescent plants, the leaves hidden by the dense hoary tomentum. On bare sand of dunes and blow-outs, locally from Lake St. John, Que., to northern Alberta, and south to the Great Lakes; also on the Atlantic Coast. Two varieties are recognized: var. *tomentosa*, with the tips of the closely crowded leaves not projecting, and var. *intermedia* Peck, with the leaf-tips more evident and the pedicels slightly longer. Our record of the species is based on a single collection, one of "a few plants growing in a sandy area near the Indiana-Ohio line" in western Williams Co. in 1929. Deam (1943) refers Indiana material to var. *intermedia*, Gleason (1952) says that this variety (which may be a hybrid of *H. tomentosa* and *H. ericoides* L.) is known only from New England and eastern Canada.

Hudsonia tomentosa

THYMELAEACEAE

Trees or shrubs; about 40 genera and 500 species, most numerous in Africa and Australia, but almost cosmopolitan; poorly represented in America.

DIRCA L. LEATHERWOOD

Two similar species, one native to a circumscribed area in California, the other widespread in the East.

1. **Dirca palustris** L. LEATHERWOOD

Tree-like shrubs with very flexible branchlets (but wood soft and brittle) and tough bark which was used by the Indians for thongs and in basket-making. Leaves entire, oval to obovate, with very short petioles enlarged at base and concealing the buds; flowers in early spring before the leaves, pendant, yellow, 6–10 mm. long, surrounded by an involucre of silky bud-scales, the calyx (or hypanthium) corolla-like, the stamens exserted; fruit a greenish to reddish berry-like drupe about 8 mm. long. In mesic or dry woods, most frequent in southern Ohio.

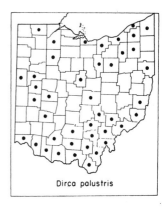

Dirca palustris

DAPHNE L.

A Eurasian genus of low shrubs, a few of which are cultivated.

1. DAPHNE MEZEREUM L.

Low shrub with alternate oblong to oblanceolate deciduous leaves, and lilac-purple or rosy-purple flowers appearing in early spring before the leaves. An escape, apparently naturalized in a swamp in Ashtabula County.

ELAEAGNACEAE

Shrubs or small trees of the north temperate zone; leaves alternate or opposite, silvery-, golden-, or reddish-scurfy; flowers axillary; fruit berry-like.

ELAEAGNUS L. OLEASTER

Shrubs with alternate leaves. Two introduced species occasionally reported as escapes.

1. ELAEAGNUS ANGUSTIFOLIA L. RUSSIAN-OLIVE

Branchlets silvery; leaves silvery beneath; fruit mealy, yellowish or silvery. From Eurasia.

2. ELAEAGNUS MULTIFLORA Thunb.

Branchlets reddish-brown; leaves silvery beneath, the scurf dotted with larger brown scales; fruit red, scaly. From eastern Asia.

SHEPHERDIA Nutt.

North American shrubs with opposite leaves; one species transcontinental, one mid-western and western, and one western.

1. **Shepherdia canadensis** (L.) Nutt. BUFFALO-BERRY
   Lepargyraea canadensis (L.) Greene
   Shrub, to 2 m. in height; leaves opposite, green and nearly glabrous above, silvery-downy and dotted with brown or rusty scales beneath; fruit yellowish-red. Trans-continental in the North, extending south into northern Ohio. Distinguished from al' other of our shrubs by the scurf and scales of the lower leaf-surface.

NYSSACEAE

This small family, often included in the CORNACEAE, is made up of only three genera. As now interpreted, it belongs in the very large order *Myrtales*, which contains the true myrtle, *Myrtus communis* L., Australian gum trees, *Eucalyptus* spp., Brazil-nut *Bertholletia excelsa*, mangroves of tropical sea-coasts, and many other plants known fo flowers or fruit.

NYSSA L. Tupelo

Trees of eastern United States and eastern Asia, with alternate, entire or few-toothe leaves, dioecious flowers on axillary peduncles, the staminate in heads, the pistillate sessile solitary or 2–8 on end of a peduncle (1.5–)3–6 cm. long, and ovoid or ellipsoid drupaceou fruit. Different authors are not in accord in the interpretation of species of *Nyssa*; th treatment here used follows Gleason (1952), who recognizes three species in the Manu range, the widespread *N. sylvatica* and two southern swamp species, *N. biflora* Wal (*N. sylvatica* var. *biflora* (Walt.) Sarg. in Fernald, 1950, and Little, 1953) and *N. uniflor* Wang. (*N. aquatica* L.).

1. **Nyssa sylvatica** Marsh. Black-gum. Sour-gum

Tree of various habitats, dry or moist, most frequent in non-calcareous soils. Bark deeply checkered, with alligator-pattern; leaves varying from elliptic to obovate, generally with short more or less acute tip, entire, or rarely few-toothed. Foliage colors brilliantly in late summer or early autumn. This species varies in shape, size, and thickness of leaves, pubescence and papillosity of leaves, characters which have been used to distinguish varieties. One of these, var. *dilatata* Fern., may be a hybrid with *N. biflora* and occurs only in the South. Within our range, two intergrading varieties occur, the two considered by Little (1953) to be synonymous. Deam (1940, 1953) maintains that they are distinct, and separates them as follows:

"Lower surface of leaves smooth, not papillose or rarely so, glabrous, glabrate or rarely densely pubescent on young specimens; leaves firm or subcoriaceous when mature, short acute or blunt at the apex, lustrous above; green branchlets usually bending when flexed to a right angle; wood difficult to split"...var. *sylvatica*
"Lower surface of leaves papillose, glabrous, glabrate or more or less pubescent, especially on the veins; leaves not firm or subcoriaceous when mature, usually acuminate at apex or some blunt; green branchlets usually breaking when flexed to a right angle; wood easy to split".........
var. *caroliniana* (Poir.) Fern.

In some specimens, the papillae are few and widely spaced, in others numerous and crowded; in some, the papillose character is found on subcoriaceous leaves, in others on thin leaves; they are more conspicuous on dry than fresh leaves. The varieties are not separated on our map, as they appear to have no geographic significance.

Shepherdia canadensis

Nyssa sylvatica

Aralia spinosa

ARALIACEAE

A chiefly tropical family, with a few representatives in temperate latitudes; among these is the well-known English Ivy, *Hedera helix* L., an evergreen trailing or climbing shrub. Trees, shrubs, or herbs; flowers small, in umbels or umbellate clusters arranged in a panicle.

ARALIA L.

A genus of North America and Asia to Australia. Leaves compound or decompound. *A. hispida* Vent., not listed below, with bristly, somewhat woody base, occurs infrequently in northeastern Ohio. Two Asiatic species, *A. chinensis* L. and *A. elata* (Miq.) Seem., are planted and may be confused with our native species, *A. spinosa*; both have the veins

$x\frac{1}{2}$

$x\frac{1}{4}$

$x\frac{1}{4}$

Aralia spinosa

ending in the teeth, while in our species the veins curve before reaching the margin. *A. elata* has the main axis of the inflorescence short, in contrast to the elongated main axis of *A. spinosa.*

1. **Aralia spinosa** L. DEVIL'S-WALKINGSTICK. HERCULES'-CLUB

A spiny small tree or shrub with stout branchlets 1–2 cm. in diameter, decompound leaves about one meter in length, and large compound panicle with elongate main axis, the whole often over one meter in length, the branchlets turning purple as the fruit ripens from red to black. Easily recognized at any season by the stout spiny branches and twigs; and by the leaf-scars (with prominent row of bundle-scars) nearly encircling the twig. Often planted for its handsome tropical aspect, and escaping from cultivation or sometimes naturalized north of its natural range.

CORNACEAE

Trees, shrubs, or rarely herbs, with mostly opposite leaves. Ten genera and nearly 100 species in temperate and subtropical latitudes of the northern hemisphere. One genus, *Cornus*, in the eastern American flora.

CORNUS L. DOGWOOD. CORNEL

Widely distributed in the northern hemisphere and represented south of the equator by one species in Peru. The name is from the Latin *cornu*, a horn, alluding to the hard wood. Flowers small, 4-merous, arranged in open naked cymes or crowded into dense heads surrounded by a corolla-like involucre. The flowering dogwood, *C. florida*, illustrates the latter arrangement; the similar western species, *C. nuttallii* Audubon, with 4–6 white petal-like bracts, and the often planted, early blooming Cornelian-cherry, *C. mas* L., of central and southern Europe and western Asia, with yellow flowers and yellowish bracts, are other trees with corolla-like bracts. A similar floral arrangement is seen in the bunchberry or dwarf cornel, *C. canadensis*, and in the circumboreal *C. suecia* L., whose range is far to the north of our area. The fruit is red in this group of dogwoods, and blue or white in those with loose cymose inflorescence. Color of fruit should be noted when collecting specimens. Among the other group of dogwoods, *C. alba* L. (several horticultural forms) and *C. sanguinea* L. are often planted for their red branches or beautifully variegated leaves.

a. Flowers and fruit in a dense cluster or head; flowers yellow or yellowish, surrounded by corolla-like involucre of 4 white or pink bracts; fruit red.
   b. Trees, sometimes shrub-like; drupes ellipsoid, about 1 cm. long, tipped with persistent calyx-lobes. . . . . .1. *C. florida*
   bb. Plants essentially herbaceous, but persisting through the winter, 0.5–2 dm. high; drupes globose. . . . . .2. *C. canadensis*
aa. Flowers and fruit in open cymes; flowers white or creamy-white; involucre absent; fruit white or blue.
   b. Leaves alternate, mostly crowded towards tips of branchlets, petioles long (to 5 cm.); pith white; drupes blue. . . . . .3. *C. alternifolia*
   bb. Leaves opposite, not crowded toward tips of branchlets, petioles shorter (to 3 cm.).
      c. Cymes paniculate, strongly convex, pedicels of fruit red; branchlets gray (light reddish brown the first year and glabrous); pith pale brown (or white in young branchlets); leaves lanceolate, elliptic, or narrowly ovate, tapering to acuminate apex and (usually) to acute base; glabrous or short appressed-pubescent above, glaucous or mealy below and with short appressed pubescence; fruit white, on red pedicels. . . . . .4. *C. racemosa*
      cc. Cymes broad, flattish to slightly convex.
         Pith of 1–2 year-old branches white; lateral veins 5–8 (–9) on each side.
           Leaves lanceolate to elliptic or ovate, 4–10 cm. long and about half (¼–⅔) as wide.

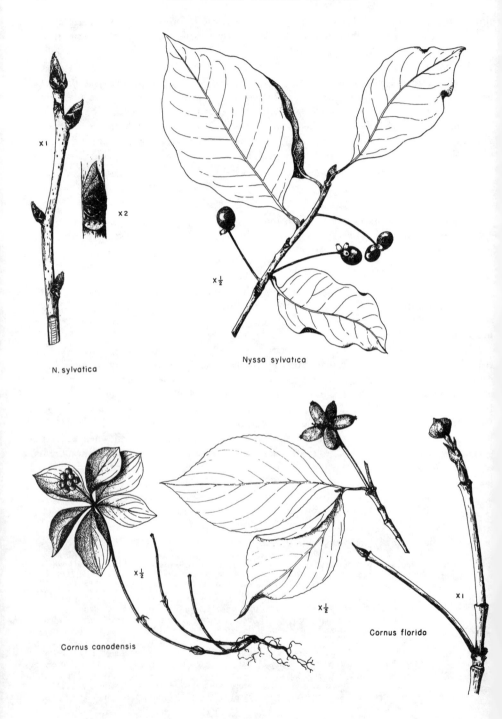

x1

x2

x½

Nyssa sylvatica

N. sylvatica

x½

x½

x1

Cornus canadensis

Cornus florida

Leaves whitened beneath, pubescent, acute or acuminate, with 5-7 lateral veins on each side; branchlets red, some often prostrate, pith ⅓ diam. of stem; fruit white..5. *C. stolonifera*
Leaves green beneath, glabrous, acuminate, with 3-4 lateral veins on each side; branchlets reddish brown to gray, pith slender; fruit white turning bluish-purple...6. *C. foemina*
Leaves broadly ovate to rotund, up to 12 cm. long and nearly as wide, abruptly contracted to a short point, woolly-pubescent beneath, with 6-9 lateral veins on each side; branchlets greenish, often with reddish blotches; fruit blue.7. *C. rugosa*
Pith of 1-2 year-old branches brown or drab, rarely white in no. 8; lateral veins 3-6 on each side.
Leaves rough above, woolly-pubescent beneath, narrowly to broadly ovate, up to 12 cm. long and about half as wide (or wider), abruptly acuminate, with 4-5 lateral veins on each side; branchlets gray (1-year reddish and rough); fruit white, on red pedicels..8. *C. drummondi*
Leaves not rough above, appressed-pubescent beneath or with some erect hairs; 3-6 lateral veins on each side; young branchlets pubescent; fruit blue.
Under surface of leaves pale, finely pubescent with appressed white or colorless hairs; leaves lanceolate to narrowly ovate-oblong, tapering to base and apex, 3-9 cm. long and ¼-⅓ as wide, with 3-5 lateral veins on each side; branches reddish to purplish...................................9. *C. obliqua*
Under surface of leaves pubescent with spreading and appressed usually brownish or reddish hairs; leaves ovate to broadly elliptic, with rounded base and abrupt short tips, 6-12 cm. long, about ½ as wide, with 4-6 lateral veins on a side; young branchlets reddish-brown, silky with brown hairs..10. *C. amomum*

WINTER KEY

The flower-buds of all species are larger than the vegetative buds, conspicuously so in those with floral involucre (*C. florida* and *C. mas*).
a. Leaf-scars irregularly alternate; twigs smooth, green or greenish, purple in sun; buds brown, ovoid or ellipsoid, with overlapping scales, glabrous......................3. *C. alternifolia*
aa. Leaf-scars opposite.
b. Low trees; spherical or oblate flower-buds usually numerous; terminal leaf-bud narrowed toward base, appearing somewhat stalked.
Lateral leaf-buds minute, covered by persistent bases of leaf-stalks........1. *C. florida*
Lateral leaf-buds ½ to as large as terminal, not hidden.... ....*C. mas*
bb. Shrubs; flower-buds absent or not greatly enlarged.
Pith white.
Twigs shining bright red to purple-red, glabrous except sometimes near tip; buds stalked, brown-silky above, terminal narrow-pyramidal, or if a flower-bud swollen toward base, the valvate leaf-like scales meeting above in long point; lateral buds long-stalked, 5-8 mm. long..5. *C. stolonifera*
Twigs greenish or greenish-purple, or blotched with purple, dull, nearly glabrous or pubescent; buds little constricted at base, pubescent at tip............7. *C. rugosa*
Pith brown (sometimes pale in young stems).
Divaricately branched.
Twigs gray, glabrous..4. *C. racemosa*
Twigs reddish brown, then gray, closely appressed-pubescent, and appearing glabrous, but rough.......................................8. *C. drummondi*
Not divaricately branched; twigs dull to bright red or purple-red, pubescent with silky pale or brown hairs; lateral buds sessile, triangular, closely appressed, 1-2 mm. long, densely brown-hairy........9. *C. obliqua* and 10. *C. amomum*, and hybrids

1. **Cornus florida** L. FLOWERING DOGWOOD

Cynoxylon floridum (L.) Raf.

Small tree, 10-12 m. tall, the branchlets spreading horizontally and upturned near tips; bark checkered, in alligator pattern (similar to, but much smaller than on *Nyssa*); wood very hard. One of our showiest trees in flower, in fruit, and in autumn coloring. Common and widespread, occurring in the open in old fields and along fence-rows, and in woods, sometimes forming a lower layer in oak and oak-pine woods.

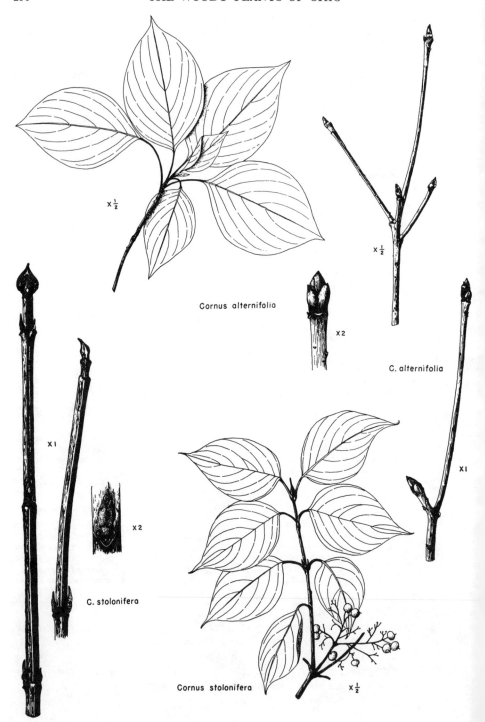

$x\frac{1}{2}$

Cornus alternifolia

$x\frac{1}{2}$

$x2$

C. alternifolia

$x1$

$x2$

C. stolonifera

Cornus stolonifera $x\frac{1}{2}$

$x1$

2. **Cornus canadensis** L. Bunchberry. Dwarf Cornel

Cynoxylon canadensis (L.) Schaffner

Essentially herbaceous, the erect stems arising from the extensively creeping and forking rhizomes. One of the characteristic species of the Hemlock-White Pine-Northern Hardwoods Forest, occurring throughout the great extent of that region, and locally beyond it in relic communities or outliers of that forest. Also in the mountains of the West.

3. **Cornus alternifolia** L. f. Alternate-leaf Dogwood. Pagoda Dogwood

The common names suggest distinctive characters of this species, one, the alternate arrangement of leaves, the other, the irregular horizontal arrangement of branches. Usually a large shrub, most frequent on mesic ravine slopes of the eastern two-thirds of the state; apparently infrequent westward except along the valleys of the Little Miami and Mad rivers.

4. **Cornus racemosa** Lam. Gray Dogwood

C. paniculata L'Her., *C. candidissima* Marsh., not Mill., *C. femina* sensu Small, not Mill. (in Schaffner)

Usually in moist soil or swamps, in open sunny situations, where it is a showy shrub because of prolific flowering and fruiting. The relatively narrow smooth leaves pale beneath, the glabrous angled branchlets, and strongly convex inflorescence aid in its recognition. Widespread in Ohio, more common northward, local southward; abundant in wet meadows of the Illinoian Till Plain, to which it is almost confined in southwestern Ohio.

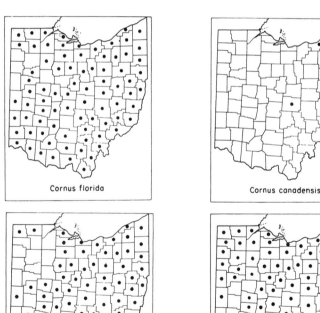

Cornus florida

Cornus canadensis

Cornus alternifolia

Cornus racemosa

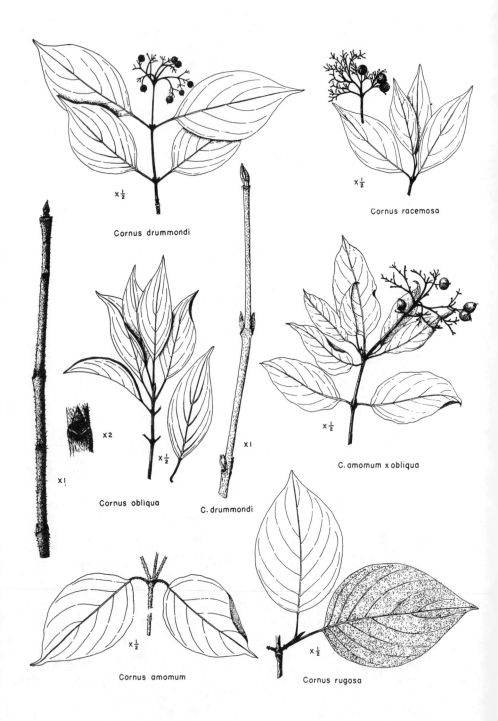

Cornus drummondi

Cornus racemosa

Cornus obliqua

C. drummondi

C. amomum x obliqua

Cornus amomum

Cornus rugosa

Hybrids between this species and *C. obliqua* (known as × *C. arnoldiana* Rehd.) are represented by specimens from Champaign, Highland, Licking, Lucas, and Medina counties. They have angled but pubescent branchlets, and leaves beneath more pubescent than those of *racemosa*.

5. **Cornus stolonifera** Michx. RED OSIER

A variable species which has been segregated into a number of species, varieties, and forms. The wide-ranging complex is transcontinental in the North, from Newfoundland to Alaska, and extends far southward in the western mountains, and into Pennsylvania, Ohio, Indiana, Illinois, and Iowa in the East. Following Gleason (1952), no varieties are recognized here; typical *C. stolonifera* and one form are in our range. The white pith, red branchlets, and long winter-buds aid in recognition.

Pubescence of lower leaf-surface sparse, strictly appressed; pubescence of younger branches minute and appressed.. *C. stolonifera* var. *stolonifera*
Pubescence partly appressed, partly distinctly spreading....................................
forma *baileyi* (Coult. and Evans) Rickett

The latter is listed by Fernald (1950) as a variety, and by Rehder as a species.

6. **Cornus foemina** Mill.
C. stricta Lam.

A coastal plain and Mississippi embayment species more or less frequent in the Wabash Valley of Indiana. One specimen from Wood County—at the northeastern end of the Wabash-Maumee valley—is referred to this species; it resembles *C. racemosa* somewhat, but has slender white pith, and leaves green beneath.

7. **Cornus rugosa** Lam. ROUNDLEAF DOGWOOD
C. circinata L'Her.

A shrub of northern distribution, most frequent in the Hemlock-White Pine-Northern Hardwoods Forest, extending southward into northern Ohio. Wide-leaved forms of *C. drummondi* are sometimes mistaken for this species, which differs in many characters, of which number of lateral veins, 6–9 pairs, is most readily observed.

8. **Cornus drummondi** Meyer ROUGHLEAF DOGWOOD
C. asperifolia of auth., not Michx.

More southern in range than our other species of *Cornus* (except *C. florida*), "its range scarcely overlapping with that of *C. rugosa*" (Gleason, 1952). A rather common species

Cornus stolonifera   Cornus rugosa   Cornus drummondi

in woods and thickets in southern Ohio. The rough upper surface of the leaves is distinctive. Very wide-leaved forms resemble *C. rugosa*, but in fruit cannot be confused with that species (fruit blue in *rugosa*, white in *drummondi*), and veins only 3–5 on a side, instead of 6–9.

9. **Cornus obliqua** Raf. SILKY DOGWOOD

C. *purpusi* Koehne, C. *amomum* in part (incl. in *amomum* in Schaffner, 1932, and other works)

Best recognized by its narrow gradually acuminate leaves, pale beneath, the pubescence closely appressed, white or colorless. Intergrades with the following species. Some specimens with the aspect of *C. obliqua* have pubescence of the midrib and principal veins beneath spreading (a character of *amomum*), white, and young stems densely whitish-pubescent. Also hybridizes with *C. racemosa*.

10. **Cornus amomum** Mill.

Leaves broad, half or more as wide as long, the pubescence in part spreading, in part appressed, usually rusty; veins beneath often reddened with rusty hairs. This and the preceding species hybridize; many of our specimens cannot be assigned to either species; because of back-crossing, some resemble *obliqua*, some resemble *amomum*. The map shows the distribution of these intermediate forms.

Cornus obliqua

Cornus amomum x obliqua

Cornus amomum

CLETHRACEAE

A family of only one small genus. Often included in the ERICACEAE, from which it differs in having the corolla of distinct petals, a 3-celled ovary, and simple pollen grains

CLETHRA L. WHITE-ALDER

Shrubs or small trees of temperate and tropical eastern America, eastern Asia, and Madeira, with fragrant white flowers in terminal solitary or panicled racemes. Stems stellate-pubescent.

1. **Clethra alnifolia** L. SWEET PEPPERBUSH

A stoloniferous shrub of swamps and moist woods from southern Maine to Florida and Texas, mostly near the coast. Often planted for its attractive fragrant white flowers in panicled racemes. Two specimens in the Oberlin College herbarium may have been taken from planted material, although this seems doubtful in view of the date of collection; one is from "northern Ohio" between 1850 and 1870, the other, 1871, " 'Akron O. B–!' This in the handwriting of Dr. H. C. Beardslee. The specimen was distributed in his 'Plantae Ohioenses Collegit et distribuit H. C. Beardslee, anno 1871.' " (Note by F. O. Grover.)

Another species, *C. acuminata* Michx., is in the Appalachian upland from Pennsylvania, West Virginia, and eastern Kentucky, southward; it is not known from nearer than about 40 miles south of the Ohio River.

PYROLACEAE

Low herbaceous or suffructicose plants, some parasitic on soil-fungi and devoid of chlorophyll, others with green, mostly evergreen leaves. A small family of boreal and temperate latitudes of the northern hemisphere, usually included in the ERICACEAE.

CHIMAPHILA Pursh PIPSISSEWA

Low evergreen plants slightly woody near the base, the leafy stems arising from horizontal subterranean stems. Leaves thick, often clustered toward top of stem; flowers 5-merous, nodding on terminal peduncles; capsules globose, splitting from the summit downward, erect, often persistent into the following year.

Leaves ovate-lanceolate to lanceolate (the lower ovate), acute, rounded or obtuse at base, remotely serrate with sharp teeth, variegated with white along principal veins. .1. *C. maculata*
Leaves oblanceolate, obtuse to acute or mucronate, cuneate at base, dentate except toward base, dark green . . . . .2. *C. umbellata*

1. **Chimaphila maculata** (L.) Pursh SPOTTED PIPSISSEWA
   One of our most beautiful flowers, ranging through much of the Central Deciduous Forest but absent westward; in acid soils, in beech, oak, pine, hemlock, or mixed woods, nowhere abundant. In Ohio, most frequent in the Allegheny Plateau, and found west of it only locally.

2. **Chimaphila umbellata** (L.) Bart. PRINCE'S-PINE
   This circumboreal species is made up of three varieties: var. *umbellata* of Eurasia, var. *occidentalis* (Rydb.) Blake of northwestern America (and northern Michigan), and var. *cisatlantica* Blake of eastern America, to which all our specimens should be referred. Readily distinguished from our commoner species by its unmarked dark green leaves.

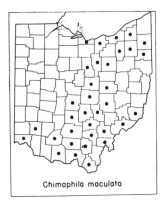

Chimaphila maculata

Chimaphila umbellata
var. cisatlantica

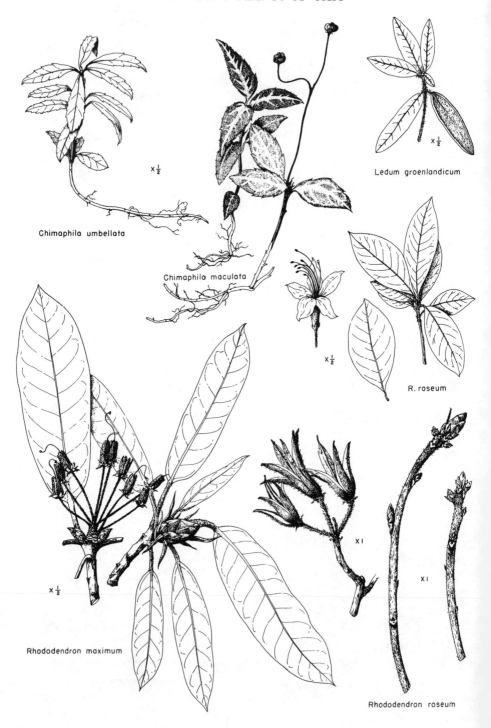

Chimaphila umbellata

Chimaphila maculata

Ledum groenlandicum

R. roseum

Rhododendron maximum

Rhododendron roseum

ERICACEAE

A large and almost cosmopolitan family of 17–18,000 species classified in about 65 genera, all trees or shrubs, often evergreen, many with handsome flowers. Although many of the genera are confined to the northern hemisphere, some to high latitudes, the largest genus, *Erica*, a genus of the eastern hemisphere with about 500 species, is best represented in South Africa; three European species are naturalized locally in the East; of these, *E. tetralix* L., cross-leaf heath, with leaves chiefly in 4's, and stiffly ciliate, is reported as "apparently adventive" under Norway spruce at Mill Creek Park, Youngstown, Mahoning County. Heather, belonging to the monotypic genus *Calluna* of Europe, is more widely naturalized; it is said to be "spontaneous and persistent" in one location in Trumbull County.

Leaves simple, usually alternate, rarely opposite or whorled; flowers 4–5–merous, regular or nearly so; fruit a capsule, berry, or drupe. Plants mostly of acid soils, therefore almost absent from western Ohio.

LEDUM L. Labrador-tea

Low shrubs of arctic and cold-temperate latitudes; only one species in the East.

1. **Ledum groenlandicum** Oeder Labrador-tea

An evergreen bog shrub with strongly revolute oblong or linear-oblong leaves, 2–5 cm. long, clothed beneath with a dense tangled mat of rusty hairs. Flowers white, about 1 cm. across, in dense terminal crowded racemes or umbel-like clusters. Bogs of the Hemlock-White Pine-Northern Hardwoods and Northern Coniferous forests; rare and local in Ohio, a glacial relic. Not reported by Detmers (1912) as growing at Buckeye Lake; a number of bog species have since been planted there, and are apparently established.

Ledum groenlandicum

RHODODENDRON L.

Evergreen or deciduous shrubs or small trees of the temperate and colder latitudes of the northern hemisphere, and into tropical latitudes in the high mountains of southern Asia and Malaysia, south into Australia. Of the 600 or more species, less than five per cent occur in North America. Among the few American species, the distinction between "Rhododendron" (subg. *Eurhododendron*) and Azalea (subg. *Anthodendron*) appears sharp; this is not so true of Asiatic species, as a number of the azaleas are evergreen or half-evergreen. A large number of hybrid Rhododendrons and hybrid azaleas are planted for their beautiful flowers; our southern Appalachian species, *R. catawbiense* Michx., is a constituent of many of the former; *R. calendulaceum* and *R. nudiflorum*, of the so-called Ghent hybrid azaleas.

Flowers large and showy, in terminal clusters, with small 5-parted calyx, 5-lobed slightly irregular corolla, 5–10 stamens, and 5-celled capsule. All of our azaleas are variable, especially in pubescence; hybridization in the wild has occurred frequently in the past and segregates and back-crosses, rather than "pure" species are common; in our area, this is especially true of *roseum* and *nudiflorum* (Skinner, 1955). Our key includes species which have been reported from Ohio, as well as those known to occur here.

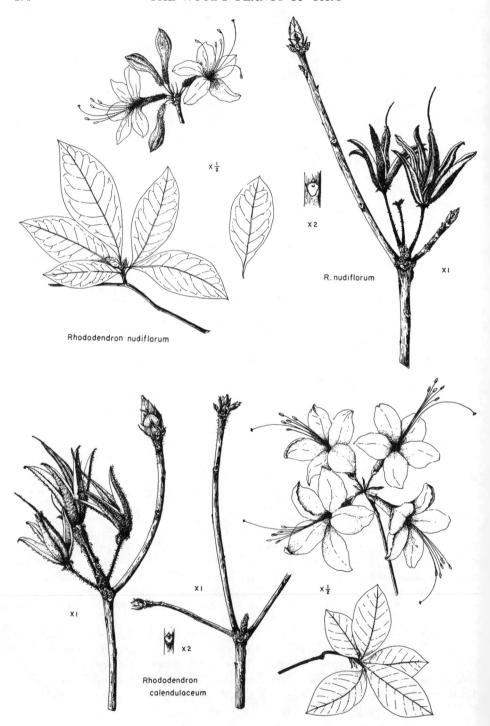

x ½

x 2

R. nudiflorum   X I

Rhododendron nudiflorum

X I

X I

x 2

Rhododendron
calendulaceum

x ½

Leaves evergreen, thick and leathery, oblong to lanceolate-oblong or sometimes oblong-obovate, 1–2 dm. long...1. *R. maximum*
Leaves deciduous, margins ciliate.
  Flowers yellow or orange, appearing with or before the leaves; leaves acute, closely short-pilose beneath; winter-buds glabrous, the scales ciliolate; pedicels and capsule beset with fine or minute bristles, the calyx-lobes 1–4 mm. long, glandular-ciliate..2. *E. calendulaceum*
  Flowers white, pale pink or deep pink; leaves acute, acuminate, or obtuse; winter-buds glabrous or pubescent.
    Leaves grayish-pubescent beneath; flowers with or before the leaves; calyx-lobes less than 1.0 mm. long; pedicels pubescent and stipitate-glandular.
      Flowers bright pink (occasionally light), fragrant; stamens twice length of corolla-tube; style 4–5 cm. long; capsule glandular-bristly, the calyx-lobes glandular-ciliate; leaves bluish-green and slightly pubescent above...........................3. *R. roseum*
      Flowers pink to almost white, slightly fragrant; stamens 3 times length of corolla-tube; capsule pubescent but not glandular, calyx-lobes usually glandless; southern.........
                                                                                    4. *C. canescens*
    Leaves glabrous beneath except on midrib; calyx-lobes 0.5–2 mm. long.
      Flowers before the leaves, pale to deep pink or white with pink tube, not fragrant; stamens 3 times length of corolla-tube; style 4.5–7 cm. long, longer than stamens; pedicels not glandular; capsule with stiff appressed hairs, not glandular; calyx-lobes ciliate with glandless hairs...................................5. *R. nudiflorum*
      Flowers opening after leaves, white to pinkish, fragrant; stamens somewhat exceeding corolla-lobes, style little longer; pedicels pubescent and glandular; capsule with stiff appressed hairs, usually glandular; eastern.........................6. *R. viscosum*

WINTER KEY

Young twigs finely and (toward tip) densely gray-pubescent, and with a few coarser ascending hairs; buds densely gray-pubescent, lowest scales light brown, apiculate, upper scales blunt, very short-ciliate; capsules thinly pubescent and glandular, pedicels villous and glandular...
                                                                                    3. *R. roseum*
Young twigs glabrous or nearly so; buds red-brown, scales glabrous except for median stripe, lowest long-apiculate, upper short-apiculate; capsules thinly pubescent with stiff spreading hairs, a few glandular; pedicels thinly pubescent with spreading and ascending hairs, rarely some glandular hairs.......................................5. *R. nudiflorum*
Young twigs somewhat pubescent with stiff upwardly-pointed hairs; buds chestnut-brown, bud-scales glabrous except for distinctly ciliate margin, strongly apiculate, the lowest with awn almost length of scale; capsules with fine short pubescence and ascending stiff non-glandular hairs; pedicels bristly hairy, some hairs gland-tipped..................2. *R. calendulaceum*

1. Rhododendron maximum L. GREAT RHODODENDRON. ROSEBAY RHODODENDRON

A large shrub or bushy tree to 10 or 12 meters in height, with crooked trunk and heavy contorted branches. This and *Kalmia latifolia* are our only broad-leaved evergreens with large leaves; distinguished from that species by the larger more leathery leaves, white to rose-color flowers spotted within with greenish-yellow or yellow, and the oblong-ovoid capsules. The leaves roll lengthwise in drought or cold weather, the more severe the cold, the tighter the roll, until the leaves become slender hollow cylinders. Widely distributed on the Appalachian upland from Alabama and Georgia northward to New England and New York, its range essentially coextensive with the area of the Mixed Meso-phytic and Oak-Chestnut Forest Regions; very local beyond this area.   In Ohio, local in deep ravines in the western, more rugged, part of the Unglaciated Allegheny Plateau.

Rhododendron maximum

2. **Rhododendron calendulaceum** (Michx.) Torr. FLAME AZALEA

R. luteum Schneid., not Sweet, *Azalea lutea* L. in part

Our only yellow or orange azalea; rare and local in Ohio. Flowers vary greatly in depth of color and in size; the bloom-period is of several weeks duration, leaves of the later-blooming shrubs almost full-grown at flowering time. The flame azalea commonly offered for sale by nurserymen is a native of the Old World.

3. **Rhododendron roseum** (Loisel.) Rehd. NORTHERN ROSESHELL AZALEA

Included in *R. canescens* in earlier manuals. A handsome pink-flowered and fragrant azalea of acid soils in oak woods of the Allegheny Plateau. Usually in patches, seldom occurring singly. Distinguished from our other pink-flowered azalea by the gray-pubescent winter-buds, and leaves more or less densely grayish pubescent beneath.

4. RHODODENDRON CANESCENS (Michx.) Sweet

Erroneously reported from Ohio; confused with *R. roseum*. *R. canescens* is a shrub of moist woods "from the South Carolina coast around the Gulf to the Trinity River in Texas and north across Arkansas and Mississippi to southern Tennessee and southern North Carolina" (Skinner, 1955).

5. **Rhododendron nudiflorum** (L.) Torr. PINXTER-FLOWER

A stoloniferous shrub of thickets, oak woods, and swamps; sometimes forming large patches. In flower, distinguished by the long-exserted stamens and noticeably longer style; in leaf, by the apically-directed coarse stiff hairs (sometimes few) on midrib and otherwise usually glabrous lower leaf-surface. This species and *R. roseum* hybridize freely; most Ohio populations of pink azaleas are mixed populations with many of the individuals intermediate.

6. RHODODENDRON VISCOSUM (L.) Torr. SWAMP-HONEYSUCKLE. SWAMP AZALEA

A late flowering white azalea, eastern in range. Although both Fernald (1950) and Gleason (1952) include Ohio in the range of this species, there are no Ohio specimens either in the Gray Herbarium or New York Botanical Garden Herbarium.

Rhododendron calendulaceum

Rhododendron roseum

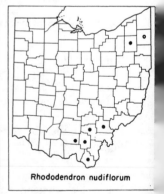
Rhododendron nudiflorum

KALMIA L.

Evergreen shrubs of North America, with entire leaves. Corolla cup-shaped, shallowly lobed, with 10 concavities in a ring about half-way between base and margin of corolla; stamens 10; capsule globose, 5-celled.

1. **Kalmia latifolia** L. Mountain-laurel

One of America's most beautiful shrubs, occasionally (in valleys of the Southern Appalachians) somewhat tree-like and reaching a height of 10–12 m., more often only 2–3 m. Confined to acid soils of bog-margins or, in our area, more common on slopes of the Allegheny Plateau. Wider ranging than Rhododendron, in much of the Appalachian Upland, and extending east and south across the Coastal Plain to southeastern Louisiana and northwest Florida, and westward on the acid soils of the Dripping Springs Escarpment (margin of western coal fields in Kentucky) and southern Knobs of Indiana; in Ohio, limited to the Allegheny Plateau, and most frequent in the unglaciated area.

The pollination mechanism of mountain-laurel is particularly interesting. In fresh flowers, the anthers lie in the concavities of the corolla; in this position, the filaments are under tension, which is suddenly released when the tongue of a bee is inserted in the crevice between ovary and stamens, changing the position of stamens and causing the pollen to be thrown onto the head of the bee, whence it is carried to the stigma of the next flower visited, thus effecting pollination (this trigger-release can be tested with a pin).

Flowers in terminal corymbs on viscid-pubescent pedicels; leaves flat (not rolling as do those of Rhododendron), elliptic to narrowly elliptic, acute at both ends.

ANDROMEDA L.

Low slender evergreen bog shrubs of northern latitudes, *A. polifolia* L. of Canada and northern Eurasia, and *A. glaucophylla*.

1. **Andromeda glaucophylla** Link. Bog-rosemary

A bog shrub of the Northern Coniferous and Hemlock-White Pine-Northern Hardwoods forests; in Ohio, confined to the northeastern counties. Distinguished by its firm evergreen narrow revolute leaves whitened beneath with dense fine puberulence; pink globose-urceolate flowers on recurved pedicels; and glaucous sub-globose or turban-shaped capsules.

LYONIA Nutt.

Deciduous or evergreen shrubs of eastern and southeastern United States, the West Indies, and eastern Asia.

1. **Lyonia ligustrina** (L.) DC. Maleberry

Xolisma ligustrina (L.) Britt.

Deciduous shrub; leaves usually obovate or oblanceolate, minutely serrulate; flowers

Kalmia latifolia   Andromeda glaucophylla   Lyonia ligustrina

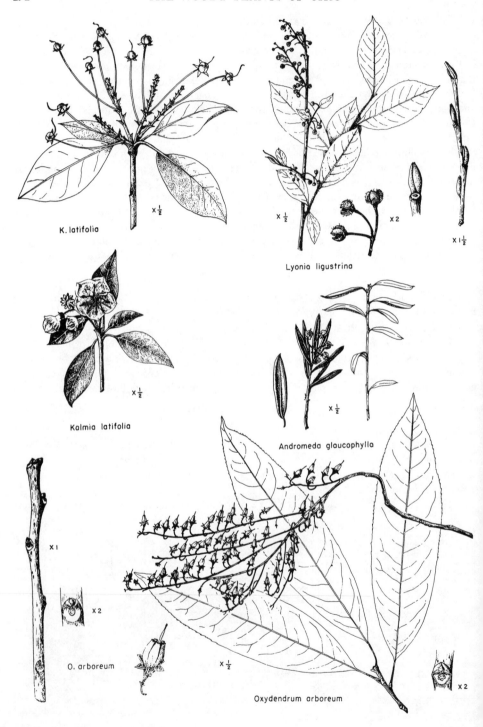

K. latifolia

Lyonia ligustrina

Kalmia latifolia

Andromeda glaucophylla

O. arboreum

Oxydendrum arboreum

numerous in loose or dense panicles terminating branches of previous year, globose or ovoid, whitish; capsules subglose, about 3 mm. in diameter. A highly variable species in which several intergrading varieties have been named. In Ohio, known only from two localities; specimens referable to the typical variety.

OXYDENDRUM DC. SOURWOOD

A monotypic genus of southeastern United States. Name from the Greek *oxys*, sour, and *dendron*, tree, referring to the sour taste of the leaves.

1. **Oxydendrum arboreum** L. SOURWOOD. SORREL-TREE

A tree with straight trunk and deeply furrowed bark, occasionally 15–20 m. in height, flowering in July, often while still small. Leaves 8–20 cm. long, bright green, glossy, acuminate, with margins finely serrulate, turning scarlet in autumn; flowers inclined downward in one-sided drooping racemes arranged in a terminal panicle, white, 6–8 mm. long, conic-ovoid, fragrant; capsules grayish- or tawny-pubescent, and frequently reddish, about as long as the flowers, erect, persistent into the winter, the panicles at length deciduous. Showy in flower and fruit, and brilliant in autumn foliage.

In acid soils of oak and oak-chestnut communities, and secondary woodlands; most abundant in the Appalachian upland; confined to unglaciated territory or rarely within areas of older glaciation.

CHAMAEDAPHNE Moench

A monotypic genus, the one species circumpolar.

1. **Chamaedaphne calyculata** (L.) Moench LEATHER-LEAF

A much-branched evergreen bog shrub, 1–1.5 m. in height, frequently forming dense thickets. Leaves thick and leathery, 1–5 cm. long, densely coated below with brown scurf, above with scattered scurf. Flowers white, nodding, in terminal leafy racemes; the buds evident the summer before; capsules depressed-globose, with thickened sutures. Three varieties distinguished by Fernald (1950): var. *calyculata* of Eurasia, var. *latifolia* (Ait.) Fern. of the far Northeast, and var. *angustifolia* (Ait.) Rehd. to which our plants are referred. This shrub was not reported by Detmers (1912) as growing at Buckeye Lake. As it could hardly be overlooked in a detailed study, it probably is one of the bog species which were later planted in this bog.

Oxydendrum arboreum

Chamaedaphne calyculata
var. angustifolia

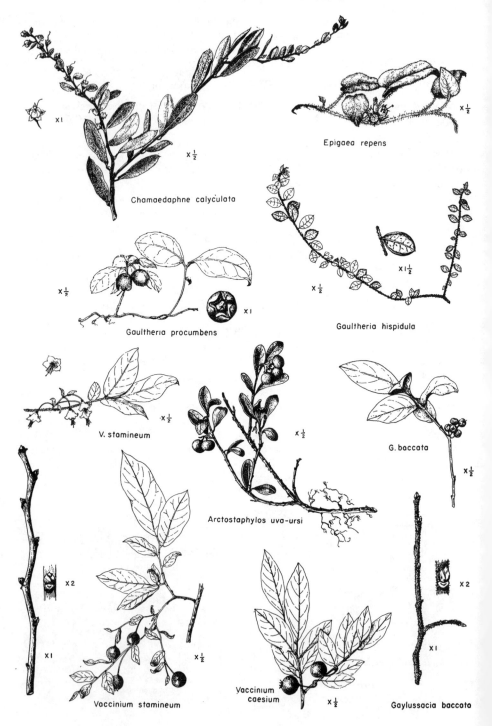

x1

x½

Chamaedaphne calyculata

x½

Epigaea repens

x½

Gaultheria procumbens

x1

x½

x1½

Gaultheria hispidula

V. stamineum

x½

Arctostaphylos uva-ursi

x½

G. baccata

x½

x2

x1

x½

Vaccinium stamineum

Vaccinium
caesium

x½

x2

x1

Gaylussacia baccata

EPIGAEA L. Trailing-arbutus

Prostrate evergreen shrubs with creeping stems. Two species, one Asiatic, limited to Japan, and our widespread eastern American species. Name from the Greek *epi*, on, and *gaea*, the earth, referring to the trailing habit.

1. **Epigaea repens** L. Trailing-arbutus. Mayflower

Distributed throughout the Allegheny Plateau in Ohio, and westward on the Lake plains to Erie County, in acid sandy or rocky soil. Leaves oval to oblong and rounded or cordate at base, 3–8 cm. long, rough and persistently pilose on both surfaces, or northward more commonly glabrous; fragrant pale to deep pink flowers with slender tube and spreading lobes, arranged in short crowded terminal and axillary spikes. Young inflorescence well developed by autumn, expanding rapidly in early spring.

Two varieties, based on differences in pubescence of leaves, are distinguished, var. *repens* with permanently pubescent and rough leaves, and var. *glabrifolia* Fern. with leaves glabrous or glabrescent. The first is said to be more southern than the latter, but both forms occur almost throughout the range of the species, and intergrade so completely that it is not desirable to separate them.

GAULTHERIA L.

A large genus of evergreen shrubs of North and South America, and eastern and southeastern Asia and Australia; ours low, almost herbaceous plants. The two species given below have been classified in different genera and even in different sub-families; the first, our common wintergreen, has erect leafy stems, and a wholly or partly superior ovary; the second has creeping stems with very small leaves, and a partly inferior ovary, which seems to ally it with the *Vaccinioideae*.

1. **Gaultheria procumbens** L. Wintergreen. Mountain-tea

The specific name refers to the extensively creeping horizontal rhizomes from which the erect leafy stems arise; the first common name, to the aromatic wintergreen flavor of leaves and berries; the second common name is often used by mountain people. Easily distinguished from other evergreens by growth-habit and by the lustrous elliptic to oblong (or in forma *suborbiculata* Fern., suborbicular), obscurely toothed leaves more or less crowded toward the top of 1–2 dm. tall stems, the wintergreen odor, the axillary or supra-axillary barrel-shaped white flowers, and subglobose red fruit. A plant of acid soils, chiefly

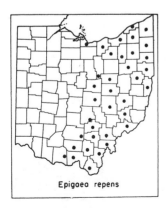

Epigaea repens

Gaultheria procumbens

of the Appalachian Highland and Lake Region; in Ohio confined to slopes of the Allegheny Plateau (and locally westward where non-calcareous Devonian shales are exposed in ravines) and sandy soils of the Lake plains.

2. **Gaultheria hispidula** (L.) Bigel. CREEPING SNOWBERRY
   Chiogenes hispidula T. & G.

In all but recent manuals, *Chiogenes* is considered a distinct genus including two species, *C. japonica* A. Gray in northern and central Japan, and *C. hispidula* of the Northern Coniferous and Hemlock-White Pine-Northern Hardwoods forest regions, and southern outliers. A creeping evergreen with slightly woody stems and small (5–10 mm.) alternate leaves, small (4 mm.) white 4-merous campanulate flowers, and white berries with mild wintergreen flavor. Moist mossy woods and bogs; probably extinct in Ohio.

ARCTOSTAPHYLOS Adans.

Evergreen shrubs, some prostrate, some tree-like. About 50 species, mostly in Central America and western North America; others in high latitudes, circumpolar or American-Asian; one in Ohio.

1. **Arctostaphylos uva-ursi** (L.) Spreng. BEARBERRY. KINNIKINICK
   Uva-ursi uva-ursi (L.) Britt.

Prostrate shrub, the long branches with papery exfoliating bark. Leaves thick, oblong-obovate, short-petioled, entire, 1–3 cm. long; flowers white to pale pink, ovoid, in short dense terminal racemes; berries globose, red, dry and mealy. Widely distributed in high latitudes of North America and Eurasia. Three varieties distinguished by differences in color and character of pubescence; our plants are referable to var. *coactilis* Fern. & Macbr., with young branchlets finely and permanently white-tomentulose, not viscid. Beaches and dunes along Lake Erie; very rare.

GAYLUSSACIA HBK. HUCKLEBERRY

Deciduous or evergreen alternate-leaved shrubs of North and South America (most numerous in the mountains of South America). Ours with resinous-dotted, deciduous leaves; flowers ovoid or tubular, 5-merous, in lateral bracted racemes; berries 5-seeded, the seeds larger than those of our common species of *Vaccinium*.

Gaultheria hispidula

Arctostaphylos uva-ursi
var. coactilis

Gaylussacia baccata

Leaves more or less copiously resinous-dotted beneath, resinous dots above usually widely scattered, yellowish-green; branchlets more or less pubescent, especially toward ends; pedicels and calyx glandular; flowers in short (1–2.5 cm.) racemes, greenish-yellow shading to dull red in corolla-lobes; pedicels less than 1 cm. long; berries black, not glaucous........1.   *G. baccata*
Leaves resinous-dotted only beneath, pale green, glaucescent; branchlets glabrous; pedicels and calyx sparsely pubescent and glandular; flowers in slender (3.5–7 cm.) racemes, greenish-purple; pedicels more than 1 cm. long; berries dark blue, glaucous............2.   *G. frondosa*

1. **Gaylussacia baccata** (Wang.) K. Koch Huckleberry
    G. resinosa T. & G.

Shrubs with the aspect of *Vaccinium*, but easily distinguished by the shining resinous globules conspicuous (especially in sun-light) on the lower leaf-surface; usually colonial. Widely distributed in acid soils, ranging from Newfoundland to Manitoba in the North, but only east of the Mississippi River southward.   In Ohio, almost confined to dry slopes of the Allegheny Plateau and adjacent exposures of non-calcareous rock, and to sandy soils of the Lake area; local elsewhere in bogs.

2. Gaylussacia frondosa (L.) T. & G.   Dangleberry

This species, whose range is mostly east of the mountains, has been reported from a number of localities in Ohio; all reports should be referred to *G. baccata*.

VACCINIUM L.

A genus of perhaps 150 species of deciduous or evergreen shrubs and small trees in the northern hemisphere from the arctic southward into the mountains of the tropics.   Leaves alternate, entire or serrate; flowers 4–5 merous, ovoid, tubular, or open campanulate; inflorescence axillary or terminal.

A complex and polymorphic genus, its distinct subgenera often elevated to generic rank; of these subgenera, representatives of *Polycodium*, *Cyanococcus*, and *Oxycoccus* in the Ohio flora.   *Cyanococcus*, to which the commercial blueberries, both wild and cultivated, belong, is the most complex.   It is made up of two groups of species, the low-bush and the high-bush blueberries, easily distinguished from one another in the field.   Diploid and tetraploid species are known in both groups, and hexaploids occur in the high-bush type. In general, tetraploids and hexaploids are larger than the species from which they are derived.   Hybrids occur in both groups, and especially north of the glacial boundary, hybrids between low-bush and high-bush species result in perplexing "half-high" types. As a result of hybridization, segregation, and back-crossing, and in some cases further crossing with other species, identification of some specimens is almost impossible.   The blueberries do not fall into distinct "pigeon-holes."   (For further information concerning the North American blueberries, see Camp, 1945.)

KEY TO SUBGENERA

Erect deciduous shrubs.
    Corolla open-campanulate in bud and at anthesis, the 5 lobes spreading; flowering and fruit-ing branchlets lateral and terminal, with leaf-like bracts usually smaller than and less acute than leaves of vegetative shoots; flowers solitary in the axils, on slender pedicels to 2 cm. long; winter-buds subglobose to round-oval, pointed, the axis of bud diverging from twig at angle of 45° or more...............................Subgen. POLYCODIUM (p. 298)
    Corolla cylindric- to subglobose-urceolate; flowers in short dense lateral or terminal clusters; pedicels short; winter-buds ovoid or oblong, appressed or ascending, the axis of bud di-verging less than 45° from twig......................Subgen. CYANOCOCCUS (p. 298)
Trailing evergreen bog shrubs with small leaves, 3–15 mm. long; corolla deeply 4-cleft or 4-parted, the narrow lobes reflexed; pedicels filiform, from axils or reduced leaves..........
                                                        Subgen. OXYCOCCUS (p. 303)

Subgen. POLYCODIUM, THE DEERBERRIES

Two well-marked features characterize this group: (1) corolla open in bud, not enclosing the stamens and pistils, but developing with them, being small and green at first, finally white and showy; (2) dimorphism of leaves, those of flowering branches (bracts) smaller and somewhat different in shape from those of vegetative branches. Differences in pubescence, glaucescence, and size of bushes are observable in each "species."

Leaf-like bracts much smaller than foliage leaves, shorter than pedicels to somewhat exceeding pedicel and flower, 2–10 mm. wide..1. *V. stamineum*
  Young branchlets pubescent; leaves pubescent on both surfaces; calyx glabrous or nearly so; fruit greenish or yellowish to purplish, not palatable...................var. *stamineum*
  Young branchlets glabrous; leaves glabrous, sometimes glaucous beneath; calyx glabrous; fruit greenish or yellowish to purplish, not palatable........................var. *neglectum*
  Young branchlets whitish-pubescent; leaves pubescent beneath; calyx white-tomentose; fruit blue to dark purple, edible..var. *melanocarpum*
Leaf-like bracts similar to, and one-half to as large as foliage leaves, exceeding the flowers, 8–15 (–20) mm. wide...2. *V. caesium*

1. **Vaccinium stamineum** L. DEERBERRY. SQUAW-HUCKLEBERRY
   Polycodium stamineum (L.) Greene
   Shrub, usually about 1 m. tall, showy when the corolla is fully developed. In Ohio, largely confined to the Allegheny Plateau, but extending westward in the pin oak flats of the Illinoian Till Plain, and locally on leached banks. A widespread polymorphic species of acid soils. Extremes in the series of variants are distinct and recognized as varieties or species; two are present in our flora: var. *stamineum*, distinguished by its pubescent twigs and foliage; var. *neglectum* (Small) Deam (*V. neglectum* (Small) Fern.), completely glabrous. The southern var. *melanocarpum* Mohr, which ranges north to eastern Kentucky, may possibly occur in southeastern Ohio.

2. **Vaccinium caesium** Greene
   Perhaps only an extreme variant of *V. stamineum*, but distinct in appearance because of the large leaves (bracts) of flowering branches. Southern in range; in Ohio, known only from open post oak—blackjack oak woodland of the Interior Low Plateau in Adams County.

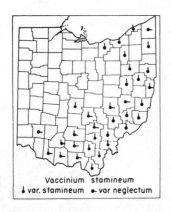

Vaccinium stamineum
↓ var. stamineum ← var. neglectum

Vaccinium caesium

Subgen. CYANOCOCCUS, THE TRUE BLUEBERRIES

Distinguished in flower by the more or less urn-shaped globose to cylindric corolla, only the small lobes spreading, and short almost leafless flowering branches. All are edible some with finer flavor than others. The student wishing to understand the complex series of variants and hybrids, and to determine intermediate specimens, should study papers by W. H. Camp (1942, 1945).

a. Leaves entire (or sometimes with a few inconspicuous teeth).
  b. Branchlets and lower leaf-surface densely pubescent; leaves thin, elliptic to sublanceolate, 1.5–3 cm. long, margin entire but sometimes appearing as if serrulate because of pubescence; corolla broadly cylindric to globose-urceolate, white or tinged with pink; fruit blue, with heavy bloom, 4–7 mm. diam . . . . .3. *V. myrtilloides*
  bb. Branchlets and leaves glabrous or glabrescent; lower leaf-surface not densely pubescent (branchlets sometimes sparsely pubescent or pubescent in lines, and leaves sometimes pubescent on veins beneath).
    Mature leaves mostly less than 4 cm. long; plants low, less than 1 m. in height.
    Leaves glaucous, broadly elliptic to oval, or the lower often suborbicular, obtuse; corolla cylindric- to globose-urceolate, greenish-white tinged with red, dull red before anthesis; fruit blue, with bloom; plants 2–5 dm. tall; twigs green . . . .
        4. *V. vacillans*
    Leaves somewhat glaucous, pale green, elliptic to elliptic-lanceolate, usually acute; corolla broadly urceolate to cylindric-campanulate, white or greenish-white, often tinged with pink; fruit dull to black, usually glaucous . . . . .5. *V. alto-montanum*
    Mature leaves mostly more than 4 (3–8) cm. long; plants tall, 1–4 m.
    Leaves glabrous or glaucous, 4–8 cm. long, ovate-lanceolate to elliptic, acuminate; corolla subglobose to cylindric, 6–10 mm. long, white or dull pinkish; fruit blue, dull, or black . . . . .12. *V. corymbosum*
    Leaves pubescent, 3–6 cm. long, narrowly to broadly elliptic, acute to acuminate; corolla ovoid to cylindric-urceolate, 5–6 mm. long, greenish or yellowish-white, tinged with pink; fruit dull black . . . . .6. *V. atrococcum*
aa. Leaves serrate or serrulate.
  b. Leaves small, 1–3 (–4) cm. long; plants low, less than 5 dm. tall.
    Leaves pubescent, elliptic to oval (or lower suborbicular), obtuse, serrate or entire . . . . .
        forms of 4. *V. vacillans*
    Leaves not pubescent.
    Plants 2 dm. or less tall, young branchlets green; leaves 4–10 mm. wide, green, shining, sharply serrulate; corolla subcylindric, 3–6 mm. long, usually white; fruit bright blue, 5–7 mm. diam . . . . .7. *V. angustifolium*
    Plants 2–4 dm. tall; leaves 10–15 (–20) mm. wide; corolla 4–8 mm. long.
    Leaf-surfaces pale green or glaucous.
    Young branchlets green; leaves variable, broadly elliptic, the lower suborbicular, irregularly serrate; corolla 4–5 mm. long; fruit blue . . . . . .forms of 4. *V. vacillans*
    Young branchlets glaucous; leaves elliptic, acuminate, sharply serrate, corolla 5–6 mm. long, white or whitish; fruit shining black, 8–12 mm. diam . . . .
        9. *V. brittonii*
    Leaf-surfaces bright green; leaves broadly elliptic; young branchlets green; corolla 4–8 mm. long, usually whitish; fruit bright blue, 8–12 mm. diam . . .8. *V. lamarckii*
  bb. Leaves larger, 3–7 cm. long, ovate-lanceolate to elliptic.
    Plants low, 3–10 dm. tall; leaves firm, pale green or sub-glaucous, elliptic to ovate-elliptic, acuminate; corolla 4–6 mm. long, greenish-white (or pink-tinged); fruit dark blue to black, sometimes glaucous, 5–7 mm. diam . . . . .10. *V. pallidum*
    Plants tall, 1–5 m. tall.
    Leaves pale green, ovate-elliptic, sometimes long-acuminate, 5–7 cm. long, sharply serrate to nearly entire; corolla 5–6 mm. long, greenish-white (or pink-tinged); fruit dark (sometimes glaucous), 6–10 mm. diam .11. *V. simulatum*
    Leaves green to subglaucous, narrowly to broadly elliptic or ovate, acuminate, 4–8 cm. long, entire or sharply serrate; corolla 6–10 mm. long, white or dull pinkish; fruit blue, dull, or black, 5–10 mm. diam . . . . .12. *V. corymbosum*

3. **Vaccinium myrtilloides** Michx. VELVET-LEAF BLUEBERRY
    V. canadense Kalm

Low shrub forming dense colonies; far northern in range and almost transcontinental (Labrador to Montana, British Columbia, and the Northwest Territory) and south into our area. The densely pubescent branchlets and small, entire thin leaves pubescent beneath distinguish this low blueberry.

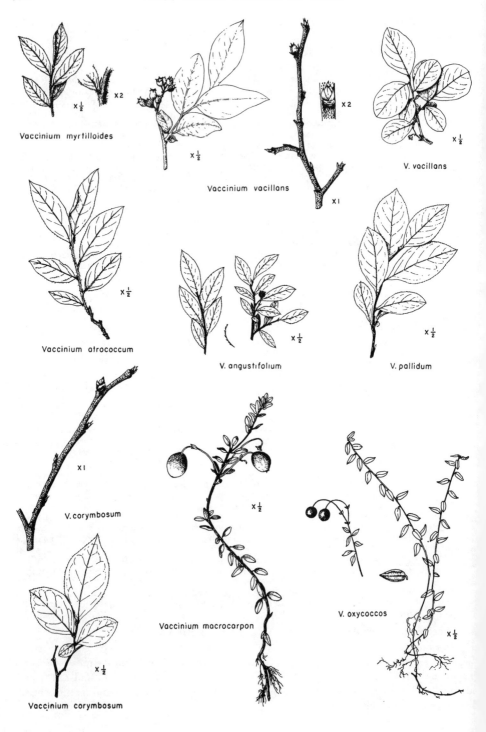

Vaccinium myrtilloides

x½   x2

Vaccinium vacillans

x½

x2

Vaccinium vacillans

x1

V. vacillans

x½

Vaccinium atrococcum

x½

V. angustifolium

x½

V. pallidum

x½

x1

V. corymbosum

Vaccinium macrocarpon

x½

V. oxycoccos

x½

Vaccinium corymbosum

x½

4. **Vaccinium vacillans** Torr. Low BLUEBERRY. DWARF DRYLAND BLUEBERRY
V. vacillans var. *crinitum* Fern., *V. torreyanum* Camp, *V. missouriense* Ashe, in part
Our most abundant low-bush blueberry, occurring almost throughout the Allegheny Plateau and Lake Plains. Usually easily recognized by its broad oval, usually entire, leaves and pale green branchlets, but quite variable because of gene interchange (in the past) with other species. The pubescent *crinitum* phase, which extends eastward from the Ozark area as far as Ohio, may be due to genes from *missouriense*, "a partially selected segregate phase involving a complicated series of crossings," then back-crossed into *V. vacillans*.

5. **Vaccinium alto-montanum** Ashe MOUNTAIN DRYLAND BLUEBERRY
This blueberry is a tetraploid derived from the diploid *V. vacillans*, and displays much the same range of variability as that species. Like most tetraploid blueberries, it is larger than its diploid ancestor. A "pubescent form is known from southern Ohio" (Camp, 1945).

6. **Vaccinium atrococcum** (Gray) Heller BLACK HIGHBUSH BLUEBERRY
This species of wide but generally southern range is with difficulty distinguished from *V. corymbosum*, and it is possible that our records (Highland, Lucas, Stark, Williams) should be transferred to that species; differs in its somewhat smaller pubescent leaves.

7. **Vaccinium angustifolium** Ait. Low SUGARBERRY
V. pennsylvanicum var. *angustifolium* Gray
This blueberry is commercially important in the Northeast; it forms dense and sometimes extensive colonies and produces bright blue fruit of excellent flavor. Distinguished from our other blueberries (except the next two which are often included in this species) by the small, sharply serrulate leaves. Northern in range: "Labrador and Newfoundland, west to Minnesota and south to New Jersey, and in the mountains to Virginia and West Virginia," in dry sandy or rocky uplands.

8. **Vaccinium lamarckii** Camp
V. angustifolium var. *laevifolium* House; *V. pennsylvanicum* Lam.
A tetraploid derivative of the diploid *V. angustifolium*, occurring in similar habitats and almost coincident in range; more susceptible to fire-damage than that species, and hence usually eliminated from commercial blueberry fields "pruned" by burning. About

Vaccinium myrtilloides

Vaccinium vacillans

Vaccinium angustifolium

twice the size of the parent species. Not known in Ohio, but to be looked for where *V. angustifolium* is common.

9. **Vaccinium brittonii** Porter

V. angustifolium var. *nigrum* (Wood) Dole

Similar to *V. lamarckii* and occupying much the same range; a tetraploid.

10. **Vaccinium pallidum** Ait. UPLAND LOW BLUEBERRY

A low-bush blueberry, coarser than but formerly included in *V. vacillans*; the line of demarcation between the two not distinct, for "they have hybridized in the past and the segregate forms today fill the gap . . . between the two basic populations" (Camp, 1945). Range: Appalachian Highland, north to West Virginia, Kentucky, Pennsylvania, and New York. Apparently local in the Allegheny Plateau area of Ohio. (All but three of the records shown on map based on specimens annotated by W. H. Camp, 1939.)

11. **Vaccinium simulatum** Small UPLAND HIGHBUSH BLUEBERRY

A high-bush blueberry resembling *V. corymbosum*, and doubtless included in that species by Fernald (1950); a tetraploid, derived from the diploid *V. pallidum*. Plants crown-forming, not in extensive patches; southern Appalachian in range, north in the mountains to Kentucky and Virginia; some southern Ohio records of *V. corymbosum* from south of the glacial boundary may belong here.

12. **Vaccinium corymbosum** L. NORTHERN HIGHBUSH BLUEBERRY

The common high-bush blueberry north of the glacial boundary, usually in bogs or swamps, or along stream or lake margins, sometimes on moist banks; southward, apparently also in dry oak woods. Plants tetraploid, forming "a highly variable complex now approaching partial stabilization." As formerly understood, *V. corymbosum* contained both diploids and tetraploids, and attempts were made to distinguish varieties and forms within the complex, which is apparently derived from hybrids involving several southern species which presumably met as they extended their ranges northward onto the land newly freed from Pleistocene ice. *V. simulatum* and *V. arkansanum* were among these, and some Ohio specimens of *V. corymbosum* from the southern counties suggest this relationship. Our map includes all specimens in this complex, whether from north of the glacial boundary (to which area *V. corymbosum* is said to be limited) or from unglaciated southeastern Ohio (where *simulatum* may occur).

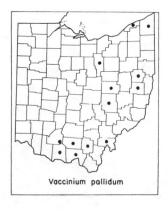

Vaccinium pallidum

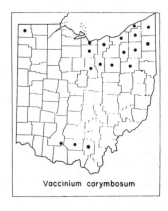

Vaccinium corymbosum

Subgen. Oxycoccus, The Cranberries

Evergreen trailing bog shrubs, with circumboreal distribution.   Corolla deeply 4-cleft or 4-parted, the narrow lobes reflexed; fruit red, acid.

Flowers 1–4 from axils of reduced leaves at tip of stem; pedicels 1–4 cm. long, glabrous or slightly pubescent, with 2 small (1–2.5 mm. long) reddish bractlets below to slightly above middle; leaves strongly revolute, whitened or glaucous beneath, ovate to ovate-oblong; berry 6–8 mm. diam..13.   *V. oxycoccos*
Flowers 2–6, from the axils of reduced leaves low on normal leafy branches; pedicels 1–3 cm. long, pubescent, with 2 green bractlets 3–5 mm. long above the middle; leaves flat or only slightly revolute, pale beneath, oblong-elliptic; berry 1–2 cm. in diam.....14.   *V. macrocarpon*

13. **Vaccinium oxycoccos** L. Small Cranberry
Oxycoccus oxycoccos (L.) MacM.
This species is a tetraploid derivative of the boreal *V. microcarpon* (Turcz.) Hook. and the more southern American cranberry, *V. macrocarpon*. It ranges across the continent in sub-arctic latitudes (also in northern Eurasia) and southward within the northern border of the United States, in the East to New Jersey and northeastern Ohio. Very rare in Ohio.

14. **Vaccinium macrocarpon** Ait. Cranberry
Oxycoccus macrocarpus (Ait.) Pers.
Extensively cultivated in the East, in Wisconsin, and to some extent in the Pacific Northwest. Natural range, Newfoundland to Minnesota, south locally to North Carolina and Tennessee. Most of the Ohio bogs where cranberries grew have been drained and vegetation destroyed or obliterated by urban growth. One, which formerly yielded residents abundant fruit for table use, was on Akron's west side (H. S. Wagner, Akron Metropolitan Park District). The best known Ohio locality is Cranberry Island in Buckeye Lake.

Vaccinium oxycoccos

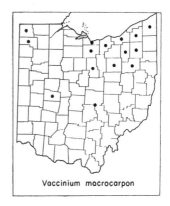

Vaccinium macrocarpon

EBENACEAE

Trees or shrubs with alternate entire leaves; six genera, including over 300 species, chiefly tropical.

DIOSPYROS L. Persimmon

The largest genus of its family, containing about two-thirds of the species, which, with few exceptions, are confined to tropical and subtropical latitudes; only our common persimmon reaches the cold-temperate belt.   One species, *D. kaki* L.f. of China, is cultivated in the Southern States and California for its large fruit.   The genus is very old, abundantly represented and widely distributed in Cretaceous time, and continuing plentiful through most of Tertiary time; later restricted by cooling climates and the advance of Pleistocene ice.   All of the continents have indigenous species.   Most of the ebony of commerce comes from species of *Diospyros*: golf-club heads and shuttles used in the textile industries are made from the hard wood of our common persimmon.

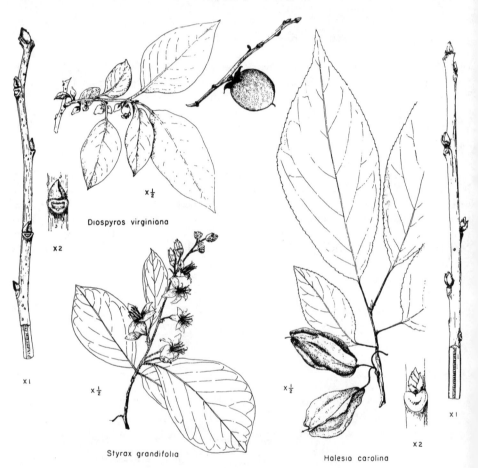

Diospyros virginiana

Styrax grandifolia

Halesia carolina

1. **Diospyros virginiana** L. PERSIMMON

Small to medium-size tree with dark bark deeply divided into more or less quadrangular blocks; twigs with chambered pith, terminal bud absent.   Leaves entire, oval or elliptic to ovate-oblong, short-acuminate.   Dioecious; corolla yellow, 4-lobed, the staminate smaller than the 10–12 mm. long pistillate flowers; calyx 4-parted, enlarging in fruit; fruit pinkish-orange, 2–4 cm. in diameter, ripening in late fall, remaining on tree after leaves have fallen. Fruit varies in flavor on different trees, astringent before ripening, sweet and edible when ripe, with high food value.   Southern in range, extending from the Gulf Coast and Florida northward through unglaciated eastern United States except at higher elevations; extending over areas of older glaciations, but rare within the limits of Wisconsin glaciation. Early planted by pioneers, and reproducing (often by suckers) from planted trees; most northern Ohio records probably are not records of indigenous trees; an Ashland County specimen bears the note that the original tree was planted, and on a Trumbull County specimen, is the note "adventive from old trees brought here . . . by first settlers." Usually in dry woodlands, thickets, or prairie patches.

Two varieties, var. *virginiana* (to which our specimens belong), glabrous or nearly so, wide-ranging and mostly east of the Mississippi River, and var. *pubescens* (Pursh) Dippel, with densely pubescent branchlets and lower leaf-surface, southern and extending northward west of the Mississippi.

STYRACACEAE

A relatively small family of trees or shrubs of the warmer parts of North and South America, Asia, and the Mediterranean.   Two genera represented in eastern United States.

HALESIA Ellis SILVERBELL

Distinguished by the pendulous white bell-shaped 4-lobed corolla, and the dry indehiscent oblong fruit, longitudinally winged.   Three or four species, two or three in eastern United States, one in eastern China.

1. **Halesia carolina** L. SILVERBELL

White bell-shaped flowers 1–2 cm. long, in clusters of 2–5, on slender pedicels.   Leaves elliptic to oblong or oblong-obovate, acute or acuminate, serrulate, 8–10 cm. long, thinly pubescent beneath; fruit 3–4 cm. long.   Young branchlets stellate-pubescent.   A tree or large shrub of rich mesophytic woodland, southern in range; one Ohio record, from southern Scioto Co.; frequently planted.

STYRAX L. STORAX

Shrubs or small trees, widely distributed in tropical and warm temperate latitudes of America, Asia, and southern Europe, extending northward into California, southeastern United States, and eastern Asia.   The aromatic resinous balsams, storax and benzoin,

Diospyros virginiana

Halesia carolina

are obtained from the Mediterranean *S. officinale* L. Distinguished by the white bell-shaped 5-parted corolla with spreading lobes, the flowers drooping on short pedicels either in racemes or on short lateral leafy branchlets, and small, more or less globular dry fruits surrounded at base by persistent calyx.

Because of the confusion which exists as to which species of *Styrax* occurs in Ohio, a key to the three southeastern species and varieties is given.

Principal lateral veins of leaf curving apically before reaching margin; branchlets densely stellate pubescent; leaves densely grayish-tomentose beneath, obovate or oval, acute, dentate with coarse teeth to entire, 9–15 (–20) cm. long; rachis of inflorescence, pedicels, and calyx stellate-tomentose; flowers several to many in elongate racemes 5–15 cm. long, leaf-like bracts only at base of raceme; corolla convolute-imbricated in bud................1. *S. grandifolia*

Principal lateral veins prolonged beyond leaf-margin, producing a toothed margin.
Branchlets nearly glabrous, sparingly stellate; leaves glabrous or slightly powdery beneath, entire or shallowly toothed, oblong to elliptic, acute at apex, cuneate at base, 2–10 cm. long; inflorescence glabrous or nearly so; flowers in 3–4-flowered leafy-bracted racemes 1–5 cm. long, or axillary; corolla valvate in bud............2. *S. americana* var. *americana*

Branchlets densely stellate pubescent; leaves sparingly puberulent above, thinly stellate-pubescent or scurfy beneath, elliptic to obovate, serrate or undulate, 2–8 cm. long; pedicels and calyx stellate-pubescent or scurfy; racemes 1–5 cm. long, bracts of inflorescence mostly leaf-like; corolla-lobes somewhat convolute in bud........2. *S. americana* var. *pulverulenta*

1. **Styrax grandifolia** Ait. Bigleaf Snowbell

Known in Ohio only from one locality in Athens County, on slopes of the preglacial Teays Valley. This is a woodland species, often on stream bottomlands, but not inundated floodplains, or on sandy banks. The curving principal lateral veins which do not reach leaf-margin clearly distinguish this species.

2. Styrax americana Lam. Snowbell

Range as stated by Fernald (1950) gives "s. O." for the typical variety. We have seen no Ohio specimens of a nearly glabrous *Styrax*, and none with the lateral veins reaching the leaf-margin. No specimens from Ohio in the Gray Herbarium. The var. *pulverulenta* (Michx.) Perkins is not known nearer than Arkansas and southern Virginia. Shrubs of bottomland woods subject to annual inundation, and borders of cypress swamps.

Styrax grandifolia

OLEACEAE

Trees or shrubs with opposite compound or simple leaves; about 20 genera and 500 species, in temperate and tropical latitudes, chiefly in the northern hemisphere. The family includes a number of plants of economic value: ornamentals, best known of which are lilac, *Syringa vulgaris* L. from Europe, and *S. persica* L. from Asia, forsythia, *Forsythia suspensa* (Thunb.) Vahl. and *F. viridissima* Lindl. from Asia, species of privet, *Ligustrum*, from Europe or Asia, and our native fringe-tree, *Chionanthus virginicus* L.; timber trees, species of *Fraxinus*; and the olive, *Olea europaea* L., of the eastern Mediterranean, cultivated for its edible fruits, olives.

FRAXINUS L. Ash

Deciduous trees with opposite, petioled, mostly pinnately compound leaves, small perfect or unisexual flowers produced before the leaves, arranged in crowded panicles or racemes from axil of leaf-scar of the previous year, and 1-seeded (rarely 2- to 3-seeded) elongate samaras. About 65 species of the northern hemisphere, in America south into Mexico, in Asia south to Java.

a. Branchlets terete; bark (except of *F. nigra*) dark, fissured; winter-buds (except of *F. nigra*) brown.
   b. Branchlets, petioles, and leaf-rachises glabrous or nearly so.
      Leaflets 5–9, stalked.
         Petiolules of lateral leaflets 3–15 mm. long; leaflets ovate to ovate- or oblong-lanceolate, usually acuminate, whitish or pale beneath, often entire; samaras linear, 3–5 cm. long, wing 4–7 mm. broad, almost entirely terminal; upper margin of leaf-scar deeply concave . . . . 1. *F. americana* var. *americana*
         Petiolules of lateral leaflets 5 mm. or less long, the blade decurrent on upper side nearly to base; leaflets elliptic-oblong to lanceolate, usually narrowed at base and decurrent, green and glabrous on both sides, or sparsely pale-villous on veins beneath, serrate except toward base; samaras linear or spatulate, 3–5 cm. long, wing 4–6 mm. broad, narrowly decurrent on body for about half or less its length; upper margin of leaf-scars convex to slightly concave . . . . 2. *F. pennsylvanica* var. *subintegerrima*
      Leaflets 7–11, the lateral sessile, rachis with thick reddish-brown tomentum at base of leaflets; leaflets oblong or oblong-lanceolate or lanceolate, tapering to acuminate apex, green and glabrous on both sides except for rusty pubescence near base of midrib beneath, coarsely and irregularly serrate; samaras narrow-oblong to oblong-obovate, 3–4 cm. long, wing decurrent to base, 7–10 mm. wide; leaf-scars vertically elliptic; winter-buds black . . . . 4. *F. nigra*
   bb. Branchlets, petioles, and leaf-rachises pubescent.
      Petiolules 5 mm. or less long, the blade decurrent nearly to base; leaflets mostly ovate, ovate-oblong to oblong-lanceolate, acuminate, entire, or entire only toward base; lower leaf-surface tawny-pubescent; wing of samara linear-oblanceolate, 3–6 mm. broad, narrowly decurrent on body for about half or less its length; upper margin of leaf-scar convex or straight . . . . 2. *F. pennsylvanica* var. *pennsylvanica*
      Petiolules mostly more than 5 mm. long (3–10–20), the leaf-blade not decurrent.
         Leaves whitened, glaucous and pubescent beneath; samaras linear, 3–5 cm. long, wing almost entirely terminal, 6–7 mm. wide; calyx in fruit 1 (–2) mm. long; upper margin of leaf-scar truncate to deeply concave . . . . 1. *F. americana* var. *biltmoreana*
         Leaves soft-pubescent beneath, not whitened; petiolules long and slender; leaflets elliptic to ovate-lanceolate, long-acuminate, usually entire; samaras linear-oblong to spatulate, 4–7 cm. long, wing of samara decurrent to below middle of body, 6–12 mm. wide; calyx in fruit 2–5 mm. long . . . . 3. *F. tomentosa*
aa. Branchlets 4-angled or narrowly 4-winged (most conspicuous on elongate internodes); bark light gray, divided into plate-like scales; winter buds gray or tawny-gray; leaves glabrous except on veins beneath (or in xeric habitats, soft-pubescent); leaflets short-stalked, lanceolate to ovate, coarsely serrate; rachis often gray-pubescent, more densely so near nodes; samaras 3–5 cm. long, elliptic to narrowly obovate, the wing 8–10 mm. wide, usually notched at apex, decurrent almost to base of body . . . . 5. *F. quadrangulata*

WINTER KEY

a. Twigs 4-angled, buds grayish or tawny, pubescent; inner bud-scales obscurely pinnate . . . .
                                         5. *F. quadrangulata*
aa. Twigs terete.
   b. Twigs more or less pubescent.
      Twigs usually velvety; buds rufous-tomentose; outer pair of bud-scales rounded on back, acute, truncate at apex; leaf-scar convex at top, indented by bud . . . .
                                       2. *F. pennsylvanica* var. *pennsylvanica*
      Buds red-brown, slightly villous; outer bud-scales keeled on back, apiculate; leaf-scar truncate at top . . . . 1. *F. americana* var. *biltmoreana*
   bb. Twigs glabrous or nearly so.
      Buds black or very dark brown; terminal bud usually as long as or longer than broad.
         Lateral buds closely appressed; terminal bud usually distinctly above last pair of laterals; twigs not prominently enlarged at nodes . . . . 4. *F. nigra*
        Lateral buds not appressed; upper pair of lateral buds often distant from terminal; twigs enlarged at nodes . . . . *F. excelsior*
      Buds brown; terminal bud usually broader than long.
        Upper margin of leaf-scar convex, but indented by bud; twigs smooth, greenish-brown
                                        2. *F. pennsylvanica* var. *subintegerrima*
        Upper margin of leaf-scar deeply concave . . . . 1. *F. americana* var. *americana*

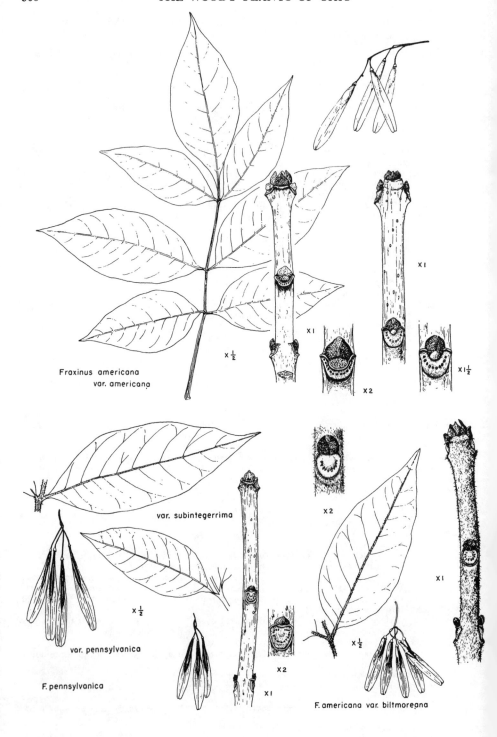

Fraxinus americana
var. americana

x ½

X I

X 2

X I½

var. subintegerrima

X 2

x ½

var. pennsylvanica

F. pennsylvanica

x ½

X 2

X I

X I

F. americana var. biltmoreana

1. Fraxinus americana L. WHITE ASH

Large forest tree with deeply fissured dark bark; a valuable timber tree. Ranging almost throughout the Deciduous Forest and Hemlock-White Pine-Northern Hardwoods region. The largest recorded Ohio tree (in Huron Co.) measures 18 ft. 5 in. in circumference (at 4½ ft.) and reaches a height of about 100 feet (Ohio Forestry Assn.).

F. americana var. americana WHITE ASH

More widely distributed than the next variety, with which it intergrades; found throughout Ohio in a variety of habitats. Leaves usually strongly whitened beneath and glabrous, or midrib beneath pubescent, petiole terete, and upper margin of leaf-scar deeply concave—characters which distinguish white ash.

F. americana var. biltmoreana (Beadle) J. Wright

F. biltmoreana Beadle

Formerly considered to be a distinct species, but experimental studies suggest that it should be included in *F. americana*. Frequent in southern Ohio, almost absent from the northern half of the state, except near Lake Erie. Similar in appearance to var. *americana*; distinguished from it by the usually pubescent branchlets, petioles, and lower leaf-surfaces (which, however, are whitened like those of var. *americana*), the truncate leaf-scar, more acute terminal bud, and more rounded lateral buds, twig characters which resemble those of red ash.

2. Fraxinus pennsylvanica Marsh.

A variable complex, representatives of which are found over much of the United

Fraxinus americana
var. americana

Fraxinus pennsylvanica
var. pennsylvanica

Fraxinus americana
var. biltmoreana

Fraxinus pennsylvanica
var subintegerrima

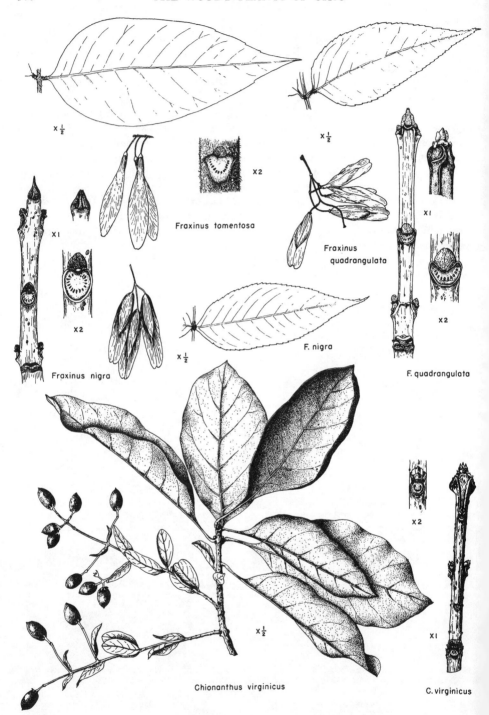

$\times \frac{1}{2}$

$\times 2$

$\times 1$

Fraxinus tomentosa

$\times 1$

Fraxinus
quadrangulata

$\times \frac{1}{2}$

$\times 2$

$\times 2$

Fraxinus nigra

F. nigra

F. quadrangulata

$\times 2$

$\times \frac{1}{2}$

$\times 1$

Chionanthus virginicus

C. virginicus

States.   Three varieties are recognized by Fernald (1950); of these, var *austini* Fern. is intermediate, and is not recognized by Gleason (1952); one specimen from Adams County, with very small samaras, may be referred to this variety.

F. pennsylvanica var. **pennsylvanica**   RED ASH

F. pubescens Lam.

Distinguished from Biltmore ash by the absence of whitened lower leaf-surface, by the grooved petioles, more slender body of the samara, and more decurrent samara-wings. Widespread in Ohio, but more frequent northward.

F. pennsylvanica var. **subintegerrima** (Vahl.) Fern.   GREEN ASH

F. lanceolata Borkh., *F. pennsylvanica* var. *lanceolata* (Borkh.) Sarg.

Wider ranging than any other ash, extending northwestward to eastern Alberta and Montana.   A streamside tree, following this habitat through the Prairie region; the state tree of North Dakota; used in shelterbelt plantings in the prairie states.   Common and widespread along streams in Ohio, where it is associated with white elm, silver maple, cottonwood, and streamside willows.   Leaves glabrous, not whitened beneath; branchlets glabrous, the upper margin of leaf-scar not concave as in white ash; body of samara slender. The samara-wings vary from almost linear to broadly spatulate; leaflets vary from lanceolate to ovate.

3. **Fraxinus tomentosa** Michx. f.   PUMPKIN ASH

F. profunda Bush

Pumpkin ash is usually a tree of deep alluvial swamps subject to long periods of inundation; in such habitats it develops a prominently swollen base.   Most abundant in hardwood and cypress swamps of the Atlantic and Gulf Coastal Plain and Mississippi embayment; very local beyond this area in wet or swampy woods.   In Ohio, known only from two tracts of swamp woods on the Illinoian Till Plain.   On technical grounds, Little (1953) rejects the name *tomentosa* for this species, retaining the name *profunda* which was used in our earlier manuals.   Fernald (1938, p. 450) discusses the nomenclature and reproduces a part of the illustration by Michaux filius.

4. **Fraxinus nigra** Marsh.   BLACK ASH

A tree of swampy woods, lake shores, swampy stream banks, and valley-flats.   Bark light gray, dividing into large plates separable into thin scales.   More northern in range than all other species of ash, its range almost coinciding with the area of the Northern Coniferious Forest (eastern half), Hemlock-White Pine-Northern Hardwoods Forest, and

Fraxinus tomentosa   Fraxinus nigra   Fraxinus quadrangulata

Beech-Maple Forest regions; very local south of these regions. The European ash, *F. excelsior* L., resembles black ash in having black winter-buds and serrate sessile leaflets; it is frequently planted. Blue ash resembles black ash in foliage and fruit; neither European ash nor blue ash has the thick pad of reddish-brown tomentum on nodes of the leaf-rachis.

5. **Fraxinus quadrangulata** Michx. BLUE ASH

A large tree of central interior range, essentially where outcrops of calcareous rock occur, or where glacial soils are calcareous; often on stream bluffs. Bark gray, fissured, with scaly and shaggy plates. Blue ash may usually be recognized by its gray- or tawny-pubescent buds with inner scales obscurely pinnate, and by its 4-angled branchlets; however, occasional trees have the branchlets almost terete, and slow-growing branchlets of tree-crowns are almost terete. Although the leaves are generally described as glabrous beneath except along midrib, specimens from the Lynx prairie (Adams Co.) have leaves soft-pubescent on lower surface, and petiole and rachis gray-pubescent throughout.

SYRINGA L. LILAC

A genus of about 30 species, in Asia and southeastern Europe.

1. SYRINGA VULGARIS L.

The broad-ovate entire leaves and upright paniculate inflorescence with 2-celled oblong capsules distinguish the common lilac, which persists about abandoned house-sites, and, northward, locally becomes abundant.

CHIONANTHUS L. FRINGE-TREE

Both scientific name (from *chion*, snow, and *anthos*, flower) and common name refer to the graceful pendant clusters of snow-white flowers. Two American species, ours and the very dwarf *C. pygmaea* Small of peninsular Florida, and one eastern Asian species.

1. **Chionanthus virginicus** L. FRINGE-TREE

Large shrub or small tree of extensive southern range, occurring in a variety of habitats: mesic woods of ravines and valley bottoms, open oak woods, high calcareous ridges, ledges on sandstone or limestone bluffs. Often planted, and sometimes spreading from cultivation north of its natural range. Easily recognized by its pendant clusters of fragrant white flowers, the corolla split almost to the base into 4 linear parts; opposite entire leaves with blades tapering at base; twigs similar to those of ash but winter-buds with more numerous scales. Functionally dioecious, although occasional individuals bear a few perfect flowers among the others. The panicles and flowers of staminate plants are larger than those of pistillate plants. Fruit a dark blue, ellipsoidal drupe 1–2 cm. long.

Chionanthus virginicus

LIGUSTRUM L. PRIVET

A dozen or more of the 50 species of privet (all eastern hemisphere) are in cultivation in the United States; several are grown for hedges. When not pruned severely, privet flowers and fruits freely; thus any of the cultivated species may occasionally appear spontaneously; only one, *L. vulgare*, is a frequent escape in Ohio. All have small white flowers (corolla 4-lobed) in terminal panicles, hard black berry-like drupes, and opposite semi-evergreen or evergreen leaves.

1. LIGUSTRUM VULGARE L. PRIVET

When found in more or less undisturbed woodland (as this species sometimes is), its identity may not be realized at once. Its long-persistent leaves, sometimes remaining green through much of the winter, make recognition easy after our deciduous shrubs have shed. Reported from more than a score of Ohio counties, well scattered over the state.

SOLANACEAE

A family of perhaps 3000 species of herbs, shrubs (rarely climbing), or small trees, widely distributed, but most abundant in the American tropics. No woody plants of this family native in Ohio; a few naturalized or escaped from cultivation.

SOLANUM L. NIGHTSHADE

The largest genus of the family, containing about two-thirds of its species. One, a partly woody or suffruticose climber, is a naturalized member of our flora.

1. SOLANUM DULCAMARA L. CLIMBING NIGHTSHADE. BITTERSWEET

A climbing plant, woody only at the base; leaves entire, ovate and acuminate, or with 1–3 pair of small but often deepcut lobes near base, or sometimes inequilateral, with only one lobe or auricle at base; flowers in long-peduncled cymes, purple (rarely white), the lobes of the 5-cleft corolla reflexed; fruit ovoid or ellipsoid, bright red, showy, but poisonous. Naturalized from Europe; widely distributed in Ohio, reported from more than 50 counties.

LYCIUM L. MATRIMONY-VINE

Shrubs of temperate and subtropical regions of both hemispheres; a few native in the Southwest and adjacent Mexico. Two species with purplish flowers and orange-red or scarlet ellipsoid berries sometimes found as escapes in our area.

Leaves lanceolate or lance-spatulate, gray-green, often fascicled; flowers on pedicels 8–20 mm. long, corolla-tube longer than corolla-lobes.............................1. *L. halimifolium*
Leaves rhombic-ovate to elliptic, deep green; flowers on pedicels 3–12 mm. long; corolla tube short and broad..2. *L. chinense*

1. LYCIUM HALIMIFOLIUM Mill. COMMON MATRIMONY-VINE

Formerly much planted; persisting in old gardens, and frequently escaped; reported from a score of counties. From southern Europe.

2. LYCIUM CHINENSE Mill. CHINESE MATRIMONY-VINE

Native of eastern Asia; sometimes planted and rarely reported as an escape.

SCROPHULARIACEAE

All native species of this very large family are herbs; one tree species is naturalized.

PAULOWNIA Sieb. & Zucc.

A genus of eastern Asia; sometimes included in the Bignoniaceae.

1. PAULOWNIA TOMENTOSA (Thunb.) Steud. ROYAL PAULOWNIA. PRINCESS-TREE

In foliage, flower shape, and coarse twigs, resembling *Catalpa*. Easily recognized by its fragrant lavender-blue flowers and large (3–4 cm.) pointed ovoid capsules in terminal panicles. Leaves large, 1–2.5 dm. long (much larger on sprouts), cordate, soft-pubescent, opposite or sometimes 3 at a node. The over-wintering panicles, with flower-buds about

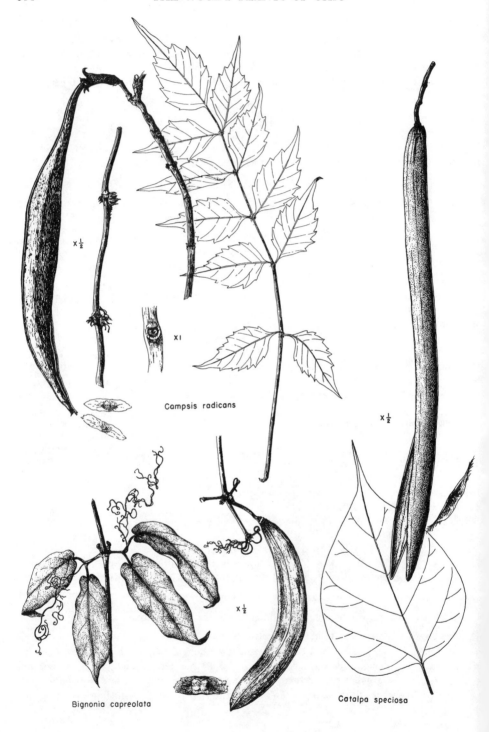

x ½

x I

Campsis radicans

x ½

x ½

Bignonia capreolata

Catalpa speciosa

1 cm. long, develop late in the season; flowers open before the leaves; buds killed in severe winters. Tree introduced from China; abundantly naturalized on mesic slopes and in valleys in the Southern Appalachians, infrequent northward. In Ohio, naturalized on slopes along the Ohio River in Adams, Scioto, and Belmont counties. Frequently planted in the southern counties.

BIGNONIACEAE

Trees, shrubs, or woody vines, chiefly of tropical and subtropical regions, a few in temperate latitudes. Flowers large and showy, corolla campanulate, funnel-form, or tubular; fruit an elongate capsule splitting into two parts; seeds transversely winged.

CAMPSIS Lour.

Woody, climbing by aerial roots; leaves opposite, pinnately compound, serrate. Flowers large, trumpet-shaped, with spreading lobes, bright orange-red; stipitate capsules cylindric, 10–15 cm. long, splitting into two valves. Two species, one American, one Chinese, the latter, *C. grandiflora* (Thunb.) Loisel. (*C. chinensis* Lam.), more showy, and sometimes planted.

1. **Campsis radicans** (L.) Seem. Trumpet-Creeper
Bignonia radicans L., *Tecoma radicans* (L.) Juss.
A woody climber whose original range was almost entirely south of the Ohio River. Commonly planted, and spreading from cultivation; now so thoroughly naturalized that it is impossible to determine in which, if any, of the Ohio counties it originally occurred. Its range as mapped merely shows the counties from which we have specimens. A handsome and well known plant, readily recognized by its large orange flowers, cigar-shaped capsules, and coarsely sharply serrate leaves; or in winter, by its straw-colored branchlets, and aerial roots in bands below the opposite, shield-shaped leaf-scars.

BIGNONIA L.
Anisostichus Bur.
A monotypic genus of eastern North America.

1. **Bignonia capreolata** L. Cross-vine
Anisostichus capreolata (L.) Bur.
A transverse section of the stem shows the wood-bundles arranged in the form of a cross. High-climbing evergreen or semi-evergreen vine, forming dense tangles in thickets

Campsis radicans

Bignonia capreolata

and over saplings in woodland, and into the crowns of trees. Flowers large, the corolla-lobes orange-yellow shading to orange-red in the throat; leaves with two petiolate entire leaflets and a terminal branched and coiling tendril, dark green in summer, dark bronzy red in winter. Ranging from Florida and Louisiana northward to eastern Maryland and the Ohio Valley; limited northward by low temperatures, the exposed flower-buds winter-killed. In Ohio, found on the hills bordering the Ohio River, rarely a few miles north of the river.

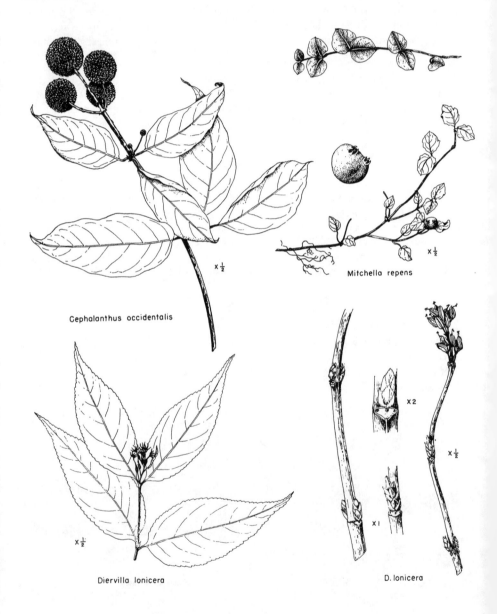

Cephalanthus occidentalis

Mitchella repens

Diervilla lonicera

D. lonicera

CATALPA Scop.

Trees of the warmer parts of eastern North America and eastern Asia. The natural ranges of our two American species, both of which are frequently planted, are circumscribed; neither is native in Ohio. Both have large white flowers with erose margins and brown-spotted within, and opposite or whorled large, broad-ovate leaves pubescent beneath. One Asiatic species, *C. ovata* G. Don, with smaller yellowish flowers striped with orange and spotted with violet, and glabrous leaves, is sometimes planted, and has been reported as an escape in Ashland County.

Leaves long-acuminate, scentless when bruised; flowers 4–6 cm. across, inside with 2 yellow stripes and inconspicuously spotted with brown-purple, in few-flowered open panicles; capsules 2–5 dm. long and 15 mm. in diam., the thick-walled valves after dehiscence remaining convex; seeds with fringe of hairs on rounded tips of wings........................1. *C. speciosa*
Leaves short-acuminate, sometimes with a pair of short lateral lobes, ill-scented when bruised; flowers 2–3 cm. across, inside with 2 yellow stripes and thickly spotted with brown-purple, in many-flowered crowded panicles; capsules 1.5–4 dm. long and 8–12 mm. in diam., thin-walled, the valves flattening after dehiscence; terminal fringe on wings of seeds narrow, connivent....
2. *C. bignonioides*

1. Catalpa speciosa Warder   Northern Catalpa.   Cigar-tree
A tree of alluvial forests of the lower Ohio Valley and central Mississippi Valley, from southwestern Indiana, southern Illinois, and Missouri, to western Tennessee and northeastern Arkansas. Widely planted for fence-posts and as an ornamental, and frequently escaped; reported (without distinction between planted and escaped) from over three-fourths of the Ohio counties. Often defoliated by caterpillars of the Catalpa sphinx.

2. Catalpa bignonioides Walt.   Southern Catalpa.   Indian-bean
Probably indigenous only on the banks of rivers from southwest Georgia and adjacent Florida westward to Louisiana; escaped northward. Often planted; a dwarf round-headed form, usually grafted high, is known as var. *nana* Bur. (erroneously referred to as *C. bungei* Hort.). Less hardy than *C. speciosa*; reported (sometimes as an escape, sometimes as cultivated) from about 15 Ohio counties. Later flowering than *C. speciosa*.

RUBIACEAE

A very large family of trees, shrubs, or herbs; nearly 400 genera and 7000 species, mostly tropical; contains a number of plants of economic importance, chief of which are coffee, *Coffea*, and Peruvian-bark trees, *Cinchona*, source of quinine; a few are grown as ornamentals, of which *Gardenia* is best known. Most of the plants of this family in cool-temperate latitudes are herbs.

MITCHELLA L.

Two evergreen almost herbaceous trailing plants, one widespread in eastern North America, the other (closely related) in Japan and Formosa.

1. **Mitchella repens** L.   Partridge-Berry
An evergreen trailing plant often forming large patches; leaves round-ovate to almost orbicular, 6–20 mm. long, dark shining green variegated with whitish lines along veins. Flowers in pairs, terminal or axillary; corolla white, funnelform, with 4 spreading or slightly recurved lobes villous on the upper (inner) face; fruit bright red, berry-like, formed from the ripened ovaries and hypanthia of the paired flowers, and crowned by the 2 persistent calyxes. Flowers always of 2 kinds; those of some plants (or clones) with short

style and exserted stamens, others with long style (stigma exserted) and included stamens; this insures cross pollination.

Partridge-berry ranges almost throughout the area of eastern forests in suitable habitats; it is generally confined to mildly acid soils, hence is infrequent in western Ohio; in wet red maple or pin oak woods on the Illinoian Till Plain, on leached banks, on shaded sandstone ledges, or even on mossy hummocks and logs.

CEPHALANTHUS L.

A small genus of deciduous or evergreen shrubs or small trees, with one Asian, one African, and five American species. Name from the Greek *kephale*, or *cephale*, head, and *anthos*, flower, alluding to the grouping of flowers in heads.

1. **Cephalanthus occidentalis** L. Buttonbush

A large summer-blooming shrub of swamps and swampy stream or pond margins; the slender tubular creamy-white flowers grouped in compact globose heads 2–3 cm. in diameter (not including the exserted styles); heads in terminal or terminal and axillary inflorescences, the peduncles 3–6 cm. long; heads of fruit 15–20 mm. in diam., composed of numerous obconic, angular nutlets, sometimes turning dark red on the sunny side in September. Leaves petioled, opposite or in whorls of 3, with minute triangular interpetiolar stipules. Very wide-ranging, throughout the eastern half of the United States, extending westward across the prairies in sloughs, and in the southwestern states and Mexico; reaching its largest size (occasionally 12–16 m.) in river-bottom swamps of southern Arkansas and eastern Texas. In Ohio, usually 1.5–2(–3) m. in height.

Two varieties are recognized:

Leaves and branchlets glabrous or essentially so..........................var. *occidentalis*
Leaves (at least the lower surface), branchlets, and peduncles pubescent.....var. *pubescens* Raf.

Most of our Ohio material is var. *occidentalis*, which probably occurs in every county; var. *pubescens* is more southern in range, and our only specimens are from Hamilton Co.

CAPRIFOLIACEAE

A family of some 400 species, mostly shrubs (some climbing), rarely small trees or herbs, chiefly north-temperate in distribution, a few in the mountains of the tropics. Many species in cultivation as ornamentals, a few of which have escaped. Leaves opposite, without interpetiolar stipules (a few species of *Viburnum* with stipules adnate to the petiole).

Mitchella repens

Cephalanthus occidentalis

Diervilla lonicera

DIERVILLA Duham.

Three species, two with sessile or subsessile leaves in the Southern Appalachians, one with petioled leaves more northern in range.

1. **Diervilla lonicera** Mill. Bush-Honeysuckle
   D. Diervilla (L.) MacM.
Shrub, about 1 m. in height, with petioled, oblong-ovate acuminate, serrate leaves. Flowers yellow, turning dull red with age, the corolla 2-lipped (similar to a honeysuckle). Northern in range; in the Hemlock-White Pine-Northern Hardwoods and Northern Coniferous Forest regions, and locally southward in the Beech-Maple region and in the mountains.

LONICERA L. Honeysuckle

About 150 species of erect or climbing shrubs; leaves opposite, entire; corolla 5-lobed, 2-lipped or almost regular. Flowers often showy and fragrant. Several Eurasian or Asian species have escaped from cultivation.

a. Branchlets hollow.
  b. One or more pairs of leaves below inflorescence connate; flowers apparently in 6-flowered sessile whorls (in opposite 3-flowered clusters) at ends of branches; fruit red; native species.
    Leaves pubescent above with straight stiffish hairs, soft-pubescent beneath, margins ciliate; young branchlets pubescent and glandular; connate disks elliptic, short-acuminate; twining and high climbing.....................................1. *L. hirsuta*
    Leaves glabrous and sometimes glaucous above, glaucous beneath (or pubescent in one var. of no. 4), margins not ciliate; branchlets glabrous or nearly so.
      Stems twining; leaves glaucous and minutely hairy beneath; corolla indistinctly 2-lipped, slenderly trumpet-shaped, 4–5 cm. long, bright red to orange-red; inflorescence a peduncled spike of few remote floral whorls...................2. *L. sempervirens*
      Stems straggling, twining, or scarcely climbing; corolla strongly 2-lipped, gaping, slightly gibbous, 1.5–3 cm. long, greenish-yellow to reddish or purplish; flowers in heads or short spikes, the whorls close together.
        Connate disk orbicular or nearly so, with rounded or retuse ends, distinctly glaucous *above* and below, much whitened; leaves obovate to suborbicular; flowers yellow, the whorls slightly separated.................................3. *L. prolifera*
        Connate disk rhombic or elliptic, usually longer than broad, bluntly pointed, green or but slightly glaucous above; whorls of flowers approximate.
          Leaves below connate disks distinctly longer than broad (1½–2 times), oblong to elliptic, narrowing to short petiole or sessile; corolla-tube gibbous at base......
          4. *L. dioica*
          Leaves below connate disks little longer than broad, obovate-elliptic to suborbicular, abruptly tapering to short petiole; corolla-tube slender, scarcely gibbous..
          5. *L. flavida*
  bb. None of the leaves connate; flowers in pairs, axillary; introduced species.
    Stems twining, high-climbing or trailing, half-evergreen; leaves ovate to oblong, pubescent on both sides when young, becoming glabrous above; flowers white later turning yellow, very fragrant; berries black............................6. *L. japonica*
    Stems upright or ascending; much branched shrubs; berries red.
      Peduncles much longer than petioles.
        Branchlets glabrous; leaves glabrous, thin, bluish-green beneath, ovate to oblong, 3–6 cm. long; peduncles 1.5–2 cm. long; corolla pink to white........7. *L. tatarica*
        Branchlets soft-pubescent to almost glabrous; leaves pubescent, at least beneath; corolla white to yellowish white (tinged with red in no. 8).
          Leaves acute, broad-cuneate to rounded at base; corolla-tube short, gibbous; bractlets about half as long as ovary........................8. *L. xylosteum*
          Leaves acute or obtusish and mucronulate, rounded at base; corolla-tube slender, gibbous; bractlets about as long as ovary.....................9. *L. morrowi*

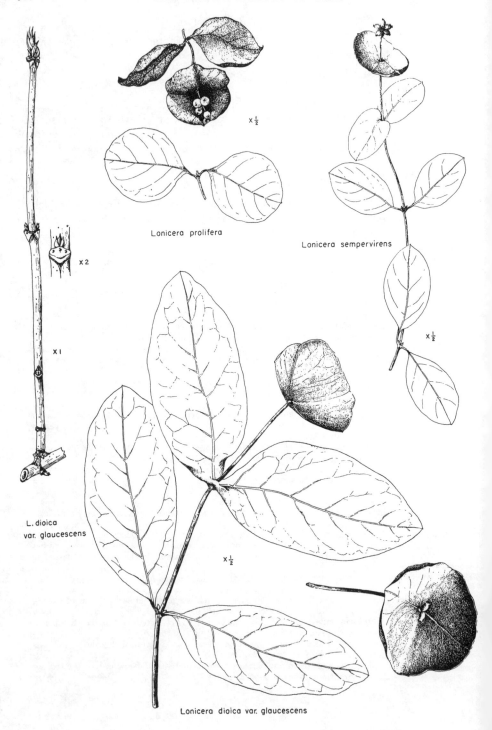

Lonicera prolifera

Lonicera sempervirens

$\times \frac{1}{2}$

$\times 2$

$\times 1$

$\times \frac{1}{2}$

L. dioica
var. glaucescens

$\times \frac{1}{2}$

Lonicera dioica var. glaucescens

Peduncles shorter than or but slightly longer than 3–5 mm. long petioles; branchlets short-pubescent; leaves dark green above, lighter beneath, pubescent on veins on both surfaces, ovate-elliptic to ovate-lanceolate, acuminate, 5–8 cm. long; corolla white changing to yellow...10. *L. maackii*

aa. Branchlets with solid white pith; none of the leaves connate; flowers in pairs on axillary peduncles.

   b. Peduncles mostly longer than flowers, 1.5–4 cm. long; leaves short-petioled; berries red.

Corolla deeply 2-lipped, yellowish-white, peduncles 2–3 cm. long; leaves minutely downy beneath, glabrescent, oblong to oblanceolate, margins and petioles not ciliate........

11. *L. oblongifolius*

Corolla nearly regular, funnelform, yellowish green or straw-color, peduncles 1.5–4 cm. long, very slender; leaves glabrous, oblong-ovate, 4–8 cm. long, margins and petioles ciliate...12. *L. canadensis*

  bb. Peduncles short, 1–7 mm. long; leaves almost sessile; berries blue...........13. *L. villosa*

Certain characters of the above key, as hollow or solid branchlets, twining habit, etc., may aid in recognition in winter condition. Growth-habit and habitat, as given in text which follows, can be used also.

1. LONICERA HIRSUTA Eat.   HAIRY HONEYSUCKLE

A twining and high-climbing plant with glandular-villous branchlets, dull oval pubescent ciliate leaves, and clammy-pubescent orange-yellow corolla 2–2.5 cm. long, with slender slightly gibbous tube. A northern species, reported for Ohio by Fernald (1950). However, the only Ohio specimen (from Painesville, 1890) in the Gray Herbarium assigned to this species has glabrous branchlets, leaves glabrous above, not ciliate, and corolla 1.5 cm. long, with strongly gibbous tube; i.e., is *L. dioica* var. *glaucescens*.

2. **Lonicera sempervirens** L.   CORAL or TRUMPET HONEYSUCKLE

A twining or sometimes high-climbing glabrous species of southern range, often planted for its showy slender trumpet-shaped coral-colored flowers. Northern Ohio reports are doubtful, specimens may be from escapes rather than from indigenous plants.

3. **Lonicera prolifera** (Kirchn.) Rehd.   GRAPE HONEYSUCKLE

A sprawling, scarcely climbing shrub, distinguished by its much-whitened glaucous foliage and almost circular floral disk which is glaucous above as well as beneath, its ends rounded or retuse. The red berries contrast strongly with the whitened floral disk of connate leaves. Dry rocky, usually calcareous banks and open woods, interior in range.

4. **Lonicera dioica** L.   WILD HONEYSUCKLE

Sprawling, reclining, or slightly climbing shrubs, the connate leaves forming oblong

Lonicera sempervirens

Lonicera prolifera

Lonicera dioica
⬥var. dioica ← var. glaucescens

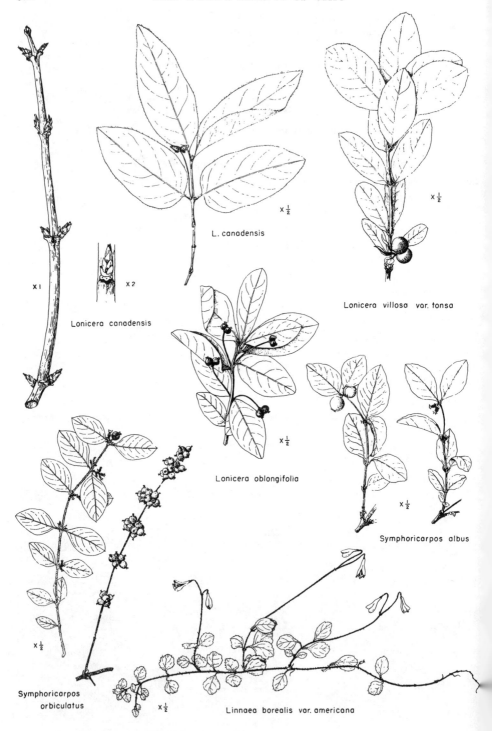

L. canadensis

$\times \frac{1}{2}$

$\times 1$ $\times 2$

Lonicera canadensis

$\times \frac{1}{2}$

Lonicera villosa var. tonsa

$\times \frac{1}{2}$

Lonicera oblongifolia

$\times \frac{1}{2}$

Symphoricarpos albus

$\times \frac{1}{2}$

Symphoricarpos orbiculatus

$\times \frac{1}{2}$

Linnaea borealis var. americana

to rhombic more or less pointed disks, green or but slightly glaucous above. Two intergrading varieties in our range:

Leaves glabrous and whitened beneath; corolla yellowish purple, glabrous outside....var. *dioica*
Leaves pubescent beneath, mostly larger; corolla pale yellow to reddish, pubescent outside; ovary often glandular or hirsute.........................var. *glaucescens* (Rydb.) Butters

The latter variety occurs more frequently. In Gleason's interpretation (1952), var. *glaucescens* has ovary glabrous, and its range is northern and western (not reaching Ohio); var. *orientalis* Gl. has ovary glandular and leaves more softly villous beneath; var. *dasygyna* (Rehd.) Gl.—forma *dasygyna* (Rehd.) Deam of var. *glaucescens* in Fernald—has ovary hirsute and leaves soft-villous beneath. Most Ohio specimens have ovary glandular (one from Erie Co. densely hirsute), but leaves vary from sparsely villous, to densely villous on veins, to softly villous.

5. **Lonicera flavida** Cockerell

Similar in habit and appearance to *L. dioica* and *L. prolifera*, and perhaps a hybrid of the latter and the southern *L. flava* Sims. It is a plant of "calcareous or circumneutral rocky shores and bluffs;" in Ohio known only from gravel bluffs (calcareous) of the Little Miami River in Clermont County.

6. Lonicera japonica Thunb. Japanese Honeysuckle

The common honeysuckle which has spread from roadside banks (where sometimes planted for erosion control) into adjacent thickets and woodlands, there festooning the trees and forming dense tangles difficult to control. Very fragrant and beautiful, but becoming a serious pest. Naturalized from eastern Asia; reported from about three-fourths of the counties.

7. Lonicera tatarica L. Tartarian Honeysuckle

Large upright glabrous shrub, 2–3 m. tall, from eastern Europe and adjacent Asia, in cultivation for about 200 years. Many horticultural forms varying in flower color (pink, rosy-pink, white, etc.) and fruit color (typically red, but yellow or orange-red in some forms). Reported from about 20 Ohio counties, mostly toward the northern part of the state.

8. Lonicera xylosteum L. European Fly Honeysuckle

Large, somewhat pubescent shrub, 1–3 m. tall, with widely spreading branches. From Eurasia; commonly planted and sometimes escaped; reported from Ashtabula, Fairfield, and Lake counties.

9. Lonicera morrowi Gray

A shrubby Eurasian honeysuckle occurring as an escape in a few northern counties.

10. Lonicera maackii Maxim.

Tall shrub, to 5 m., introduced from China; showy in fall because of its abundant red berries. Reported only from Hamilton County, where it is becoming abundant in pastures and woodlands.

11. **Lonicera oblongifolia** (Goldie) Hook. Swamp Fly Honeysuckle

An upright shrub (1–2 m. tall) of bogs and wet woods of the Hemlock-White Pine-Northern Hardwoods region, hence restricted to extreme northeastern Ohio.

12. **Lonicera canadensis** Bartr. Fly Honeysuckle

A loosely branched and straggling woodland shrub (0.5–1.5 m. tall) of the Hemlock-White Pine-Northern Hardwoods region and adjacent areas southward. Confined to northeastern Ohio.   Distinguished by its glabrous leaves, with margin and petiole ciliate, and filiform peduncles.

13.   **Lonicera villosa** (Michx.) R. & S.   MOUNTAIN FLY HONEYSUCKLE

A low shrub of the North, distinguished by its almost sessile blunt oblong leaves, very short peduncles, and blue berries.   Several varieties are recognized; our only specimen, from Ashtabula County, is referred to var. **tonsa** Fern., which has branchlets glabrous and leaves glabrous to sparingly pilose.

Lonicera oblongifolia   Lonicera canadensis   Lonicera villosa var. tonsa

SYMPHORICARPOS Duham.

A small genus, 15–16 species, all in North America except one in China.   Low, much-branched shrubs, the rather small flowers and white or coral-red berries in tight clusters in axils of the opposite leaves, or in leafy spikes.   Name alluding to these clusters, from the Greek *symphorein*, to bear together, and *karpos*, fruit.

Berries coral-red, 4–6 mm. in diam.; corolla greenish or purplish, 3–4 mm. long; style bearded; leaves oval to suborbicular, 1.5–3 cm. long, pubescent beneath..............1.   *S. orbiculatus*
Berries white, 6–10 (–15) mm. in diam.; corolla pink, 6 mm. long; styles glabrous; leaves oval to suborbicular or elliptic-oblong, 2–5 cm. long, pilose or glabrous beneath........2.   *S. albus*

1.   **Symphoricarpos orbiculatus** Moench   CORALBERRY

A shrub of dry over-grazed pastureland, open slopes, and roadsides, sometimes forming large patches; usually in disturbed places.   Coral-colored berries conspicuous after the leaves have fallen, and persisting long into the winter.   Ranging through much of the United States westward to the Great Plains.

2.   **Symphoricarpos albus** (L.) Blake   SNOWBERRY

S. racemosa Michx.

A low shrub (to about 1 m.) with slender branches and shreddy bark.   Two varieties

Branchlets usually finely pubescent; leaves pubescent beneath; berries 6–10 mm. in diam., mostly in leaf-axils, sometimes in terminal spikes.................................var. *albu*
Branchlets usually glabrous; leaves glabrous beneath; berries 12–15 mm. in diam., usually in elongate terminal racemes or spikes ....var. *laevigatu*

S. albus var. *albus*, of "calcareous ledges, barrens, and gravels," is northern, and occurs in Ohio in the Lake area.   *S. albus* var. *laevigatus* (Fern.) Blake (*S. racemosus* var. *laevigatu* Fern., *S. rivularis* Suksd.) of the Pacific Coast states, is commonly planted and locall escaped from cultivation.

LINNAEA Gronov. TWINFLOWER

A genus dedicated to Carl Linnaeus, the great Swedish botanist and founder of the modern system of nomenclature; Linnaeus seems to have been very fond of this delicate and beautiful circumpolar plant which Gronovius, at his request, named in his honor. Usually considered to contain but one variable species, of which the typical variety ranges through northern Eurasia and Alaska, the American variety, through the northern forests from Greenland and Labrador to Alaska, southward locally into our area.

1. **Linnaea borealis** L. var. **americana** (Forbes) Rehd. TWINFLOWER

Low evergreen trailing somewhat hairy plants with slender, scarcely woody stems with reddish brown finally exfoliating bark. Leaves opposite, short-petioled, roundish to obovate, usually obtuse, with a few crenate teeth. Flowers very fragrant, white to pink, shaded and striped with deep pink, nodding on short pedicels paired at the summit of filiform 4–7 cm. long peduncles. A plant of the Northern Coniferous and Hemlock-White Pine-Northern Hardwoods forests; very local in northeastern Ohio, and probably now extinct.

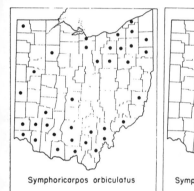

Symphoricarpos orbiculatus   Symphoricarpos albus var. albus   Linnaea borealis var. americana

VIBURNUM L.

Shrubs or small trees of the northern hemisphere, and southward to Java and South America; many species, mostly Asian, in cultivation. Winter-buds naked or with paired scales; leaves opposite, without stipules or sometimes with adnate stipules, entire, toothed, or lobed; flowers small, white (pink in some planted species), in compound terminal cymes, the corolla 5-lobed, some of the flowers in certain species enlarged and sterile; fruit a 1-seeded, red, blue, or black drupe. The genus is made up of several well defined groups of species, of which the dentatum group (*Odontotinus*) is most confusing.

a. Winter-buds naked, stalked, 2–3 cm. long; young branchlets, buds, and petioles reddish-brown to grayish stellate-scurfy; leaves round-ovate to suborbicular, 10–20 cm. long, serrate, stellate scurfy beneath (at least when young) and on veins; cymes sessile, stellate pubescent, 8–12 cm. across, bordered with large flat sterile flowers 2–2.5 cm. wide; fruit red, turning purple-black
1. *V. alnifolium*

aa. Winter-buds with 1 or 2 pairs of scales visible; pubescence, if present, not stellate-scurfy; some fascicled hairs may be present.
    b. Leaves 3–5 lobed, palmately 3–5 nerved at base, rarely with some unlobed leaves, or lobes suppressed.
       c. Leaves 3-lobed, broad-ovate, 3.5–10 cm. long, the lobes long-acuminate; petioles with 2 large glands near blade; cymes 7–10 cm. across, on peduncles 1.5–3 cm. long, with reddish glands; marginal sterile flowers present, 2 cm. across; fruit red, very acid.

Petiolar glands columnar or clavate, dome-shaped; stipules long, slender, with thickened tips...2. *V. trilobum*
Petiolar glands concave-topped or saucer-like disks; stipules slender with attenuate tips...3. *V. opulus*
   cc. Leaves 3–5–lobed, ovate, 5–14 cm. long, brown- to black-dotted beneath, lobes acute to acuminate, rarely almost absent; petioles without glands; cymes 4–7 cm. broad, on peduncles 1.5–7 cm. long, usually pubescent; no marginal sterile flowers; fruit black without bloom...4. *V. acerifolium*
  bb. Leaves not lobed, pinnately veined (no. 9 almost appearing palmately veined); fruit blue-black or purple-black.
   c. Leaves entire, crenulate, or serrulate, lateral veins curving and anastomosing before reaching margin; winter-buds with 1 pair of scales.
Leaf-margin crenulate; cymes on peduncles 1–3 cm. long; winter-buds covered with golden- to yellow-brown scurf; young branchlets scurfy-punctate; leaves dull above, elliptic, oval, or oblong, acute, short-acuminate or rounded at apex; fruit while ripening yellowish white, pink, then blue-black, with bloom; northern............
                                                               5. *V. cassinoides*
Leaf-margin serrate or serrulate; cymes sessile or nearly so.
Winter-buds and broad, short petiole rusty-red tomentose; leaves elliptic, oval, to obovate, or, on short branches broad, the lowest pair much reduced, thick, glossy-green above, usually with some rusty tomentum beneath; fruit black, with bloom; southern...6. *V. rufidulum*
Not rusty-red tomentose as above.
Winter-buds short-pointed, ashy-brown; leaves oval or ovate, or rarely sub-orbicular, acute or obtuse, dull above, scarcely if at all scurfy; fruit blue-black, with bloom; wide-ranging...............................7. *V. prunifolium*
Winter-buds brown to lead-color, the terminal long-pointed; leaves ovate, oval, obovate, or suborbicular, abruptly acuminate, glabrous or reddish-scurfy beneath, petiole margins undulate; fruit blue-black, with bloom...8. *V. lentago*
   cc. Leaves coarsely toothed, lateral veins straight, simple or 1–2 forked, ending in the teeth.
Leaves subsessile or very short-petioled, with linear stipules; blades ovate to sub-orbicular, truncate or subcordate at base 3–7 cm. long, with 4–6 pairs of veins, and 4–10 coarse teeth on each side; cymes 2–6 cm. broad.........9. *V. rafinesquianum*
Leaves usually long-petioled, petioles 1–5 cm. long; blades with 6–10 pairs of veins; with or without stipules.
Leaves cordate at base, broadly oval or suborbicular, 3 lowest pairs of lateral veins approximate; veins beneath glandular-dotted; teeth 18–30 on each side; leaves 5–13 cm. long, slender petioled, petioles 1.5–5 cm. long; glabrous except for minute stipitate glands; stipules linear..........................10. *V. molle*
Leaves not cordate at base, base narrowed or rounded, shape variable; lowest veins not approximate, all more or less parallel to one another; 4–22 teeth on each side; petiole 0.5–2.5 cm. long, with or without stipules..11. the *dentatum* complex

WINTER KEY

a. Buds naked, scurfy-hairy, the lateral veins of bud-leaves clearly visible; scurf red-brown to grayish.
   b. Leaf-scars broad; scurf rusty to tawny...............................1. *V. alnifolium*
   bb. Leaf-scars narrow; scurf pale tawny-gray....................................*V. lantana*
aa. Buds with scales.
   b. Buds scurfy.
   c. Buds densely red-brown scurfy, blunt; terminal flower-bud about 1 cm. long, swollen near base; lateral buds less than twice as long as wide...............6. *V. rufidulum*
   cc. Buds not densely red-brown scurfy.
Buds loosely brown-scurfy with small peltate scales; upper internode often sparsely scurfy; terminal flower-bud 15–25 (–30) mm. long, swollen about ⅓ way above base, slender above; lateral buds slender, curved.......................5. *V. cassinoides*
Buds dull brown, grayish brown, or lead-color, smooth but with minute scales.
Terminal flower-bud very long, 20–25 mm., long-acuminate, swollen near base, lateral buds slender, to 10 (–20) mm. long, curved; twigs long and flexuous.....
                                                               8. *V. lentago*
Terminal flower-bud shorter, 10–12 mm., acute or acuminate, swollen toward middle; lateral buds shorter; twigs shorter, rigidly spreading...7. *V. prunifolium*

bb. Buds glabrous, or pubescent toward tip; scales sometimes ciliate.
 c. Outer bud-scales connate, as long as bud, splitting across apex; buds ovoid to globose,
    stalked, green to red; twigs and buds lustrous. . . . .
                                                                2. *V. trilobum* and 3. *V. opulus*
 cc. Outer bud-scales shorter than inner.
    Outer pair of bud-scales blunt, forming a horizontally striate "collar" about ⅓
    length of bud, inner scales acute; low spreading shrub. . . . . . . .9. *V. rafinesquianum*
    Buds without distinct "collar," bud-scales distinctly paired.
    Bark of 2-year old twigs silvery gray, on 2-year and older branches exfoliating in
    thin flakes, exposing reddish-brown surface. . . . . .10. *V. molle*
    Bark not exfoliating.
    Twigs glabrous or nearly so; outer bud-scales ½ length of bud, strongly keeled;
    tall erect shrubs. .glabrous forms of the *dentatum* complex
    Twigs pubescent.
    Twigs finely soft-pubescent (rarely glabrous); outer bud-scales short, about
    ¼ length of bud. . . . . .4. *V. acerifolium*
    Twigs pubescent with coarse and fascicled hairs; outer bud-scales about ½
    length of bud. . . . . .12. *V. dentatum* var. *scabrellum*

1. **Viburnum alnifolium** Marsh. HOBBLEBUSH
   V. grandifolium Ait.
   A straggling shrub to 3 m. tall, in mesic woods and ravines of the eastern or Northern
Appalachian Highland division of the Hemlock-White Pine-Northern Hardwoods forest
and high mountains southward. Cymes large, expanding with the leaves or later, and
bordered with sterile flowers; leaves large, at first densely pubescent, later nearly glabrous
above but stellate-scurfy below, especially on veins; winter-buds conspicuous in late
summer, naked, consisting of tightly folded undeveloped leaves densely covered by rusty
to golden-brown pubescence, the veins of the outer pair clearly visible. The European
Wayfaring-tree, *V. lantana* L., sometimes planted, has similar winter-buds; its inflorescence
does not have marginal sterile flowers.

2. **Viburnum trilobum** Marsh. HIGHBUSH-CRANBERRY
   V. opulus L. var. *americanum* Ait.
   A tall shrub or small tree, 1–3 (–4) m. tall, of wet woods and borders of lakes and
streams, in northern regions from Newfoundland to British Columbia, southward into
the northern tier of states. The red, acid fruit used for preserves. This and the preceding
are the only native species of *Viburnum* which have the outer flowers of the cyme enlarged
and sterile; this and *V. acerifolium* are the only Ohio species with lobed leaves.

Viburnum alnifolium

Viburnum trilobum

3. VIBURNUM OPULUS L. EUROPEAN CRANBERRY-BUSH

The European *V. opulus* is very similar to *V. trilobum*, and best distinguished by the disk-like glands on petiole, slender, usually attenuate-tipped stipules, glabrous groove of petiole, and somewhat bitter fruit. It is frequently planted and is an occasional escape (reported from a few counties); the common snowball is a horticultural variety of *V. opulus* in which all the flowers are enlarged and sterile.

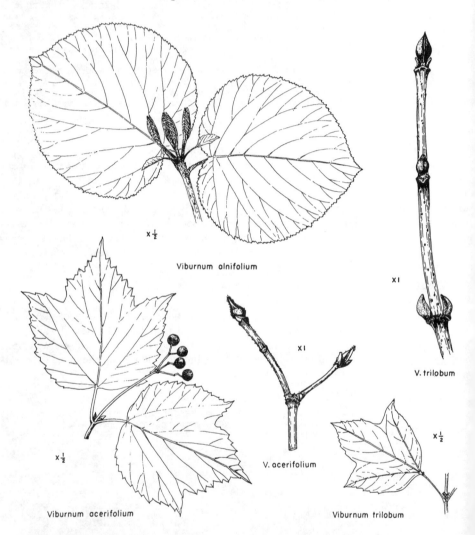

Viburnum alnifolium

x ½

x 1

V. trilobum

V. acerifolium

x 1

Viburnum acerifolium

x ½

Viburnum trilobum

x ½

4. **Viburnum acerifolium** L. MAPLE-LEAF VIBURNUM. ARROW-WOOD

Erect shrub, 1–2 m. tall, of dry or moist but well-drained soil in woodlands; often with beech, but not confined to any forest type. Ranging through much of the Deciduous Forest east of the Mississippi River. Widespread in Ohio, but less frequent in the western counties. Foliage turns carmine to purple in fall. Two intergrading varieties are recognized:

Leaves soft-downy beneath; young branchlets, petioles, and inflorescence finely pubescent....
  var. *acerifolium*
Leaves glabrous or glabrescent beneath except on veins; young branchlets glabrous.........
  var. *glabrescens* Rehd.

The latter is more southern and Appalachian in range; specimens from Adams Co.; specimen approaching this variety from Fairfield Co. cited by McAtee (1956). In var. *acerifolium* forma *ovatum* Rehd., the lateral lobes of leaves are very short or almost wanting; specimens from Coshocton, Lake, Lorain, and Scioto counties, and cited by McAtee (1956) from Jackson Co. The more or less copious dotting of the lower leaf-surface, and the usually lobed leaves are diagnostic characters of the species.

5. **Viburnum cassinoides** L. WITHE-ROD

A shrub of swamps and wet woods, or locally, of sandy oak woods; about 2 m. in height (or up to 4 m.); northern in range, and in the higher mountains southward. The crenulate leaf-margin and golden or yellow-brown scurf of winter-buds and somewhat scurfy-punctate young branchlets distinguish this from other similar species. Beautiful in late summer with its varicolored fruits.

6. **Viburnum rufidulum** Raf. SOUTHERN BLACK-HAW

V. rufotomentosum Small

Easily recognized by its thick lustrous leaves and rusty-red tomentose blunt buds and wide petioles. A southern species, often tree-like (10–12 m.), the dark bark deeply fur-

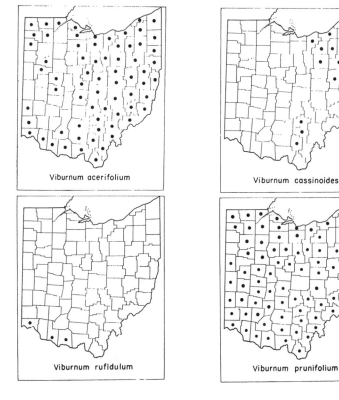

Viburnum acerifolium

Viburnum cassinoides

Viburnum rufidulum

Viburnum prunifolium

rowed and cut crosswise. Found in Ohio in a few southern counties, usually on open rocky calcareous slopes.

7. **Viburnum prunifolium** L. BLACK-HAW

A large shrub, sometimes tree-like and 3–5 m. tall, growing singly or in patches, in dry or wet situations. The leaves vary from narrowly elliptic to almost suborbicular, the drupes from ellipsoid to globose. More extensive in range than any of our other species, occurring throughout Ohio. Specimens of this species are sometimes mistaken for *V. lentago*, from which they differ in having usually blunt-tipped leaves (acute to obtuse), inconspicuously wavy-margined and usually more slender petioles.

8. **Viburnum lentago** L. NANNYBERRY

A large shrub of northern range, usually in swamps or wet woods; most frequent in

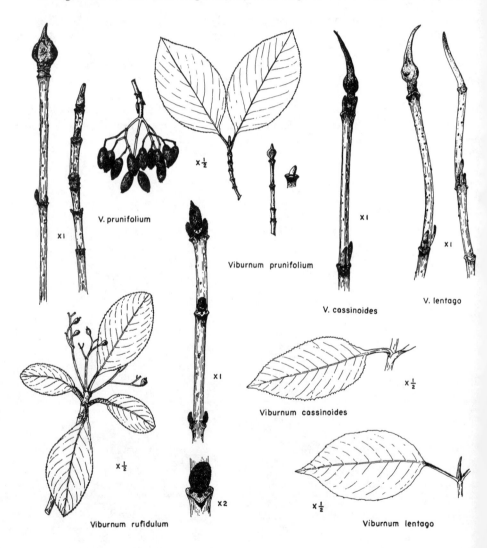

V. prunifolium

Viburnum prunifolium

V. cassinoides

V. lentago

Viburnum cassinoides

Viburnum rufidulum

Viburnum lentago

northern Ohio, local southward and on the flats of the Illinoian Till Plain. Similar in appearance to *V. prunifolium*; winter-buds longer, laterals often curved or flattened, 1–2 cm. long, terminal flower-bud longer, swollen near base (instead of toward middle), and leaves, at least some, long-acuminate.

9. **Viburnum rafinesquianum** Schultes DOWNY ARROW-WOOD
   V. pubescens of ed. 7, not Ait., *V. affine* Bush, var. *hypomalacum* Blake
   A low and spreading shrub distinguished from others of its group by the small sessile or subsessile, or short-petioled, often cordate leaves with relatively few teeth and linear-subulate stipules, and by the small cymes. Two varieties occur:

   Leaves soft-pubescent beneath with dense fascicled hairs, sessile or subsessile............
   var. *rafinesquianum*
   Leaves glabrous beneath except on veins; petioles about 1 (0.5–2) cm. long..............
   var. *affine* (Bush) House

   The two varieties intergrade; the typical is more frequent in Ohio.

10. **Viburnum molle** Michx.
    Erect shrub, 1–3 m. tall; "bark of young shrubs exfoliating like that of a birch" (Deam, 1924). Recognized by the ovate or nearly orbicular, cordate leaves, densely soft-pubescent beneath, petioles minutely red-glandular, usually with filiform stipules, and by the characteristic venation: 3 lowest pairs of lateral veins approximate, thus suggesting palmate venation; the lowest pair small, recurved, and ending in first marginal tooth; the second pair diverging perpendicularly, and branching on lower side, these branches ending in teeth; the third pair ascending and branching, similar to other veins arising from midrib above. Western-interior in range: Indiana to Missouri, south to Kentucky and Arkansas, and into Ohio.

11. The **dentatum** complex
    Two interpretations of this complex are made. Fernald (1941, 1950) recognizes two species, *V. dentatum* which is southern or southeastern ("Fla. to Tex., n. to se. Mass., s. R.I., L.I., N.J., Pa., W.Va. and Tenn."), and *V. recognitum* which is northern ("N.B. to s. Ont., s. to s. N.E., L.I., S.C., n. O. and Mich."). These differ in pubescence of stems, petioles, and rays of cymes, the former usually pubescent, the latter glabrous or pubescent only on veins beneath; both have comparable variation in leaf-outline. The only morphologic difference given is in shape of stone and depth of ventral groove of stone, characters which can be found in few specimens. The presence or absence of stipules is not a satisfactory diagnostic character. Fernald distinguishes only one variety, var. *deamii* (in

Viburnum lentago

Viburnum rafinesquianum
↓ var. rafinesquianum ← var. affine

Viburnum molle

addition to the typical) under his *V. dentatum*; this variety has "branchlets and petioles glabrous or essentially so," and is intermediate in range, "w. Pa. to e. Mo. and Tenn." Gleason (1952) recognizes one species, *V. dentatum*, which is made up of six varieties, differing in pubescence (amount and kind) and in stipules. One of these, var. *lucidum*, is equivalent to Fernald's *V. recognitum*; two, var. *indianense* and var. *deamii* are united under var. *deamii* by Fernald; the other three, two almost confined to the Coastal Plain

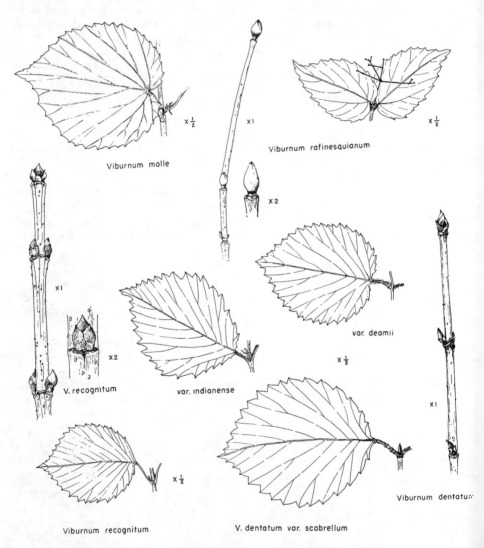

Viburnum molle

Viburnum rafinesquianum

var. deamii

V. recognitum

var. indianense

Viburnum dentatum

Viburnum recognitum

V. dentatum var. scabrellum

and one (var. *scabrellum*) to the Gulf Coastal Plain and southern Ohio, are included in *V. dentatum* (typical) as interpreted by Fernald.

It is evident from the published ranges, that Ohio material of the *dentatum* complex has not been studied. We have specimens of some form of the complex from about half of the counties, well distributed over the state except in the western part. If *recognitum* and *dentatum* (sensu Fernald) are distinct from one another where geographically isolated, such is not the case in Ohio where the supposed interval between ranges is occupied by many variants, some impossible to classify satisfactorily. The recombination of characters which can be observed suggests that our Ohio population is the result of interbreeding and backcrossing. Extensive study and collecting are necessary to determine the true relationship of plants of the *dentatum* complex. In our treatment, the essentially glabrous *V. recognitum* Fern. is segregated, and all other forms included in *V. dentatum*. *V. recognitum* is almost confined to the northeastern third of the state; the few specimens from the southern counties are not quite typical. *V. dentatum*, highly variable, is more southern in range. Two of its varieties, *deamii* and *indianense*, are in general interior in range; three, *dentatum*, *scabrellum*, and *venosum*, are largely confined to the Coastal Plain. The occurrence, in the rugged western part of the Appalachian Plateau, of taxa of southeastern range is not unusual. Our maps show the range of *V. recognitum* and of *V. dentatum* (all forms); the occurrence of varieties (when sufficiently representative) is noted in the text.

KEY TO OHIO REPRESENTATIVES OF THE *DENTATUM* COMPLEX

Petioles glabrous; leaves glabrous beneath except for tufts in vein-axils; cymes glabrous; stone "globose-ovoid, with shallow and broad trough-like furrow"..............12. *V. recognitum*
Petioles more or less pubescent, at least in furrow, with simple or fascicled hairs; stone ellipsoid-ovoid, with deep furrow-like ventral groove...........................13. *V. dentatum*
   Petioles pubescent only in furrow; leaves glabrous or nearly so, pubescent beneath only in vein-axils, and above in vein-furrows.
      Cymes glabrous except for bracts and nodes, where some fascicled hairs.......var. *dentatum*
      Cymes glabrous or nearly so (hairs not fascicled), but with some minute stipitate glands..
                                                                                var. *indianense*
   Petioles pubescent on back as well as in furrow, with simple or fascicled hairs.
      Leaves more or less scabrous on upper surface, pubescent beneath with fascicled hispid hairs; petioles densely pubescent with fascicled hairs.
         Leaves scabrous above, relatively thin; corolla (outer surface) and young fruit hispid..
                                                                                var. *scabrellum*
         Leaves scarcely scabrous above, thick; veins beneath prominently elevated; young fruit glabrous or glandular..var. *venosum*
      Leaves but slightly scabrous above, and more or less hispid pubescent, pubescent beneath at least on veins, the pubescence soft; hairs of petioles mostly separate or sometimes 2 together...var. *deamii*

12. **Viburnum recognitum** Fern. ARROW-WOOD
   V. dentatum of ed. 7, not L., *V. dentatum* L. var. *lucidum* Ait. (sensu Gleason)
   Very similar to *V. dentatum* (sensu Fernald); essentially glabrous—branchlets, petioles,

Viburnum recognitum

Viburnum dentatum

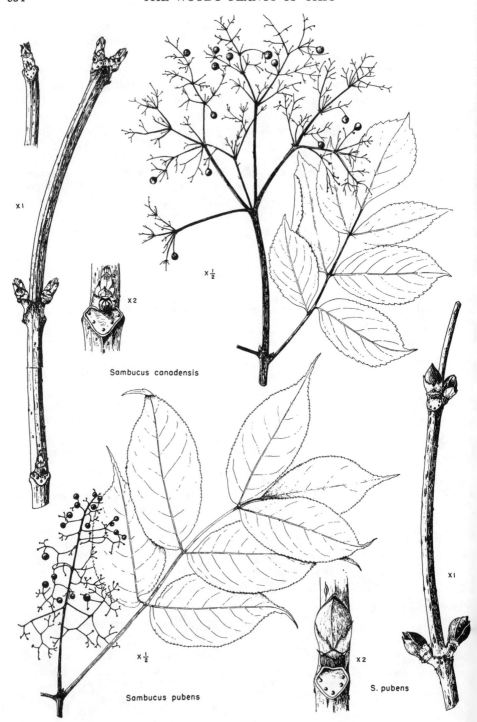

Sambucus canadensis

Sambucus pubens

S. pubens

leaves (except sometimes veins and vein-axils beneath), and cymes. Northeastern in range, extending into Ohio from the northeast.

13. **Viburnum dentatum** L. (sensu Fernald)

A polymorphic species including all variations of the complex except the essentially glabrous *V. recognitum*. As interpreted here, five more or less distinct varieties are recognized.

V. dentatum var. **dentatum**

A northern form of this southern variety, formerly known as *V. venosum* Britt. var. *canbyi* Rehd., is represented by one specimen from Scioto Co. and is also reported by McAtee (1956) from Scioto Co.

V. dentatum var. **indianense** (Rehd.) Gleason

V. pubescens var. *indianense* Rehd.

Rather frequent on the Illinoian Till Plain of southwestern Ohio in open wet woods and meadows. An almost glabrous taxon, with petioles pubescent only along the furrow; leaves beneath pubescent in vein-axils and on main veins toward base. Brown, Clermont, Coshocton, Franklin, Licking, and Perry counties.

V. dentatum var. **deamii** (Rehd.) Fern. (sensu Gleason)

A more or less pubescent form, with petioles pubescent on back as well as in furrow. Adams, Delaware, Highland, Franklin, and Muskingum counties.

V. dentatum var. **scabrellum** T. & G.

V. scabrellum (T. & G.) Chapm.

More or less frequent in southern Ohio in the western border of the Allegheny Plateau. Intermediates occur which suggest var. *deamii* or var. *dentatum*. Adams, Highland, Jackson, Lawrence, Ross, Scioto.

V. dentatum var. **venosum** (Britt.) Gleason

A form with more or less orbicular leaves with veins beneath elevated and prominent. Jackson Co.

SAMBUCUS L. ELDER

Deciduous shrubs or small trees, or rarely herbs; about 40 species in the north temperate zone and mountains of the tropics. Stems stout, pithy; leaves opposite, pinnate, leaflets serrate. Flowers in compound cymes, saucer-shaped or open urn-shaped, deeply 5-lobed; fruit berry-like, juicy.

Branchlets yellowish gray, pith white; young branchlets and leaves glabrous; flowers white, fragrant, in flat or slightly convex cymes; berries usually purple-black.......1. *S. canadensis*
Branchlets pale yellow-brown, pith brown; young branchlets and leaves finely pubescent; flowers white (buds often red), in ovoid or pyramidal cymes; berries usually red..2. *S. pubens*

WINTER KEY

Pith white; buds broad-based, conical, light brown and green, with tendency to grow in winter; 4–5 pair of bud-scales visible...1. *S. canadensis*
Pith brown.
  Buds large, 10 mm. long, solitary, almost globular, red, stalked; 2–3 pair of bud-scales visible, outermost very small..2. *S. pubens*
  Buds small, often multiple...*S. racemosa*

1. **Sambucus canadensis** L. COMMON ELDER

Stoloniferous shrub, to 3 m. tall, the stems simple or few-branched; pith large and wood-cylinder thin. Berries purple-black (or red in forma *rubra*, yellow or orange in forma *atroflavula*, greenish in forma *chlorocarpa*). Plants of more western range with leaves grayish-green and soft pubescent beneath are distinguished as var. *submollis* Rehd. Our plants belong to the typical variety. Ranges almost throughout the Deciduous Forest; probably in every county in Ohio. Common in rich moist soil, in woods, fields, and on roadsides (where they are being destroyed by poison spray).

2. **Sambucus pubens** Michx. RED-BERRiED ELDER

S. racemosa of Am. auth., not L.

A large shrub, 3–4 m. tall, or occasionally as much as 8 m. Variations in foliage and in fruit color are the basis of form names. A northern and mountain species, ranging across the continent in the North—Newfoundland to Alaska, southward into New England and the northern states, and in the Appalachian Highland to Georgia and Tennessee, in the Rocky Mountains and Cascade Range to New Mexico and Oregon. Most frequent in northeastern Ohio, local southward; the Clifton Gorge locality is definitely disjunct. The European red elder, *S. racemosa* L., is sometimes planted and may be distinguished from the American species by its more compact cymes, reflexed lower branches, and glabrous branchlets and leaves.

Sambucus canadensis

Sambucus pubens

GLOSSARY

Abruptly pinnate. Pinnate, without a terminal leaflet. See Fig. 2 I.

Accessory buds. Buds accompanying the axillary or lateral bud; see *collateral* and *superposed.*

Achene. A small dry one-seeded indehiscent fruit.

Acuminate. Tapering gradually to a point, the sides usually slightly concave. See Fig. 3.

Acute. Pointed; tapering to a point, the sides straight or slightly convex.

Adnate. United to, as stipules attached to petiole. See *Rosa*, p. 212.

Adventitious. Not in the usual place.

Adventive. Imperfectly naturalized; not fully established.

Aerial roots. Roots produced above ground; seen on climbing stems of poison ivy and trumpet-creeper (p. 314).

Aggregate fruit. Made up of matured ovaries of a number of pistils of a single flower. See *Magnolia*, pp. 140, 142, 144.

Alternate. Of leaves, leaf-scars, etc.—one at a node. See Fig. 1 A.

Ament. See catkin.

Anther. The pollen-bearing part of a stamen. See Fig. 4.

Apetalous. Without petals.

Apiculate. Ending in a short pointed tip. See Fig. 3.

Appressed. Lying flat against; of hairs lying flat against leaf-surface.

Arborescent. Tree-like; becoming a small tree.

Aril. An appendage, often fleshy, attached at or about the hilum (the point of attachment of stalk to seed) of a seed and tending to envelop it.

Aristate. Tipped with an awn; tapering to a bristle-like point. See Fig. 3.

Attenuate. Gradually tapering to a slender point.

Auricle. Ear-shaped appendage or lobe, as at base of leaf. See Fig. 3.

Awl-shaped. Tapering from base to a slender stiff point.

Awn. A bristle-like appendage.

Axil. The angle above a leaf, between leaf and stem, or the point above a leaf-scar.

Back-cross. Cross between hybrid and one of the parent species.

Berry. A fleshy, usually small, fruit developed from a single ovary, the seeds (not stones) embedded in the pulp.

Bipinnate. Twice pinnate. See Fig. 2 I.

Blade. The expanded portion of a leaf. See Fig. 2 A, B.

Bloom. A whitish powdery, usually waxy, coating on leaves, stems, fruits (as on a blue plum), and easily rubbed off.

Bract. A modified leaf, often small, in flower-cluster, or sometimes on stem. See *Rubus frondosus*, p. 208; *Vaccinium stamineum*, p. 294.

Bractlet. A bract borne on a secondary axis; a bract on a flower-pedicel.

Branchlet. Ultimate division of a branch, here used for leafy branchlet; see also *twig.*

Bristle. A stiff strong hair.

Bud-scales. The scale-like structures enveloping a winter-bud.

Bundle-scars. Small scars or dots within the leaf-scar, formed at broken ends of vascular bundles passing from stem to leaf. See Fig. 1.

Calyx. The outer of two floral envelopes or series of floral leaves, sometimes the inner series (corolla) lacking; composed of sepals, which may be distinct or united. See Fig. 4.

Campanulate. Bell-shaped.

Canescent. Hoary with a gray pubescence.

Capsule. The dry dehiscent fruit resulting from maturing of a compound ovary, i.e., one with 2 or more carpels. See p. 18.

Carpel. One of the highly modified foliar units of a compound pistil, or the one unit of a simple pistil.

Catkin. A dense scaly-bracted flexuous spike or spike-like inflorescence, often pendant, and usually unisexual. See *Salix, Corylus, Castanea.* An ament.

Caudate. With a slender tail-like appendage or tip.

Cauaex. The thickened and persistent base or stem of a perennial plant.

Chambered pith. Pith with cavities separated by plates or disks. See walnut, p. 92.

Ciliate. With marginal hairs; fringed. See Fig. 3.

Ciliolate. Minutely ciliate.

Clavate. Club-shaped; gradually increasing in diameter upward.

337

Clone. A group of individuals that have descended from one plant by vegetative propagation.

Collateral buds. Buds side-by-side; often, flower-buds on either side of a vegative lateral bud. See *Lindera.* p. 148, and *Prunus angustifolia,* p. 218.

Colonial. Forming colonies, usually from root-shoots.

Compound leaf. A leaf composed of two or more leaflets. See Fig. 2 G, H, I, J.

Cone. A structure composed of central axis to which are attached bracts and seed- or pollen-bearing structures, the whole detachable and fruit-like; characteristic of conifers, p. 64.

Connate. Joined or united; grown together, as upper leaves of some species of *Lonicera.* See Fig. 3 and p. 320.

Cordate. Heart-shaped, with the point apical. See Fig. 3. Term applied to whole leaf or to base only.

Coriaceous. Firm and leathery in texture.

Corolla. The inner of the two floral envelopes, consisting of petals, distinct or united. See Fig. 4.

Corymb. A flat or round-topped inflorescence with relatively short axis, flowering from the margin upward and inward. (p. 15).

Cotyledon. A seed leaf; a leaf of embryonic plant within seed.

Crenate. Scalloped with rounded teeth. See Fig. 3.

Crenulate. Finely crenate.

Cuneate. Wedge-shaped, narrowly triangular, and attached at the sharp angle. See Fig. 3.

Cuspidate. Tipped with a sharp firm point.

Cyme. A broad flat-topped or slightly convex inflorescence in which center or terminal flower of each branch opens before lower flowers; p. 15.

Deciduous. Falling away at the close of the growing season; said of trees which lose their leaves in winter.

Decumbent. Stems reclining, but with ends ascending.

Decurrent. Extending downward on the stem and attached as a ridge or wing.

Dehiscence. The opening of a dry fruit or anther at maturity.

Deltoid. Triangular, with more or less equal sides.

Dentate. Toothed, with teeth projecting outward or perpendicular to margin. See Fig. 3.

Denticulate. Finely dentate.

Diaphragm. A partition or dividing membrane, as seen at the nodes of grape (p. 264) or in the pith of walnut (p. 92).

Dichotomous. Forking regularly into two nearly equal branches.

Diffuse. Widely or loosely branched or spreading.

Digitate. Compound, with the members arising from one point, as the fingers of a hand. Fig. 2 J.

Dimorphic. Having 2 forms.

Dioecious. Unisexual, with the two kinds of flowers on separate plants.

Diploid. An organism containing two sets of chromosomes derived from egg and sperm; the basic number of chromosomes in the nuclei.

Divaricate. Widely divergent.

Downy. Covered with short soft spreading hairs.

Drupe. A stone-fruit; a fleshy fruit with seed enclosed in a hard covering (the inner layer of the fruit-wall), as in a plum or peach. Many drupes are small and look like berries.

Drupelet. A small or diminutive drupe, as one of the divisions of a blackberry.

Echinate. Prickly, the prickles often stout.

E-. A prefix denoting absence.

Eciliate. Without cilia.

Eglandular. Without glands.

Ellipsoid. A solid which is elliptical in section.

Elliptic. With the form of an ellipse, longer than broad and rounded about equally to both ends. See Fig. 3.

Emarginate. With a shallow notch at the apex. See Fig. 3.

Endemic. A native species confined, geographically, to a single, often restricted, natural area.

Entire. With the margin continuous, not toothed or lobed. See Fig. 3.

Epigynous. With perianth and stamens appearing to arise from summit of ovary. See Fig. 4 D.

Erose. With margin appearing as if gnawed or eroded, but not regularly toothed.

Escape. A plant which is usually seen in cultivation, but which occasionally grows as if wild, i.e., as an escape.

Excurrent. Extending out or projecting beyond the margin, as the midrib and principal veins of members of the red oak group (pp. 124 126, 128).

Exfoliating. Peeling or separating in thin plates or shreds.

Falcate. Scythe-shaped; curved and flat, tapering to a point.

Fascicle. A dense cluster.

Filament. A thread or thread-like structure; the stalk of a stamen which supports the anther. See Fig. 4 A.

Filiform. Thread-like.

Fimbriate. Fringed.

Floccose. Clothed with tufts of soft woolly, easily detachable, hairs.

Flocculent. Diminutive of floccose.

Floricane. The flowering cane in *Rubus*, usually a 2-year old stem.

Flower-bud. A winter-bud containing an undeveloped flower or flower-cluster.

Follicle. A fruit with a single carpel and dehiscing along one suture.

Forma (pl. *formae*). Often expressed by the English word form. A minor category under species. The species may include subspecies, varieties, and forms, in descending rank.

Fruit. The ripened ovary, with or without accessory parts; the seed-bearing organ.

Fruit-scar. The scar left by the falling of a fruit.

Fusiform. Spindle-shaped; narrowed to both ends from the middle portion.

Gene. A unit concerned with the determination of hereditary characters; an element of the germ-plasm.

Gibbous. Swollen or enlarged on one side.

Glabrate. Becoming glabrous with age, or nearly glabrous.

Glabrous. Without hairs, although not necessarily smooth.

Gland. A secreting structure; glands (on plants) may be external and sessile (appearing as dots) or stalked, or internal.

Glandular. With glands, or gland-like.

Glandular-pubescent. With hairs terminated by glands (small glandular knobs) or with hairs and glands intermixed.

Glaucescent. Slightly glaucous; tending to be glaucous.

Glaucous. Covered with a fine bluish or whitish bloom; whitened with a thin often removable coating, usually waxy in nature.

Glutinous. Sticky; having a sticky exudation.

Habitat. The kind of place in which a plant (or other organism) lives; *not* the locality.

Hastate. Shaped like an arrow-head, but with basal lobes pointing outward. See Fig. 3.

Head. A dense, often spherical flower-cluster, the flowers sessile or nearly so on a short axis. See *Cephalanthus*, p. 316, *Platanus*, p. 160.

Hexaploid. With three times the chromosome complement of a diploid.

Hirsute. Pubescent with rather coarse and stiff hairs.

Hirtellous. Minutely hirsute.

Hispid. Beset with rigid or bristly hairs.

Hybrid. A plant resulting from a cross between unlike parents, usually, of two different species.

Hypanthium. An expansion of the receptacle, usually derived from fusion of floral series. See p. 17.

Hypogynous. With stamens and petals attached to the receptacle below the ovary. See Fig. 4 A.

Imbricate. Overlapping vertically, as the shingles of a roof, or spirally, then successively inwrapping.

Imperfect. Said of flowers having either stamens or pistils, but not both.

Impressed. With the leaf-surface bent inward at veins.

Incised. Cut sharply and irregularly, more or less deeply.

Indehiscent. Not opening at maturity; a term usually applied to fruits.

Indigenous. Native to area.

Inflorescence. The flowering part of a plant, including axis, bracts, and pedicels, as well as flowers; more correctly, the mode of arrangement. See p. 15.

Infrastipular. Below the stipules; *infra-*, below.

Internode. The part of a stem between two nodes. See Fig. 1.

Introgression. The result of infiltration of germ-plasm of one species into another; *introgressive hybridization*, the infiltration of genes of one species into germ-plasm of another as a result of hybridization and repeated back-crossing.

Introduced. Brought into an area, intentionally or accidentally, from another region, often from another continent.

Involucel. A secondary involucre about a portion of a flower-cluster.

Involucre. A set of bracts surrounding a flower-cluster or a single flower. See enlarged involucre of a hazelnut, p. 102.

Lanceolate. Much longer than wide; tapering from above base to apex. See Fig. 3.

Lateral bud. A bud situated on the side of a twig, usually just above leaf-scar. See Fig. 1.

Lateral veins. Veins on either side of the midrib.

Leaf-scar. The scar left on twig by the falling of a leaf. See Fig. 1.

Leaflet. One part of a compound leaf.

Legume. A dry dehiscent fruit resulting from ripening of a simple ovary, and splitting along both sutures; a pod; the characteristic fruit of the *Leguminosae.*

Lenticel. Corky spots or prominences breaking through the bark; often of contrasting color.

Linear. Long and narrow, the sides parallel. See Fig. 3.

Lobed. Divided rather deeply. See Fig. 2 E, F.

Lobulate. Divided into small lobes.

Loculicidal. Dehiscing on the back, into the cavities of the capsule.

-merous. In composition, refers to number of parts, as a 4-merous flower (some species of *Euonymus,* p. 242).

Mesic. A mesic habitat is one with moderate or intermediate moisture conditions.

Mesophyte. A plant growing in a habitat with moderate moisture conditions, neither dry nor wet.

Midrib. The central or main vein of a leaf or leaf-like structure.

Monoecious. Having staminate and pistillate flowers on the same plant.

Monotypic. Used of a genus containing only one species.

Mucronate. With a short stiff point. See Fig. 3.

Multiple fruit. A fruit formed from several flowers, as a mulberry or osage-orange.

Naked bud. A bud without specialized bud-scales. See *Hamamelis,* p. 157, and *Viburnum alnifolium,* p. 328.

Native. A plant belonging to an area, originally present, not brought in accidentally or intentionally.

Naturalized. Thoroughly established and reproducing, thus appearing as if native, but originally from some other region.

Node. The part of a stem from which leaves arise or may arise. See Fig. 1.

Nut. A one-celled and one-seeded indehiscent hard and bony fruit.

Nutlet. A small nut, distinguished from a nut only by size, and from an achene by hardness and thickness of wall. Also used for the hard bony seeds in pome of *Crataegus.*

Ob-. A prefix signifying inversion.

Obconic. Inversely conical, with attachment at the point.

Obcordate. Inverted heart-shaped; deeply lobed at apex. See Fig. 3.

Oblanceolate. Much longer than wide and tapering from near apex to base; the opposite of lanceolate.

Oblique. Said of leaf when unequal-sized and slanted at base. See Fig. 3.

Oblong. Longer than broad, and with sides nearly parallel. See Fig. 3.

Obovate. The reverse of ovate, broadest above middle.

Obovoid. Egg-shaped, but attached at smaller end.

Obtuse. Blunt or rounded at apex. See Fig. 3.

Odd-pinnate. Pinnate, with a terminal (odd) leaflet, in contrast with abruptly pinnate. See Fig. 2 H.

Opposite. Two at a node, on opposite sides of stem, applied to leaves and leaf-scars. See Fig. 1 B.

Orbicular. Circular. See Fig. 3.

Oval. Broadly elliptic.

Ovary. The ovule-bearing part of a pistil (also called *ovulary*). See Fig. 4.

Ovate. Widest below the middle, and shaped like section of an egg. See Fig. 3.

Ovoid. Egg-shaped.

Ovule. The body which, after fertilization, becomes the seed.

Palmate, palmately lobed, palmately veined. Diverging radiately from a common point. See Fig. 2 D, F, J.

Panicle. A branched inflorescence, longer than broad, the branches arranged as in a raceme, and at least the lower secondarily branched; flowering from base toward summit. See p. 16.

Papilionaceous. Corolla as in a sweet pea, with standard, wings, and keel; see subfamily III of Leguminosae, p. 227.

Papillose (*papillate*). With minute blunt projections or papillae.

Pectinate. With narrow closely-set segments; Comb-like.

Pedicel. The stalk of an individual flower in a flower-cluster.

Pedicellate. Having pedicels.

Peduncle. The stalk of a flower-cluster, or of a flower if a solitary-flowered inflorescence.

Pellucid. Clear or translucent.

Pellucid-dotted. With clear or translucent spots, as seen when held to the light.

Peltate. Attached to stalk inside of margin. See Fig. 3.

Perfect. Said of flowers having both stamens and pistils.

Perfoliate. Said of a leaf or bract when base of blade surrounds stem, so that stem appears to pass through leaf. See Fig. 3.

Perianth. The two floral envelopes, calyx and corolla, considered together.

Perigynous. Said of a flower with calyx-tube or hypanthium surrounding ovary but free from it or only partly adnate, the calyx-lobes, petals, and stamens then arising from margin of cup. See Fig. 4 B, C.

Persistent. Long-continuing, remaining attached.

Petal. One unit of the inner floral envelope or corolla. See Fig. 4 A.

Petiole. The stalk of a leaf. See Fig. 2 A, B.

Petiolule. The stalk of a leaflet; *petiolulate,* having the leaflets stalked.

Pilose. With long soft hair.

Pinna (pl. *pinnae*). One of the divisions of a pinnate leaf.

Pinnate. A compound leaf with the leaflets arranged on either side of a central axis or rachis. Fig. 2 H.

Pinnately veined. With the principal lateral veins arising from either side of the midrib. Fig. 2 C.

Pinnatifid. Pinnately cleft or parted. See *Comptonia,* p. 90.

Pistil. The organ in a flower which bears ovules and later seeds; composed of ovary (the ovule-bearing part), style (which may be wanting), and stigma. See Fig. 4 A.

Pistillate. Having pistils, but no stamens (at least not functional stamens).

Pith. The central soft spongy cylinder of angiosperm stems. See Fig. 1 A.

Pod. A legume, the fruit of members of the Leguminosae. Loosely used for any dehiscent dry fruit.

Polymorphic. Having various forms or characters.

Polyploid. Having more chromosomes than the basic number of a diploid.

Pome. The apple-type of fruit, more or less fleshy, with a papery or bony core at center, and sepals (or their scars) crowning the fruit.

Prickle. A sharp pungent outgrowth from epidermis or cortex; not connected with the vascular system.

Primocane. The first year's cane of certain woody plants, especially *Rubus;* not flower-bearing until second year, when stem is a *floricane.*

Procumbent. Trailing, lying on the ground, but not rooting at nodes.

Pruinose. Having a waxy coating or bloom on surface.

Puberulent. Minutely pubescent, the hairs soft and straight, but not evident to the naked eye.

Pubescent. Covered with hairs, usually short and soft.

Pulvinus. An enlargement or swelling of petiole or petiolule at point of attachment.

Punctate. Dotted with depressions or with translucent glands or colored dots.

Pustulate. Having slight blister-like elevations.

Pyriform. Pear-shaped.

Raceme. A more or less elongate unbranched inflorescence with the flowers stalked; lowest flowers open first.

Rachis. Axis along which leaflets of a pinnate leaf are attached, or axis of an inflorescence.

Rank. A term applied in connection with leaf (or leaf-scar) arrangement, and referring to the number of vertical rows; 2-ranked, leaves in 2 vertical rows; 3-ranked, in 3 vertical rows. See p. 9.

Receptacle. The more or less enlarged or elongated end of stem to which the floral parts are attached. See Fig. 4 A.

Regular flower. One in which the members of each whorl of parts are alike in size and shape.

Relic. A term applied to localized plants or plant communities which are survivors of an earlier geologic time.

Reniform. Kidney-shaped. See Fig. 3.

Repand. With a slightly wavy margin. See Fig. 3.

Resinous. With conspicuous resin; appearing shiny and gummy.

Reticulate. In the form of a network, netted.

Retuse. With a shallow notch at a rounded end.

Revolute. Rolled back; said of leaves when margin is rolled under and therefore visible from beneath. See *Andromeda,* p. 292.

Rhombic. Shaped more or less like a parallelogram with its angles oblique.

Rostrate. Having a beak.

Rotate. Wheel-shaped.

Rugose. Wrinkled, usually because of impressed veins.

Rugulose. Diminutive of rugose.

Samara. An indehiscent winged fruit, as of elm (p. 134), maple (p. 246), or ash (p. 308).

Scabrous. Rough to the touch.

Scarious. Thin, dry and papery, not green; *scarious-margined,* said of leaves with narrow colorless (or straw-colored) dry margin.

Scurfy. With scale-like particles, scaly rather than hairy.

Second-growth forest. One which has grown after disturbance, by fire, cutting, grazing, etc.

Seed. A ripened ovule; should not be confused with small dry indehiscent fruits.

Sepal. One member of the outer floral envelope or calyx. See Fig. 4 A.

Septicidal. A mode of dehiscence in which the splits occur along the partitions; contrasted with *loculicidal.*

Serrate. Saw-toothed, with the teeth sharp and pointing forward (toward apex). See Fig. 3.

Serrulate. Finely serrate.

Sessile. Not stalked.

Setose. Beset with bristles.

Simple. Said of a leaf with only one part, as opposed to compound. See Fig. 2, A–F.

Sinus. The space or recess between two lobes of a leaf.

Spatulate. Somewhat spoon-shaped; oblong, with basal end attenuate. See Fig. 3.

Spike. A simple (unbranched) inflorescence with the flowers sessile or nearly so, and opening progressively from bottom toward apex.

Spine. A sharp woody outgrowth on a stem which arises from or contains vascular tissue. Same as thorn, and contrasted (as to origin) with prickle.

Spinescent. Becoming spine-like; ending in a spine. See *Pyrus coronaria*, p. 163.

Spontaneous. Appearing unexpectedly, not planted.

Stalked bud. A bud definitely narrowed near base, with outer scales above base.

Stamen. A pollen-bearing organ of a seed-plant, typically composed of filament (sometimes wanting) and anther. See Fig. 4 A.

Staminate. Having stamens, but no pistils.

Stellate. Star-shaped; said of hairs with radiating branches; often applied to groups of hairs in which the individual hairs radiate from center.

Stigma. The part of a pistil which receives the pollen. See Fig. 4 A.

Stipellate. Having stipellae; said of leaflets which have stipule-like structures.

Stipitate glands. Glands with stalk-like base.

Stipule. An appendage at the base of a petiole, and often adnate to it; sometimes modified into tendrils or spines. Fig. 2 A, B.

Stipule-scar. Scar left on twig after stipules fall. See *Magnolia macrophylla*, p. 142.

Stoloniferous. Producing stolons, i.e. shoots which arch over and take root, or, which extend horizontally at or below ground-level and produce upright stems.

Striate. With fine longitudinal lines, ridges, or channels.

Strobile. A cone; an inflorescence with imbricated bracts or scales.

Style. The usually slender and more or less elongate part of a pistil between ovary and stigma. Fig. 4 A.

Sub-. As a prefix, denotes somewhat, slightly, or nearly.

Subulate. Awl-shaped; tapering from base to apex.

Suffruticose. Low and somewhat woody at base.

Superposed buds. One bud above another; an axillary bud and one or more other buds above it.

Supra-axillary. Above the axil.

Syncarp. A fruit made by the union or crowding of carpels of one or of several flowers.

Synonym. A name no longer in use, but formerly applied to the taxon in question; a name subsequently given to an already-described taxon.

Taxon (pl. *taxa*). A general term which may be applied to a taxonomic unit of any rank, to a genus, a species, a variety, etc.

Tendril. A slender clasping or coiling structure modified from leaf or leaflet (*Bignonia*, p. 314), stipule (*Smilax*, Fig. 2 B and p. 70); or stem (*Vitis*, p. 262).

Terete. Circular in cross-section.

Terminal bud. The bud which terminates a stem; should be distinguished from uppermost lateral on those twigs where there is no true terminal bud, but instead a tip-scar. See Fig. 1.

Ternate. In threes; a compound leaf with three leaflets. Fig. 2 G.

Tetraploid. With twice the chromosome complement of a diploid.

Thorn. A spine.

Tip-scar. A scar left at end of twig by the early falling of terminal bud or shoot-tip (before winter); often small and inconspicuous, and close to uppermost lateral bud. Fig. 1 A.

Tomentose. Densely woolly, with a matted or woolly pubescence.

Tomentulose. Finely tomentose.

Tomentum. Close woolly hair.

Trichome. A hair or hair-like outgrowth of the epidermis.

Tri-. In composition, three.

Trifoliate. With 3 leaflets.

Trilobate. Three-lobed.

Truncate. Appearing as if cut off; said of apex or base of leaf when nearly straight across. See Fig. 3.

Tuberculate. Warty, or with minute protuberances.

Turbinate. Top-shaped, or inversely conical.

Twig. A young woody stem, the terminal, usually one-year old, part of a branch. Here used for the winter-twig, the leafless branchlet.

Type. The specimen from which the original description was made.

Typical. Referring to the originally described variation of a species.

Umbel. An inflorescence in which the peduncles or pedicels of the cluster arise from the same point; usually a more or less flat-topped cluster, the outer flowers opening first.

Umbellate. In or like an umbel.

Umbo. A stout or conical projection arising from the surface, as the umbo on a pine-cone scale.

Undulate. With wavy margin, sinuate.

Unisexual. Of one sex; either staminate or pistillate.

Urceolate. Urn-shaped; hollow and cylindric or ovoid, somewhat contracted at the throat.

Valvate. Meeting at the edges but not overlapping.

Variety. A taxon of lesser rank than the species.

Vascular. Furnished with vessels or ducts; *vascular flora,* that part of the flora whose members have vascular tissue, i.e., the ferns and fern-allies, the gymnosperms, and the angiosperms.

Vascular bundle. A strand of woody fibers together with associated tissues; a conducting strand.

Veins. The vascular or woody bundles of a leaf.

Vein-eyelet. The small area enclosed by veinlets, the ultimate divisions of veins.

Verticillate. Arranged in whorls, with 3 or more leaves or branchlets at a node.

Villous. With long soft but not matted hairs.

Virgin forest. One which has reached maturity through natural processes of development and has not been influenced by human activity.

Viscid. Sticky or glutinous.

Whorled. Arranged in a circle around the stem; three or more leaves or leaf-scars at a node.

Winter-bud. An overwintering bud (leaf or shoot bud, flower bud, or mixed bud) which begins to develop early in summer and reaches full size at about the time of or before leaf-fall.

Wood-cylinder. The cylinder of wood surrounding the pith.

×-. Used before a scientific name denotes a hybrid, as it also does when between two names.

Xeric. A xeric habitat is a dry habitat, a habitat with low soil water content and often with high evaporation rate.

Xerophyte. A plant that normally grows in a dry habitat.

LIST OF ABBREVIATIONS OF AUTHORS' NAMES

Adans.—ADANSON, M.
Ait.—AITON, W.
Anderss.—ANDERSSON, N. J.
Arnold—ARNOLD
Arn.—ARNOTT, G. A. W.
Ashe—ASHE, W. W.
Audubon—AUDUBON, J. J.

Bailey—BAILEY, L. H.
Ball—BALL, C. R.
Bart.—BARTON, B. S.
Bartr.—BARTRAM, W.
Batsch—BATSCH, A. J. G. K.
Beadle—BEADLE, C. D.
Bebb—BEBB, M. S.
Benth.—BENTHAM, G.
Best—BEST, G. N.
Bge.—BUNGE, A. VON
Bigel.—BIGELOW, J.
Blake—BLAKE, S. F.
Blanch.—BLANCHARD, W. H.
Bl.—BLUME, K. L.
Borkh.—BORKHAUSEN, M. B.
Britt.—BRITTON, N. L.
Buckl.—BUCKLEY, S. B.
Bur.—BUREAU, E.
Butters—BUTTERS, F. K.

Camp—CAMP, W. H.
Carr.—CARRIÈRE, É. A.
Chapm.—CHAPMAN, A. W.
Cockerell—COCKERELL, T. D. A.
Coult.—COULTER, J. M.
Cov. & Britt.—COVILLE, F. V. & BRITTON, N. L.
Crépin—CRÉPIN, F.

DC.—DeCANDOLLE, A. P.
DC., A.—DeCANDOLLE, A.
Dcne.—DECAISNE, J.
Desf.—DESFONTAINES, R. L.
Desmarais—DESMARAIS, Y.
Dipp.—DIPPEL, L.
Dode—DODE, L. A.
Dole—DOLE, E. J.
Donn—DONN, J.

Duham.—DUHAMEL DU MONCEAU, H. L.
Dunal—DUNAL, M. F.
Durazz.—DURAZZINI, A.
Du Roi—DU ROI, J. P.

Eat.—EATON, A.
Egglest.—EGGLESTON, W. W.
Ehrh.—EHRHART, F.
Ehrh. B.—EHRHART, B.
Ell.—ELLIOTT, S.
Ellis—ELLIS, J. E.
Endl.—ENDLICHER, S. L.
Engelm.—ENGELMANN, G.
Engl.—ENGLER, H. G. A.
Engl. V.—ENGLER, V.

Farw.—FARWELL, O. A.
Fern.—FERNALD, M. L.
Fluegge—FLUEGGE, J.
Focke—FOCKE, W. O.
Forbes—FORBES, J.
Fritsch, K.—FRITSCH, K.

Gaertn.—GAERTNER, J.
Gleason—GLEASON, H. A.
Gmel.—GMELIN, S. G.
Goldie—GOLDIE, J. G.
Graebn.—GRAEBNER, P.
Gray—GRAY, ASA
Greene—GREENE, E. L.
Griggs—GRIGGS, R. F.
Gronov.—GRONOVIUS, J. F.

Hand-Mazz.—HANDEL-MAZZETTI, H. VON
HBK.—HUMBOLDT, F. W. H. A. VON, BONPLAND, A., & KUNTH, C. S.
Hedrick—HEDRICK, U. P.
Heller—HELLER, A. A.
Henry—HENRY, A. H.
Hill—HILL, J. H.
Hill, E. J.—HILL, E. J.
Hitchc.—HITCHCOCK, A. S.
Hook.—HOOKER, W. J.

344

Hort.—Hortorum or hortulanorum; of gardens or gardeners.
House—House, H. D.
Houtt.—Houttuyn, M.
Huds.—Hudson, W.
Hyland.—Hylander, N.

Jacq.—Jacquin, N. J.
Jaeg.—Jaeger, H.
Juss.—Jussieu, A. L. de

Kalm—Kalm, P.
Karst.—Karsten, H.
Kerner—Kerner, J. S. von
Kirchn.—Kirchner, G.
Knerr—Knerr, E. B.
Koch, K.—Koch, K.
Koehne—Koehne, E.
Ktze.—Kuntze, O.

L.—Linnaeus, Carolus, or Linné, Carl von
L. f.—Linné, Carl von (the son)
Lam.—Lamarck, J. B. P. A. Monet
Lamb.—Lambert, A. B.
Lange—Lange, J. M. C.
Lauche—Lauche, W.
Laxm.—Laxmann, E.
Le Conte—Le Conte, J. E.
L'Her.—L'Héritier de Brutelle, C. L.
Lieblein—Lieblein, F. K.
Lindl.—Lindley, J.
Link—Link, H. F.
Lodd.—Loddiges, C.
Loisel.—Loiseleur-Deslongchamps, J. L. A.
Loud.—Loudon, J. C.
Lour.—Loureiro, J.

MacM.—MacMillan, C.
Mak.—Makino, T.
Marsh.—Marshall, H.
Mart.—Martfeld, J.
Maxim.—Maximowicz, C. J.
Med, Medic.—Medicus, F. C.
Meyer—Meyer, E. H. F.
Michx.—Michaux, A.
Michx. f.—Michaux, F. A.
Mill.—Miller, P.
Miq.—Miqel, F. A. W.

Moench—Moench, C.
Muell., P. J.—Mueller, P. J.
Muench, Muenchh.—Muenchhausen, O. von
Muhl.—Muhlenberg, G. H. E.
Munson—Munson, T. V.

Neck.—Necker, N. J. de
Nees—Nees von Esenbeck, C. G.
Nutt.—Nuttall, T.

Oeder—Oeder, G. C.
Ohwi—Ohwi, J.
Oliver—Oliver, D.

Palmer—Palmer, E. J.
Palmer & Steyerm.—Palmer, E. J. &. Steyermark, J. A.
Pax—Pax, F.
Peck—Peck, C. H.
Perkins—Perkins, J. R.
Pers.—Persoon, C. H.
Piper—Piper, C. V.
Planch.—Planchon, J. E.
Poir.—Poiret, J. L. M.
Porter—Porter, T. C.
Pursh—Pursh, F.

R. & S.—Roemer, J. J. & Schultes, J. A.
Raf.—Rafinesque-Schmaltz, C. S.
Regel—Regel, E. A. von
Rehd.—Rehder, A.
Richard—Richard, L. C. M.
Richards.—Richardson, J.
Robins.—Robinson, B. L.
Roem.—Roemer, M. J.
Rowlee—Rowlee, W. W.
Rydb.—Rydberg, P. A.

St. John—St. John, H.
Salisb.—Salisbury, R. A.
Sarg.—Sargent, C. S.
Schaffner—Schaffner, J. H.
Schelle—Schelle, Ernst
Schneid.—Schneider, C. K.
Schub.—Schubert, B. G.
Schultes—Schultes, J. A.

Scop.—Scopoli, J. A.
Seem.—Seemann, B. C.
Ser.—Seringe, N. C.
Shuttlw.—Schuttleworth, R.
Sieb.—Siebold, P. F. von
Simon-Louis—Simon-Louis
Sims—Sims, J.
Small—Small, J.
Sm.—Smith, J. E.
Soper—Soper, J. H.
Soul.—Soulange-Bodin, E.
Spach—Spach, E.
Spreng.—Sprengel, K.
Steud.—Steudel, E. G.
Steyerm.—Steyermark, J. A.
Sudw.—Sudworth, G. B.
Suksd.—Suksdorf, W. N.
Sweet—Sweet, R.
Swingle—Swingle, W. T.

T. & G.—Torrey, J. & Gray, A.
Thomas—Thomas, F. A. W.
Thunb.—Thunberg, C. P.
Torr.—Torrey, J.
Tratt.—Trattinnick, L.
Trautv.—Trautvetter, E. R. von

Trel.—Trelease, W.
Turcz.—Turczaninow, N. S.

Vahl—Vahl, M.
Vent.—Ventenat, E. P.
Voss—Voss, A.

Walt.—Walter, T.
Wang.—Wangenheim, F. A. J. von
Warder—Warder, J. A.
Wats. S.—Watson, S.
Wendl.—Wendland, J. C.
Wendland f.—Wendland, H. L.
Wieg.—Wiegand, K. M.
Wight & Hedrick—Wight, W. F., & Hedrick, U. P.
Willd.—Willdenow, K. L.
With.—Withering, W.
Wood—Wood, A.
Wright, J.—Wright, J. S.

Zucc.—Zuccarini, J. G.

LITERATURE CITED

ANDERSON, E. 1949. Introgressive Hybridization. John Wiley & Sons. New York.
ANDERSON, E. 1953. Introgressive hybridization. Biol. Rev. **28**: 280–307.
Atlas of American Agriculture. 1918, 1922. Pt. II. Climate. Frost and the growing season (1918); Precipitation and humidity (1922). Gov. Print. Office. Washington.
BAILEY, L. H. 1941–45. A monograph of the genus *Rubus* in North America. Gentes Herb. **5**: 1–932.
BARTLETT, H. H. 1951. Regression of × *Quercus Deamii* toward *Quercus macrocarpa* and *Quercus muhlenbergii.* Rhodora **53**: 249–264.
BEATLEY, J. 1960. Distribution of buckeyes in Ohio. In preparation.
BERRY, E. W. 1923. Tree Ancestors. Williams & Wilkins. Baltimore.
BRAUN, E. L. 1928. The vegetation of the Mineral Springs region of Adams County, Ohio. Ohio Biol. Surv. Bull. 15.
BRAUN, E. L. 1936. Forests of the Illinoian Till Plain of southwestern Ohio. Ecol. Monog. **6**: 89–149.
BRAUN, E. L. 1941. A new station for *Pachystima Canbyi.* Castanea **6**: 52.
BRAUN, E. L. 1950. Deciduous Forests of Eastern North America. Blakiston. Philadelphia.
BRAUN, E. L. 1955. The phytogeography of unglaciated eastern United States and its interpretation. Bot. Review **21**: 297–375.
BRAUN, E. L. 1960. The genus *Tilia* in Ohio. Ohio Jour. Sci. **60** (5): 257–261.
CAMP, W. H. 1942. The structure of populations in the genus *Vaccinium.* Brittonia **4**: 189–204.
CAMP, W. H. 1945. The North American blueberries with notes on other groups of Vacciniaceae. Brittonia **5**: 203–275.
CAMP, W. H. 1950. A biogeographic and paragenetic analysis of the American beech (*Fagus*). Amer. Phil. Soc. Yearbook: 166–169.
CLARKSON, R. B. 1958. The genus *Robinia* in West Virginia. Castanea **23**: 56–58.
CLAUSEN, R. T. 1951. *Smilax hispida* versus *S. tamnoides.* Rhodora **53**: 109–111.
DANSEREAU, P. and DESMARAIS, Y. 1947. Introgression in sugar maples, II. Amer. Midl. Nat. **37**: 146–161.
DEAM, C. C. 1924. Shrubs of Indiana. Indiana Dept. Conservation.
DEAM, C. C. 1940. Flora of Indiana. Indiana Dept. Conservation.
DEAM, C. C. 1953. Trees of Indiana. Indiana Dept. Conservation.
DEAN, F. W. and CHADWICK, L. C. Ohio Trees. Ohio State Univ. Press.
DESMARAIS, Y. 1952. Dynamics of leaf variation in the sugar maples. Brittonia **7**: 347–388.
DETMERS, F. 1912. An ecological study of Buckeye Lake. Proc. Ohio Acad. Sci. Spec. Paper 19.
Ecological Society of America. 1951. Symposium: The glacial border—climatic, soil, and biotic features. Ohio Jour. Sci. **51**: 105–146.
FENNEMAN, N. M. 1938. Physiography of Eastern United States. McGraw-Hill. New York.
FERNALD, M. L. 1938. Noteworthy plants of southeastern Virginia [*Fraxinus tomentosa*, p. 450–2]. Rhodora **40**: 434–459.
FERNALD, M. L. 1941. *Viburnum recognitum,* nom. nov., pp. 647–652, in "Another century of additions to the flora of Virginia." Rhodora **43**: 635–657.
FERNALD, M. L. 1950. Gray's Manual of Botany, ed. 8. American Book Company.
FERNALD, M. L. and KINSEY, A. C. 1943. Edible Wild Plants of Eastern North America. Idlewild Press.
FLINT, R. F. 1947. Glacial Geology and the Pleistocene Epoch. John Wiley & Sons. New York.
GLEASON, H. A. 1952. Illustrated Flora of the Northeastern United States and Adjacent Canada. 3 vols. N. Y. Botanical Garden.
HALL, M. T. 1952. Variation and hybridization in *Juniperus.* Ann. Mo. Bot. Gard. **39**: 1–64.
HARDIN, J. W. 1957a. Studies in the Hippocastanaceae, IV. Hybridization in *Aesculus.* Rhodora **59**: 185–203.

347

HARDIN, J. W. 1957b. A revision of the American Hippocastanaceae. Brittonia 9: 145–195.

JILLSON, W. R. 1930. Filson's Kentucke—a facsimile reproduction of the original Wilmington Edition of 1784. Louisville.

JONES, G. N. 1946. American species of *Amelanchier*. Ill. Biol. Mon. 20, no. 2.

KRIEBEL, HOWARD B. 1957. Patterns of genetic variation in sugar maple. Ohio Agr. Exp. Sta. Res. Bull. 791.

LITTLE, E. L., JR. 1949. To know the trees: important forest trees of the United States. In Trees, Yearbook of Agriculture, U. S. Dept. Agr.

LITTLE, E. L., JR. 1953. Checklist of native and naturalized trees of the United States. U. S. Dept. Agr., Forest Service, Agr. Handbook no. 41.

MANNING, W. E. 1950. A key to the hickories north of Virginia with notes on the two pignuts, *Carya glabra* and *C. ovalis*. Rhodora 52: 188–199.

MASSEY, A. B. 1940. Discovery and distribution of *Pachystima Canbyi* Gray. Castanea 5: 8–11.

McATEE, W. L. 1956. A review of the Nearctic *Viburnum*. Publ. by author, Chapel Hill, N. C.

MICHAUX, F. ANDRÉ. 1859. North American Sylva, vol. IV: Nuttall, Vol. I. [*Quercus leana*, pp. 25–27 and Plate V.]

MILLER, G. N. 1955. The genus *Fraxinus*, the ashes, in North America, north of Mexico. Cornell Univ. Agr. Exp. Sta., Memoir 335.

NIELSEN, E. L. 1939. A taxonomic study of the genus *Amelanchier* in Minnesota. Am. Midl. Nat. 22: 160–206.

Ohio Department of Natural Resources, Division of Water. 1950. The climatic factors of Ohio's water resources.

Ohio Forestry Association. Ohio's Big Trees. Columbus.

PALMER, E. J. 1948. Hybrid oaks of North America. Jour. Arnold Arboretum 29: 1–48.

PALMER, E. J. 1956. *Crateagus* in Ohio with description of one new species [*C. Horseyi*]. Ohio Jour. Sci. 56: 205–216.

REHDER, A. 1940. Manual of Cultivated Trees and Shrubs. Macmillan Co., New York.

SALAMUM, P. J. 1951. A population study of the variation in the inflorescence of *Spiraea tomentosa*. Rhodora 53: 280–292.

SARGENT, C. S. 1922. Manual of the Trees of North America. Houghton Mifflin Co. New York.

SCHAFFNER, J. H. 1932. Revised catalog of Ohio vascular plants. Ohio Biol. Surv. Bull. 25.

SEARS, P. B. 1925. The natural vegetation of Ohio. I. A map of the virgin forest. Ohio Jour. Sci. 25: 139–149.

SEARS, P. B. 1926. The natural vegetation of Ohio. II. The prairies. Ohio Jour. Sci. 26: 128–146.

SEARS, P. B. 1953. What worth Wilderness? Bulletin to the Schools of the Univ. of the State of New York. Reprinted by Nature Conservancy, Washington.

SEYMOUR, F. C. 1952. The type of *Ulmus americana* L. Rhodora 54: 138–139.

SKINNER, H. T. 1955. In search of native azaleas. Bull. Morris Arboretum 6: 3–10; 15–22.

STEBBINS, G. L., JR., MATZKE, E. B. and EPLING, C. 1947. Hybridization in a population of *Quercus marilandica* and *Quercus ilicifolia*. Evol. 1: 79–88.

STOUT, WILBER. 1943. Rock Water. Bull. 44, Geol. Surv. Ohio.

TRANSEAU, E. N. 1935. The prairie peninsula. Ecol. 16: 423–437.

TRANSEAU, E. N. 1941. Prehistoric factors in the development of the vegetation of Ohio. Ohio Jour. Sci. 41: 207–211.

TRELEASE, W. 1918. Winter Botany. Publ. by auth., Urbana.

TRELEASE, W. 1924. The American Oaks. Mem. Nat. Acad. Sci. 20: 225 pp., pl. 1–420.

WIEGAND, K. M. 1912. The genus *Amelanchier* in eastern North America. Rhodora 14: 117–161.

WOLFE, J. N. 1942. Species isolation and a proglacial lake in southern Ohio. Ohio Jour. Sci. 42: 1–12.

WRIGHT, J. W. 1944. Genotypic variation in white ash. Jour. Forestry 42: 489–495.

WRIGHT, J. W. 1944. Ecotypic differentiation in red ash. Jour. Forestry 42: 591–597.

GENERAL INDEX

See Glossary, also.

349

INDEX TO PLANT NAMES

The page on which a species is illustrated is in **boldface**. Varieties and forms are not indexed unless name occurs in a heading. Synonyms are in *italics*. Names occurring in introductory chapters are not indexed, unless not occurring elsewhere. For generic names in KEYS TO GENERA see INDEX TO GENERA IN KEYS, p. 55.

350